T0224596

The Customer Centric Enterprise

Springer
Berlin
Heidelberg
New York
Hong Kong
London
Milan
Paris
Tokyo

Mitchell M. Tseng · Frank T. Piller
(Editors)

The Customer Centric Enterprise

Advances in Mass Customization and Personalization

With 121 Figures and 47 Tables

 Springer

Professor Mitchell M. Tseng, Ph. D.
The Hong Kong University of Science & Technology
Department of Industrial Engineering &
Engineering Management
Clear Water Bay, Knowloon
Hong Kong
tseng@ust.hk
http://iesu5.ust.hk

Dr. Frank T. Piller
Technische Universität München
TUM Business School, Lst. AIB
Leopoldstraße 139
80804 Munich
Germany
piller@ws.tum.de
http://www.mass-customization.de

ISBN 3-540-02492-1 Springer-Verlag Berlin Heidelberg New York

Cataloging-in-Publication Data applied for
A catalog record for this book is available from the Library of Congress.
Bibliographic information published by Die Deutsche Bibliothek
Die Deutsche Bibliothek lists this publication in the Deutsche Nationalbibliografie; detailed bibliographic data is available in the Internet at <http://dnb.ddb.de>.

Springer-Verlag Berlin Heidelberg New York
a member of BertelsmannSpringer Science+Business Media GmbH

http://www.springer.de

© Springer-Verlag Berlin · Heidelberg 2003
Printed in Germany

Softcover-Design: Erich Kirchner, Heidelberg

SPIN 10926178   42/3130-5 4 3 2 1 0 – Printed on acid-free paper

Acknowledgments

We wish to acknowledge the work of many people making this book finally happen. First of all we have to thank our authors for contributing to this book. More than 60 researchers from all parts of the world share their knowledge and thoughts on customer centric enterprises and mass customization with us making this book a most comprehensive reference for future work in the field. We acknowledge delivering their contributions on time and apologize for taking so long to bring all the pieces together.

We gratefully acknowledge the support by many scholars and practitioners in our field which contributed both to the World Congress on Mass Customization and Personalization and this book: Prof. Claudio Roberto Boer (National Research Council Italy ITIA), Prof. Hans-Jörg Bullinger (Fraunhofer Gesellschaft), Ing. Sergio Dulio (National Research Council Italy ITIA), Dr. Ravindra S. Goonetilleke (The Hong Kong University of Science & Technology), Prof. Jochen Gros (Academy of Art & Design Offenbach), Mr. Poul Kyvsgaard Hansen (Aallborg University), Prof. Martin G. Helander (Nanyang Technological University), Dr. Roger Jianxin Jiao (Nanyang Technological University Singapore), Prof. Halimahtun M. Khalid (Universiti Malaysia Sarawak), Mrs. Eva Kuehn (form:format Berlin), Dr. Bart MacCarthy (University of Nottingham), Prof. Farokh Mistree (Georgia Institute of Technology), Prof. Ralf Reichwald (Technische Universitaet Muenchen), Mr. Falk-Hayo Sanders (MSR Consulting Group), Mr. Ralf Seelmann-Eggebert (Fraunhofer Institute IFF Magdeburg), Prof. Nam P. Suh (MIT) and Mr. Peter Tredwin (Independent Consultant).

To finish this book, we got great help by our colleagues and teams in Hong Kong and Munich. At The Hong Kong University of Science and Technology, Ms. Rebecca Tsang and Ms. Sri Hartati Kurniawan contributed to editing this book. Same did Ms. Helen Burton, Ms. Silvia Meyer, and Mr. Joachim Wimmer at the TUM Business School of the Technische Universität München. The publishing of this book was partly supported by a grant of the Deutsche Forschungsgemeinschaft (DFG) within its National Research Cluster on Mass Customization (SFB 582, see www.sfb582.de) at the Technische Universität München, Germany. We are also indebted to the Hong Kong Research Grant Council, Innovation Commission of Hong Kong SAR Government and several industrial grants including Rockwell Foundation, Esquel Enterprises Limited., Artesyn Technologies Asia Pacific Limited, Yusan Industry Limited, Effect Group, Honeywell Consumer Product Group and Sterling Products Limited. They have generously provided not only financial supports, but also problem context for us to address the broad issues of customer centric enterprises.

Finally we thank Dr. Werner Müller, Publishing Director at Springer, and his team for their patience with the editors and for helping to get this book in print.

Preface: All yours

„All yours – mass customization transforms manufacturing in the 21st century,"
wrote *The Economist* in a 2001 feature article. Enterprises in all branches of
industry are being required to become more customer centric, yet, at the same
time, increasing competitive pressure dictates that costs must also continue to
decrease. Mass Customization and Personalization are strategies developed to
address this challenge by producing goods and services meeting individual
customer's needs with near mass production efficiency. However, while mass
customization and personalization have already been discussed in the literature for
almost two decades, reports on practical implementation of the principles of mass
customization in businesses can been found only within the last years.

Also, academic research and development of the theoretical and managerial
aspects of mass customization and personalization is increasing rapidly. While an
internet search of the term mass customization got about 350 results in 1995,
nowadays there are more than 75,000 hits. According to a recent literature
research, there are more than 2700 articles in English language published about
the topic since the term was coined in 1989, about 60% of them within in the last
two years.

The 2001 World Congress on Mass Customization and Personalization wanted
to distinguish the buzz from the facts and to provide the first international, multidisciplinary and broad platform for exchange and sharing best practices and
innovative ideas in the field. The congress, jointly organized by the Advanced
Manufacturing Institute and the Department of Industrial Engineering and
Engineering Management (IEEM) of the Hong Kong University of Science and
Technology (HKUST) and the Department of General and Industrial Management
of the Technsiche Univesität München (TUM), was held at the HKUST in
October 2001. As Co-Chairs of the conference, we were very pleased by the
enormous feedback on our call for papers and the participations of the conference.
Scholars from various academic disciplines, corporate executives from all over the
world and other interested audience discussed the many faces of mass customization and personalization intensively and with much personal involvement.

The objective of this book is to share the results from the conference with a
larger audience. We selected 29 papers from the original conference proceedings
of more than 70 papers [1]. Selection of the papers was based first of all on the
rankings of the reviewers' evaluation. In addition, we tried to select papers
discussing specific topics or papers that provide a perspective on the broad scope
of contemporary mass customization research and applications. After the
selections and based on the discussion in the conference, authors were invited to
revise, extend and update their original conference contribution. The idea of the
book is to give the reader an introduction into the field, to show the scope of mass

customization research, and to present recent research findings and the state of the art in selected perspectives of this subject. We hope that our selection may fit your personal interests.

Mitchell M. Tseng and Frank T. Piller

References

[1] Tseng, M.M. and Piller, F.T. (Eds.): Proceedings of the 2001 World Congress on Mass Customization & Personalization, Hong Kong University of Science and Technology, Hong Kong 2001 (available on CD-ROM at ami.ust.hk/MC01/MCPC.htm).

Contact to the editors:

Dr. Frank T. Piller
Technische Universität München
TUM Business School, Lst. AIB
Leopoldstraße 139, 80804 Munich, Germany
Tel.: +49 89 289 24820
E-Mail: piller@ws.tum.de
www.aib.ws.tum.de/piller
www.mass-customization.de

Prof. Mitchell M. Tseng, Ph.D.
The Hong Kong University of Science & Technology
Department of Industrial Engineering & Engineering Management,
Clear Water Bay, Kowloon, Hong Kong
Tel.: +852 2358 7091
E-Mail: tseng@ust.hk
http://iesu5.ust.hk/
http://ami.ust.hk/

Further information about the World Congresses on Mass Customization and Personalization:

Hong Kong, 2001:   http://ami.ust.hk/MC01
Munich, 2003:   http://www.mcpc2003.com

Contents

Part I: Heading Towards Customer Centric Enterprises

An introduction

Enterprises in all branches of industry are becoming more customer centric. The increasing interest and effort of business practices heading towards mass customization and personalization is met by an intensified and ongoing study of these approaches in research and academia. Though the oxymoron 'mass customization' was coined in the mid 1980's, research has started to pick up pace only in recent years. The number of papers published on mass customization and personalization has increased threefold in the last decade. With this in mind, the intention of this book is not only to discuss the state of the art of methods and approaches of more customer centric manufacturing, but also to show the obstacles and challenges of mass customization, and to analyze its potentials and capabilities. To open the discussion, the first part of this book gives a brief introduction into mass customization and personalization as key strategies of customer centric enterprises. *Tseng and Piller* comment on their understanding of both terms and illustrate the levels of a mass customization system from a generic perspective. Chapter 1 also presents a framework of the flow of activities in an extended mass customization system and integrates mass customization in the larger framework of supply chain management.

Part I: Heading Towards Customer Centric Enterprises

An Introduction

1 The Customer Centric Enterprise

An integrative overview on this book

Mitchell M. Tseng[1] and Frank Piller[2]
[1] Department of Industrial Engineering & Engineering Management,
Hong Kong University of Science & Technology
[2] TUM Business School, Department of General and Industrial
Management, Technische Universitaet Muenchen, Germany

> *The most creative thing a person will do 20 years from now is to be a*
> *very creative consumer... Namely, you'll be sitting there doing things like*
> *designing a suit of clothes for yourself or making modifications to a stan-*
> *dard design, so the computers can cut one for you by laser and sew it to-*
> *gether for you by NC machine...*

Robert H. Anderson, Head Information Systems, RAND Corporation,
quoted in Alvin Toffler *"Third Wave"* [1], p. 274

1.1 Open questions and increasing implementation

More than two decades later, in 2003, this prophecy is still a vision not only in the
clothing business but also in most other industries. What causes the renowned
futurist miss the mark? Though we have most, if not all, the necessary hardware,
software, powerful computing and communication systems, including laser
cutting, high performance sewing etc, we are still not really able to meet the
special yearning of human beings, that very important feature that sets us apart
from animals, i.e. *creativity*. We believe the missing gap is the capacity to put the
systems, including organization, process and business models together and make
them customer centric. Building a customer centric enterprise that places the
demands and wishes of each single customer in the center of value creation
implies much more than investing in advanced technologies. Firms have to build
not organizations and structures to produce customized services, but organizations
and structures for customers. With the customers at the center, human beings can
then focus on being creative and be isolated from mundane tasks in order to
concentrate on expressing themselves more freely.

Although few enterprises can be truly considered as customer centric today,
successful companies have entered their particular market with initiatives and
products that break with the paradigm of mass production. In order to be
responsive to customers, companies have shortened the product development time
dramatically. Taking the automotive industry as an example, the time to market

has been reduced from six years to three months. One of the side effects of time competition in new product development is product proliferation. We witness thousands of new products vying for customers' attention in supermarkets. Even for commodity items such as milk, we have different flavors, fat contents, and types of feeds (organic or ordinary) as factors for differentiation. The main driving force behind the wide spread product proliferation is the expectation of many product managers that somehow one variety will fit with the buying decision of a high enough number of customers. However, variety does not come cheap. The cost of the explosive increase of product variety is tremendous, just consider inventory costs, customer returns (because of buying the wrong items), sales force training, post sale services, and confusion (complexity costs) inside and outside a company.

For these reasons economic forces have led companies to build to order instead of building to stock. This development is made possible by the availability of flexible manufacturing and information technology. The rapid advent of computer and communication technology has enhanced the flexibility not only of shop floor machinery but also on the information system level such as MRP (Manufacturing Resource Planning), ERP (Enterprise Resource Planning) and CRM (Customer Relationship Management). Furthermore, with a more educated workforce, there is an inherent flexibility within companies that has not been fully deployed by the firms. In fact, it has been widely reported that a lot of this new flexibility has been under-utilized in terms of enterprise business. Leading managers have identified the opportunity of applying these under-utilized resources (of flexibility) as an option to better serve customers as individuals instead of just providing variety to the mass. Instead of designing and producing goods to stock for an anonymous market, their companies provide goods and services that are customized and assembled on-demand for each individual customer. Sanjay Choudpouri, former director of mass customization at Levi Strauss, is one example. He foresees that customization in his industry "will become a competitive necessity rather then a nice to have fringe offering" [2].

Thus, Anderson's prediction quoted at the beginning of this chapter may not be accomplished entirely. Putting customers at the center of the enterprise and building its processes and systems towards serving customers as an individual best *and* efficiently has emerged as a profitable business proposition. The basic idea of being customer centric can be considered as achieving the ultimate goal of quality management: meeting individual customers' requirements exactly without a significant increase in production and distribution cost. There are already significant steps towards a new, more customer centric economy. For example, a listed Chinese company reported that by using mass customization technology, it produced more than 500 customized uniforms a day for lawyers, service workers, and policemen. Most people would think that a uniform would have to be uniformly produced. However, due to differences in body measurements, styles in different cities and replacement cycles, production orders require significant amounts of customization. Today's industrial move towards mass customization is an answer to the growing demands in competition. Enterprises in all branches of industry are being forced to react to the growing individualization of demand and to find ways to manage the increasing product variety. Yet, at the same time, increasing competitive pressure dictates that costs must also continue to decrease. Companies have to adopt strategies that embrace both cost efficiency and a closer reaction to customers' needs. Mass customization and the corresponding approach of personalization have the potential to address competitive market requirements while improving a firm's profitability.

The increasing interest and effort of business practices towards mass customization is met by an intensified and ongoing study of mass customization in research and academia. Though the oxymoron 'mass customization' was coined in the mid 1980's, research has started to pick up pace only in recent years. The number of papers published on mass customization has increased threefold in the last decade. With this in mind, the intention of this book is firstly to discuss the state of the art of methods and approaches of more customer centric manufacturing, and secondly to show the obstacles and challenges of mass customization and personalization, and to analyze their potentials and capabilities.

1.2   What is mass customization?

There is a wide variety of understandings and meanings of mass customization and personalization: „Extant literature has not established good conceptual boundaries for mass customization", state Duray et al. [3, p. 606] after a literature review. The same is true for managers and practitioners who use the term mass customization for many forms of being more customer centric. Davis, who coined the phrase in 1987, refers to mass customization when "the same large number of customers can be reached as in mass markets of the industrial economy, and simultaneously they can be treated individually as in the customized markets of pre-industrial economies" [4, p. 169]. In order to address the implementation issues of mass customization, a working definition of mass customization was adopted as "the technologies and systems to deliver goods and services that meet *individual* customers' needs with *near* mass production efficiency" [5]. This definition implies that the goal is to detect customers needs first and then to fulfill these needs with efficiency that almost equals that of mass production. Often this definition is supplemented by the requirement that the individualized goods do not carry the price premiums connected traditionally with (craft) customization [4, 6, 7, 8, 9]. However, mass customization practice shows that consumers are frequently willing to pay a price premium for customization to reflect the added value of customer satisfaction due to individualized solutions, i.e. the increment of utility customers gain from a product that bette fits to their needs than the best standard product attainable [10, 11, 12]. We consider the value of a solution for the individual customer as the defining element of mass customization. A customer centric enterprise recognizes that customers have alternatives of choice which are reflected through their purchase decisions [13]: Customers can either choose mass customized goods which provide better fit, compromise and buy a standard product of lesser fit (and price), or purchase a truly customized product with excess features but also at a higher price. Thus, value reflects the price customers are willing to pay for the increase in satisfaction resulting from the better fit of a (customized) solution for their requirements. Mass customization is only applicable to those products for which the value of customization, to the extent that customers are willing to pay for it, exceeds the cost of customizing.

The competitive advantage of mass customization is based on combining the efficiency of mass production with the differentiation possibilities of customization. Mass customization is performed on four levels (see Figure 1). While the *differentiation level* of mass customization is based on the additional utility customers gain from a product or service that corresponds better to their needs, the *cost level* demands that this can be done at total costs that will not lead to such a price increase that the customization process implies a switch of market segments [14]. The information collected in the course of individualization serves to build up a lasting individual relationship with each customer and, thus, to increase customer loyalty (*relationship level*). While the first three levels have a customer centric perspective, a fourth level takes an internal view and relates to the fulfillment system of a mass customizing company: Mass customization operations are performed in a fixed *solution space* that represents "the pre-existing capability and degrees of freedom built into a given manufacturer's production system" [15]. Correspondingly, a successful mass customization system is characterized by *stable* but still flexible and responsive processes that provide a dynamic flow of products [16]. While a traditional (craft) customizer re-invents not only its products but also its processes for each individual customer, a mass customizer uses stable processes to deliver high variety goods. A main enabler of *stable processes* is to modularize goods and services [17, 18]. This provides the capability to efficiently deliver individual modules of customer value within the structure of the modular architecture. Setting the solution space becomes one of the foremost competitive challenges of a mass customization company, as this space determines what universe of benefits an offer is intended to provide to customers, and then within that universe what specific permutations of functionality can be provided [16].

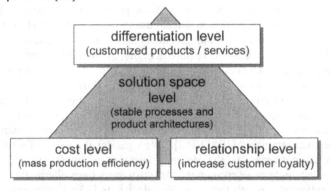

Figure 1:   The four levels of mass customization

1.3   Personalization versus customization

Personalization must not be mixed up with customization. While customization relates to changing, assembling or modifying *product* or *service* components according to customers' needs and desires, personalization involves intense *communication* and *interaction* between two parties, namely customer and supplier. Personalization in general is about selecting or filtering information objects for an individual by using information about the individual (the customer profile) and then negotiating the selection with the individual. Thus, personalization compares strongly to recommendation: From a large set of possibilities, customer specific recommendations are selected [13, 19, 20]. From a technical point of view, automatic personalization or recommendation means matching meta-information of products or information objects against meta-information of customers (stored in the customer profile). Personalization is increasingly considered to be an important ingredient of Web applications. In most cases personalization techniques are used for tailoring information services to personal user needs. In marketing, personalization supports one-to-one marketing [21] which should increase the customer share over a lifetime.

A good example of both customization and personalization is provided by *Land's End*, a catalog retailer. The company has been using a virtual model and recommendation service on its web site since 1999. The system recommends a customized bundle of standard mass products matching each other and the customers' style profile. This service provides customers with a set of coherent outfits rather than with isolated articles of clothing. But each product is still a standard product. In 2001, Land's End also introduced mass customization. Customers can order made-to-measure trousers and shirts. All products are made to order in a specially assigned factory. The company offers a substantial number of design options and varieties. However, this customization process is not supported by personalization. A consumer has to know by herself which style, waistline and length suits her best. The configuration toolkit used by Land's End does not provide any information or consultancy. For this company, personalization as performed for (almost ironically) standard products would only provide real additional benefit for the mass customization operations. Combining personalization with customization would empower a customer without the knowledge of a tailor to customize a product more easily.

Thus, while mass customization and personalization may have different methodological backgrounds and use different instruments, personalization can support mass customization. By presenting, for example, a personalized pre-configuration, the co-design or design by the customer [11] process could be shortened, streamlined and focused on providing real customer value. Instead of starting to combine the core product from scratch with hundreds or millions of options, the customer could concentrate on designing a solution that fits her known needs. Thus, during the configuration process the complexity and burden of choice could be reduced heavily by only presenting options identified as relevant from a customer's profile [22].

1.4   Placing customers at the centre of the enterprise

While mass customization has been addressed in literature as a promising approach to meeting today's market demands, some authors have recently discussed its limits and concerns (e.g., [23, 24, 25, 26, 27]). One limitation of mass customization is that excess variety may result in an external complexity that Pine defined as "mass confusion" [28]. The number of choices could overwhelm customers during product configuration [24, 29]. However, this challenge of mass customization can also serve as the source of its competitive advantage. The interaction with each single customer and their integration in the supplier's value chain is the prerequisite for delivering a customized solution. Essentially, by making customers the center of an enterprise, every transaction implies information gathering about each customer's specific product design requirement. Thus, a *two stage process of development of product offerings* bestows the foundation for customer centric enterprise and is the basis of mass customization:

(1) Product architectures and the range of possible variety are fixed during a preliminary design stage linking the overall company strategy to the available capabilities including manufacturing, logistics and service capacity. During this step, the 'solution space' of a mass customization system is set.

(2) The second design and development stage takes place in close interaction between the customer and the enterprise. Here, the capabilities of the solution space from stage 1 are turned through adequate configuration tools into a specific customer order. This process is called the *elicitation* of a mass customization system [27]. The enterprise has to interact with its customers to obtain specific information in order to define and translate the customers' needs and desires into a concrete product specification.

Elicitation during the course of configuration and the process of co-development results in customer integration. The customer is integrated into the firm's value creating activities. Customer integration can be defined as a form of industrial value creation where "the consumers take part in activities and processes which used to be seen as the domain of the companies" [30, p. 360]. The customer becomes a 'co-producer' respectively 'prosumer' [31]. The result is a system of co-production, i.e. an interaction of supplier and customer for the purpose of attaining added value [32, 33]. The customer can be seen from the firm's perspective as a production factor fulfilling tasks that in a mass production system are done internally [34]. As in a mass customization system the main part of the interaction with the customer takes place during the configuration and therefore the design of a customer specific product, it seems appropriate to call the customer more a *co-designer* than a co-producer. Customer co-design describes a process that allows customers to express their product requirements and carry out product realization processes by mapping the requirements into the physical domain of the product [35, 36, 37]. While a personalization system is often able to provide individualized communication with a client without any explicit interaction between both parties (if the information for personalization can be based on existing information about this client), mass customization requires the active participation of each customer. Here, the organization's primary thrust is to identify and fulfill the individual wants and needs of each and every customer [16, 38].

1.5   Linking mass customization and supply chain management

Interacting with customers and enabling them to become co-designers and part of value creation is an important capability of a customer centric firm. Another is to deliver the results of the configuration process, to finally produce customized products and services. While specific means and approaches of flexible manufacturing are addressed in many chapters of this book, we want to place mass customization here in the larger framework of *supply chain management*. In many ways supply chain management and mass customization are important building blocks for a customer centric enterprise. However, supply chain management addresses production management from the inside out, particularly from the manufacturers' point of view, while mass customization approaches from the outside (i.e. customer requirements) towards the production and distribution systems. Finding synergies between these two approaches may open up new avenues to better serve the end customers, particularly the creative ones. Here are areas where we believe the synergy between supply chain management and mass customization can be built upon [39]:

Product design as an integral part of the supply chain: In supply chain management, the product design is often taken as given. By considering products at the SKU level, the emphasis of supply chain management is on the synchronization of material, financial and information flows. In mass customization, customers can actively participate in the value chain by making informed trade offs not only regarding cost and delivery schedules, but also with regard to product design features and functionality. Thus, expanding the three flows in supply chain management to include the customers' choices on product design could provide a new dimension for synchronization. For example, customers may be persuaded to buy a specialized design to alleviate the shortage of certain components.

Bring customers into the value chain: In supply chain management, customers are an integral part of the value chain. In many cases, they are at the initiation point of the value chain. Consequently, mass customization has been working on issues such as getting customers to explore the possibilities, identify what they want and getting information about the capability of producers, and hence the delivery schedule, without laborious searches. The idea of helping customers to articulate their needs, negotiate and make informed decisions could be useful to leverage on supply chain management information flow. Efficiency of the supply chain can be improved because the customers and producers are able to share information about material availability and customer preference in a direct, non-intrusive, and *in situ* mode. For example, customers and enterprise can negotiate the price, features, and delivery simultaneously so that best value can be provided to the customers at that time.

Measure value chain performance directly: By considering each customer as an individual, value contribution of the supply chain moves beyond traditional metrics of inventory and delivery performance. Direct links between customer preference and the supply chain provide the possibility of measuring value contribution directly including metrics such as customer loyalty and gross margin. Furthermore, pushing accountability closer to the value contributors could create flexibility to utilize more economical production capabilities such as electronic cottage industry and global division of works.

Structure product family to achieve efficiency in high-variety low-volume production: Recognizing the individual differences among customers while still achieving economies of scale as well as product family, modularity and commonality are important techniques for implementing mass customization. Much has been learned in mass customization about the importance of structuring product families that are both easy for customers to comprehend and conducive for organizing repetition in high variety production. These techniques will be very useful for supply chain management to streamline not only various processes but also increase the scale of economy in operation.

1.6   The knowledge circle of mass customization: A survey of this book

Mass customization and customer integration create a customer centric enterprise system that transcends the traditional manufacturing enterprise. In essence, the scope of a customer centric enterprise is not limited to producing its products, it also applies the capabilities to detect customers' needs, to proactively satisfy them, and to strategically position the enterprise capabilities around the customers' future requirements. Companies no longer design, make and then sell products; instead, companies will sell capabilities, get orders, and then fulfill these requests. Consequently, their success depends very much on the ability to manage knowledge – that not only covers one transaction but uses information gathered during the fulfillment of a customer-specific order to improve the knowledge base of the whole company [14, 27, 40, 41]. The representation of these processes in a knowledge loop model stresses the importance of an interconnected and integrated flow of information (Figure 2). The four premier activities of a mass customization system [14, 42, 43] link together to form a loop that serves as the backbone of a customer centric enterprise. These four activity bundles are the design of the solution space, configuration and customer interaction, fulfillment of customized manufacturing, and customer relationship management. These activities have to be delivered in an integrated, streamlined way, starting from the point of interaction where information required for customization is surveyed, then on to processing this information for fulfillment and providing the customized offer, and finally activities that deepen the customer relationships and create customer loyalty. The knowledge generated during serving one customer has to be stored to serve this customer even better, faster, and more efficiently when a second order is placed. Furthermore, information acquired during the output process must be saved, assessed and employed to continuously improve the solution space and foster efficiency and quality enhancement in follow-up business. By doing so, both new and old customers can be served better. Supply chain management can also be improved continuously (e.g. with a given amount of inventory, the customer service level can be optimized).

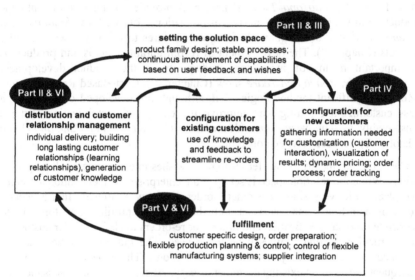

Figure 2: The enterprise knowledge loop and the structure of this book

We will use the knowledge loop model to briefly give an overview of the subsequent parts and chapters of this book. However, before starting with the first activity of the knowledge loop, readers will be introduced with the fundamentals of being customer centric. Thus, Part II of this book provides an introduction to principles, concepts, demarcations, and business models for mass customization and personalization. In Chapter 2, *MacCarthy, Brabazon and Branham* present an introduction into the scope of mass customization principles in industry. Correspondingly, *Franke and Mertens* discuss the use of personalization in industry and public administration (Chapter 6). Does mass customization and personalization pay? Both *Riemer and Totz* (Chapter 3) and *Reichwald, Piller, Jäger and Zanner* (Chapter 4) evaluate this question from an economic perspective in two different settings. *Thoben* contributes to the understanding of the nature of mass customization by comparing this approach with (traditional) customer driven manufacturing (Chapter 5). Part II concludes with a new perspective: individualization and personalization are characteristics of art, as *Gros* discusses in Chapter 7. After this introduction, the structure of the book follows the activity sequence of the knowledge loop (see Figure 2 above):

Design of the solution space: Before a single customer can interact within a mass customization system, the solution space has to be defined. As described above, building the solution space consists of developing the basic product architectures and families, setting the number of customization options for one component, and defining the pricing schemes for each selection. This activity cluster further includes the development and implementation of the fulfillment processes in manufacturing. These design activities are addressed in Part III of our book. Here, *Du, Jiao and Tseng* (Chapter 8) provide an introduction into the product family approach for mass customization, followed by *Siddique and Rosen's* discussion of common platform architectures to identify a set of similar products (Chapter 9). The development of reconfigurable models and products is an important means of increasing productivity in the product development process, as *Cox, Roach and Teare* show (Chapter 10). Cost-Based Reasoning and TRIZ are significant methodologies that, although not developed originally in a mass customization setting, can improve the design of customization systems considerably, as *Wongvasu, Kamarthi and Zeid* (Chapter 11) and *Mann and Dmob* (Chapter 12), respectively, demonstrate.

Configuration means to transfer customers' wishes into concrete product specifications. While the solution space is set up at the enterprise level, elicitation activities take place with every single customer's order. For new customers, first a general profile of their desires and wishes has to be built up. This profile is transformed into a concrete product specification and order. For re-orders made by regular customers their particular existing profiles have to be used. The old configuration may be presented and customers just asked for variations. The objective is to make subsequent orders of an existing customer as easy, efficient and fast as possible – an important means of increasing customer loyalty. Part IV comprises the discussion of configuration methodologies and modes for customer interaction. *Khalid and Helander* give an introduction into web-based configuration approaches (Chapter 13), an issue that is elaborated by *Oon and Khalid* in Chapter 15. How consumers behave in such an environment is discussed by *Kurniawan, Tseng and So* (Chapter 14). Afterwards, *Porcar, Such, Alcantara, Garcia and Page* show how consumer expectations can be captured by the Kansei Engineering methodology (Chapter 16). *Hvam and Malis* present a documentation tool for configuration processes to foster knowledge based product configuration (Chapter 17). Part IV concludes with *Svensson and Jensen's* plea that the customer should always be at the final frontier of mass customization (Chapter 18).

Fulfillment: After an order has been placed, it is transferred into specific manufacturing tasks. Scheduling activities follow, whereby suppliers may be integrated in the customization of some parts, too. Often in a segmented production layout, the order is fulfilled. Part V analyses how such manufacturing operations are set up and performed. *Urbani, Molinari-Tosatti, Bosani and Pierpaoli* present an analytical approach on system flexibility and reconfigurability for mass customization (Chapter 19). Flow systems can enable mass customization, as *Tsigkas, de Jongh, Papantoniou and Loumos* explore in Chapter 20. An adaptive system for production planning and control for customized manufacturing is discussed by *Lopitsch and Wiendahl* (Chapter 21). *Schenk and Seelmann-Eggebert* take a broader view on customized manufacturing including training activities in Chapter 22. *Hansen, Jensen and Mortensen* explain how modularization in manufacturing is implemented in industry (Chapter 23). Finally, *Mchunu, de Alwis and Efstathiou* introduce a framework to select the best fitting mass customization strategy for a specific manufacturing task (Chapter 24).

Customer relationship management: Increasing customer loyalty is one of the foremost objectives of companies going into mass customization. Mass customization has especially pronounced capabilities to intensify the relationship between a supplier and its customers and, thus, to increase customer loyalty. Customer loyalty can be seen as a result of switching costs, opportunity costs, and sunk costs based on technological, contractual, and psychological obligations faced by a customer [44]. All sources of switching, sunk, and opportunity costs are positively influenced in a mass customization system during the course of customer integration, as *Riemer/Totz* present in Chapter 3. Corresponding activities are also discussed in various other chapters of this book (e.g., *Reichwald et al.* in Chapter 4, *Franke and Mertens* in Chapter 6, *Svensson and Jensen* in Chapter 18, *Gurtzki and Hinderer* in Chapter 27, and *Piller and Tseng* in Chapter 30).

How the different activities of the knowledge loop of mass customization are connected and applied in an industry model is the theme of Part VI. Five chapters present lessons learned from the apparel and footwear industry and give an insight into some specific challenges of mass customization. We have chosen this industry as a premier example as here many experiments, pilot studies and business implementations of mass customization principles have already taken place. *Bullinger, Wagner, Kürümlüoglu and Bröckner* open this part and substantiate the knowledge loop model for the footwear industry by presenting a framework for an extended user oriented shoe enterprise (Chapter 25). Issues of retail and customer interaction are addressed by *Taylor, Harwood, Wyatt and Rouse* in Chapter 26 and *Gurtzki and Hinderer* in Chapter 27. Special issues of customization in the footwear sector are illustrated with regard to measurement and fit by *Luximon, Goonetilleke and Tsui* (Chapter 28), and with regard to design and manufacturing by *Sacco, Vigano and Paris* (Chapter 29).

Even with all the references to the state-of-the-art of mass customization research in Parts II to VI, there are still many unanswered questions. Thus, we share our view about the future of the customer centric enterprise, mass customization, customer integration, and personalization in Part VII. We also comment on business opportunities and fields for further research where it is necessary to develop new processes, tools and programs for integrating the customer into value creating activities, both on the technological and the operational process side.

But despite open questions and demand for further research, many companies have proved that it is already possible today to identify specific (perhaps sometimes still narrow) customer centric niches as a successful business opportunity. These are the companies which took serious investments and business risks to identify major factors for differentiation in order to create economic value and precipitate enablers for increasing customer loyalty.

References

[1]   Toffler, A.: The Third Wave, New York: Bantam Books 1980.

[2]   Choudpouri, S.: Mass Customization at Levi Strauss, Speech at the "Think Custom" Conference, New York, February 2000.

[3]   Duray, R. et al.: Approaches to Mass Customization: Configurations and Empirical Validation, in: Journal of Operations Managements, 18 (2000), pp. 605-625.

[4]   Davis, S.: Future Perfect, Reading: Addison-Wesley 1987.

[5]   Tseng, M.M. and Jiao, J.: Mass Customization, in G. Salvendy (Ed.) Handbook of Industrial Engineering, 3rd edition, New York: Wiley, 2001, pp. 684-709.

[6]   Hart, C.: Mass Customization: conceptual underpinnings, opportunities and limits, in: International Journal of Service Industry Management, 6 (1995) 2, pp. 36-45.

[7]   Schenk, M. and Seelmann-Eggebert, R.: Mass Customization Facing Logistics Challenges, in: C. Rautenstrauch et al. (Eds.) Moving towards mass customization, Heidelberg: Springer 2002, pp. 41-57.

[8]   Victor, B. and Boynton, A.C.: Invented Here, Boston: HBSP 1998.

[9]   Westbrook, R. and Williamson, P.: Mass customization, in: European Management Journal, 11 (1993) 1, pp. 38-45.

[10]   Chamberlin, E.H.: The Theory of Monopolistic Competition, 8. ed., Cambridge: Harvard University Press 1962.

[11]   Du, X. and Tseng, M.M.: Characterizing Customer Value for Product Customization, in: Proceedings of the 1999 ASME Design Engineering Technical Conference, Las Vegas 1999.

[12]   Porter, M.E.: Competitive Strategy, New York, 1980.

[13]   Imhoff, C.; Loftis, L.; Geiger, J.G.: Building the Customer-Centric Enterprise: Data Warehousing Techniques for Supporting Customer Relationship Management, London: Wiley 2001.

[14]   Piller, F.: Mass Customization, 3rd edition, Wiesbaden: Gabler: 2003.

[15]   Von Hippel, E.: Perspective: User Toolkits for Innovation, in: The Journal of Product Innovation Management, 18 (2001), pp. 247-257.

[16]   Pine , B.J.: Challenges to Total Quality Management in Manufacturing, in: J.W. Cortada and J.A. Woods: The Quality Yearbook 1995, New York: McGraw-Hill, 1995, pp. 69-75.

[17]   Du, X., Jiao, J. and Tseng, M.M.: Architecture of Product Family: Fundamentals and Methodology, in: Concurrent Engineering: Research and Application, 9 (2001) 4, pp. 309-325.

[18]   Tseng, M.M and Jiao, J.: Design for Mass Customization, in: CIRP Annals, 45 (1996) 1, pp. 153-156.

[19]   Elofson, G. and Robinson, W.N.: Creating a custom mass-production channel on the Internet, in: Communications of the ACM, 41 (1998) 3, pp. 56-62.

[20] Resnick, P. and Varian, H.: Recommender systems, in: Communications of the ACM, 40 (1997) 3, pp. 56-58.

[21] Peppers, D. and Rogers, M.: Enterprise one to one: Tools for competing in the interactive age, New York 1997.

[22] Piller, F.; Koch, M.; Möslein, K. and Schubert, P.: Managing high variety, Proceedings of the European Academy of Management (EURAM) Conference 2003, Milan: SDA Bocconi, 2003.

[23] Agrawal, M.; Kumaresh, T.V.; Mercer, G. A.: The false promise of mass customization, in: The McKinsey Quarterly, 38 (2001) 3, pp. 62-71.

[24] Huffman, C. and Kahn, B.: Variety for Sale: Mass Customization or Mass Confusion, in: Journal of Retailing, 74 (1998), pp. 491-513.

[25] Piller, F. and Ihl, Ch.: Mythos Mass Customization, in: New Management, 71 (2002) 10, pp. 16-30 (in German language).

[26] Sahin, F.: Manufacturing competitiveness: Different systems to achieve the same results, in: Production and Inventory Management Journal, 42 (2000) 1, pp. 56-65.

[27] Zipkin, P.: The Limits of Mass Customization, in: Sloan Management Review, 42 (2001) 12, pp. 81-87.

[28] Teresko, J.: Mass Customization or Mass Confusion, in: Industry Week, 243 (1994) 12, pp. 45-48.

[29] Friesen, G.B.: Co-creation: When 1 and 1 make 11, in: Consulting to Management, 12 (2001) 1, pp. 28-31.

[30] Wikström, S.: Value Creation by Company-Consumer Interaction, in: Journal of Marketing Management, 12 (1996), pp. 359-374.

[31] Toffler, A.: Future Shock, New York: Bantam Books 1970.

[32] Milgrom, P. and Roberts, J.: The Economics of Modern Manufacturing: Technology, Strategy, and Organization, in: The American Economic Review, 80 (1990) 6, pp. 511-528.

[33] Normann, R. and Ramirez, R.: From value chain to value constellation, in: Harvard Business Review, 71 (1994) 4, pp. 65-77

[34] Ramirez, R.: Value co-production: intellectual origins and implications for practice and research, in: Strategic Management Journal, 20 (1999) 1, pp. 49-65.

[35] Helander, M. and Khalid, H.: Customer Needs in Web-Based Do-It-Yourself Product Design in: Proceedings of the M.Sc. Ergonomics International Conference, Lulea, Sweden, 29-30 October 1999, Lulea University of Technology, Department of Human Work Sciences, Division of Industrial Ergonomics, Lulea, Sweden 1999, pp. 9-14.

[36] Tseng, M.M. and Du, X.: Design by Customers for Mass Customization Products, in: CIRP Annals, 47 (1998) 1, pp.103-106.

[37] Von Hippel, E.: Economics of Product Development by Users: The Impact of "Sticky" Local Information, in: Management Science, 44 (1998) 5, p. 629-644.

[38] Pine, B.J.: Mass Customization, Boston: Harvard Business School Press, 1993.

[39] Tseng, M. M.: Are You Ready to Serve Creative Customers?, in: The Supply Chain Connection, Stanford Supply Chain Forum, Spring, 2002, pp4-5.

[40] Anderson, D.M.: Agile product development for mass customization, Chicago: Wiley 1997.

[41] Gilmore, J.H. and Pine, B.J.: Customization that counts, in: J.H. Gilmore / B.J. Pine (Eds.): Markets of one: creating customer-unique value through mass customization, Boston: Harvard Business School Press 2000, pp. vii-xxv.

[42]   Reichwald, R., Piller, F, and Moeslein, K.: Information as a critical success factor for mass customization, in: Proceedings of the ASAC-IFSAM 2000 Conference, Montreal, 2000.

[43]   Rautenstrauch, C., Tangermann, H. and Turowski, K.: Manufacturing Planning and Control Content Management in Virtual Enterprises Pursuing Mass Customization, in: C. Rautenstrauch et al. (eds.) Moving towards mass customization, Heidelberg: Springer 2002, pp. 103-118.

[44]   Jackson, B. B.: Build customer relationships, in: Harvard Business Review, 63 (1985) 6, pp. 120-128.

Contacts:

Prof. Mitchell M. Tseng, Ph.D.
Department of Industrial Engineering & Engineering Management,
Hong Kong University of Science & Technology
E-mail: tseng@ust.hk

Dr. Frank T. Piller
TUM Business School, Department of General and Industrial Management,
Technische Universitaet Muenchen, Germany
E-mail: piller@ws.tum.de

Part II: Mass Customization and Personalization

Key Strategies for Customer Centric Enterprises

Being customer centric includes a wide range of strategies, approaches and ideas. Agile manufacturing, focused factories, flexible specialization, lean manufacturing, customer relationship management, and mass customization are strategies that emerged from the literature in the last decades. Despite different backgrounds and focus, the major objective of these new concepts is to improve the ability of enterprises to react faster to changing customers' needs and to address the heterogeneity of demand more efficiently. This book's emphasis is placed on mass customization and personalization which can be seen as key strategies for making firms more customer centric. Thus, Part II provides an introduction into principles, concepts, demarcations, and business models for mass customization and personalization. The scope of the contributions in this part is relatively broad. The intention is to sharpen the reader's view on customization and personalization and to give an overview into the reach and scale of these concepts.

Part II starts with an introduction into the extent of mass customization principles in industry. In Chapter 2 *MacCarthy, Brabazon and Branham* contribute to our understanding of both the potential of mass customization and the constraints under which real mass customizers may operate. The authors show that there is not one mass customization strategy. They present five case studies from a range of sectors – bicycles, computer assembly, communications components, mobile phones and commercial vehicles – and analyze their approaches to customization as well as their modes of operations. The scope of being customer centric is also the topic of Chapter 3 by *Riemer and Totz* on the many faces of personalization and mass customization. Their focus in on the emergence of internet technology enabling cost-effective one-to-one relationships with customers and, thus, new ways of doing business. Personalization (individual (one-to-one) communication) and mass customization (efficient product individualization) are discussed and set in relation to each other. The authors conclude that customization has to be accompanied by personalization of communication and customer interaction. They integrate customization and personalization into the online marketing mix. By doing so, the chapter provides a thoughtful discussion of the economic motivation of personalization and mass customization based on the capability of individualization to increase switching costs for the customers – resulting in deeper and more profitable customer relationships.

Does mass customization and personalization pay? *Reichwald, Piller, Jäger and Zanner* (Chapter 4) evaluate this question from an economic perspective. They apply a general framework for the economic evaluation of mass customization on a special setting of decentralized, customer centric production units (so-called mini-plants) located in close proximity to a particular local market. The chapter examines whether such a decentralized scenario of value creation could provide a suitable framework for the efficient production of individualized goods. The authors discuss whether the additional costs and hurdles of mass customization in mini-plants could be counterbalanced by the advantages of such a decentralized setting (compared to both mass production and centralized mass customization). Advantages could arise from new cost saving potentials and a higher consumers' willingness to pay for a customized solution. However, at the bottom line there is no generic rule as to when mass customization does pay. Only by evaluating the influencing factors of a particular situation can an answer be provided. With this in mind, *Thoben* contributes in Chapter 5 to the understanding of the nature of mass customization by comparing its system design principles with (traditional) customer driven manufacturing. Especially in Europe there is a long tradition of designing and manufacturing customer specific products such as machinery, ships and cars. The author evaluates synergies, similarities as well as limitations and potentials of both mass customization and (traditional) customer driven manufacturing. Bringing the discussion back to life experiences and case studies, *Franke and Mertens* discuss in Chapter 6 the use of personalization approaches in industry and public administration. While the theoretical foundations of user modeling and personalization techniques have been discussed in literature for several years, their practical implementation has been neglected for a long time. The authors share their experiences from a couple of cases of computer-assisted information, consulting, decision support and offering systems. These systems use personalization technologies to individualize the dialogue between man and machine pragmatically by user modeling based on content based filtering as well as social filtering.

Part II concludes with a new perspective: individualization and personalization are characteristics of art, as *Gros* discusses in Chapter 7. As Chapter 1 of this book has already shown, using the creativity of consumers may lead not only to better fitting products but also demands a new way of performing – and evaluating – value creation in industry. *Gros* sharpens our view of being customer centric by approaching customization as art. Applied art was once an important field of industry. However, as a result of industrialization and mass production, the link between art and consumer goods has been broken for almost a century. Now it could be assumed that new mass customization technologies may favor a rebirth of the association between art and consumer goods, a relationship coined 'art customization' by the author.

2 Examination of Mass Customization Through Field Evidence

Bart MacCarthy, Philip G. Brabazon and Johanna Bramham
Mass Customization Research Centre, School of Mechanical, Materials,
Manufacturing Engineering & Management, Nottingham University, UK

Mass customization excites interest across both the research community and business and industry. However there are issues and question marks over what it means and how it may be realized. More evidence of practice is required to understand the implications of adopting a mass customization strategy. This chapter presents five case studies from a range of sectors – bicycles, computer assembly, communications components, mobile phones and commercial vehicles – and analyzes their approaches to customization as well as their modes of operations. The type of the customization practiced by these different businesses is identified in terms of dimensionality (fit/size), hardware functionality, software functionality, properties of the whole product, grade, quality level, aesthetics and style, personalization, literature and packaging. All five businesses offer more than one type of customization. The implications of customizing different product attributes are discussed. The operational modes observed in the case studies are analyzed with respect to a typology of five modes of mass customization presented elsewhere. The reasons why different operational modes occur in different environments are speculated on. The chapter contributes to understanding both the potential for mass customization and the constraints under which real mass customizers may operate.

2.1 Introduction

This work presents an examination and discussion of five manufacturing enterprises in which customization is a significant component of their business strategy. The companies are from the consumer electronic, automotive, telecommunications and consumer products sectors and are either selling to other businesses or direct to consumers. All of the companies are customizing products for at least some of their customers. All of them are flexible to some extent in what they customize, when they fulfill customization orders and to where they deliver them, and how many they customize for a customer. They are all interested in the mass customization (MC) concept. However only one company, a computer assembler, thinks of itself as an all-out mass customizer and that is due to its similarities with competitors acknowledged as leaders in mass customization rather than by reference to objective quantitative criteria.

Mass customization has been defined and described at a high level (e.g. [1,2]). Certainly the concept of MC has been conveyed well, but descriptions of how mass customization is put into practice are infrequent, as has been noted by Da Silviera [3]. A starting point for understanding how it is implemented and a basis for the subsequent development of operations templates is the case study and this is the approach taken here [4]. The case studies are used to explore the factors that influence or dictate how mass customization is put into practice. A summary is given of the market and business environment of each company to aid the reader. However it is not an aim of this chapter to look at the factors that make mass customization an attractive strategy. Furthermore, the focus is on the customization of products, hence the customization of the selling process or of services that accompany a product are not examined.

Each case study follows the same format, with the market and business environment described first, followed by a summary of each company's customization strategy, a description of their product and lastly details of the operations to fulfill orders. The case studies are followed by an analysis and discussion of the type of product customization and the mode of operation observed in each of the five organizations. We also speculate why different environments lead a company to follow one mode rather than another.

2.2   Case study summaries

2.2.1   European bicycle company

Market and business environment: There is an increasing diversity in bicycle types and components. There are bicycles designed for comfort, for road racing, for off-road activities, for stunts. There are bicycles without suspension, with front suspension and those with front and rear suspension. They can have steel, aluminum or carbon fiber components. Component choice has grown, such as with brakes where there are several types of rim brakes and now there are disc brakes also as used on motorcycles. Furthermore there are many manufacturers vying to be the leading brand for one or more component types.

The number of brands has grown with many being imported into Europe from Southeast Asia as finished bicycles. The SE Asia region dominates component manufacture. The market for bicycles is stratified with the significance of styling features, componentry and the bicycle's properties differing across layers. In general, color and style are important for lower price segments, componentry becomes increasingly important as price rises, as does dimensional fit and properties of the bicycle (e.g. weight, durability) which are often key concerns at the upper end of the market. The majority of sales are through large retail chains or via small independent retailers where there has been a tradition for bicycles to be customized at the point of sale to a small degree, such as changing the saddle and adding accessories. Mail order companies are a smaller but significant channel. At present there is an appetite for differentiated products, especially from the mail order companies and larger retailers, who wish to distinguish themselves from other channels. The market is strongly seasonal with approaching a half of all sales coming in the pre-Christmas period.

Strategy: The company's strategy is to offer to resellers a high level of product diversity to cover the many market segments, and to provide a high level of service in terms of replenishing their stock. Consequently the company is constantly balancing component variety, inventory levels and service performance. It develops a catalogue of products for each segment that is revised annually, though launch periods differ across segments. The company is prepared to customize 'specials' if expected volumes are sufficient, resulting in customization being limited to larger buyers. No customization is undertaken for individual small retailers. It customizes a product for a period of time, not for a fixed volume. Once the product is specified and designed, that customer can order it as if it were a catalogue model – any time and in any volume. The model is not available to other customers.

Product: There are approaching 350 products divided across nearly 20 product families. Each bicycle is constructed from around 100 types of components. The number of bought out component lines changes each year, but is in the range of 2000 – 2500. The bicycle has evolved into a modular product. The frame is the defining component, but a bicycle is an ensemble of sub-systems with the three main systems being the frame and suspension, braking system and drive system. Other than for graphic transfers that are used as part of the color scheme, the company does not source new components for a special product. A typical customization involves selecting one of the standard catalogue products and changing the mix of components and the color/graphics. Short run specials provide an opportunity for consuming obsolete components.

Operations: Throughput flexibility is insufficient to tolerate seasonal demand undulations, hence the company operates a make-to-stock policy even with seasonal staff being brought in for peak periods. The extent to which temporary staff can be used is limited by many assembly tasks requiring skill and experience to achieve the quality at the fast production rate. Manufacturing is a batch process, with batches from 30 to 100. The process is divided into areas, including component warehouses, wheel assembly, frame and fork painting, final assembly, finished stock warehouse. A synchronous line is used for assembly with tasks of equal cycle time. Throughput is of the order of 500k units per year. There are two main barriers to lowering the batch size - the time to 'pick' components and the painting operation. It takes over an hour for a person to pick the components for a batch of 100, and only a little less time to pick for a smaller batch. For painting, the frames are hung on a moving line that passes through one of two painting booths. The time lost in changing color precludes small batches.

Initial inquiries for specials are not handled through the routine ordering mechanism, but via a Product Manager or a member of the sales team. Product development, sourcing, manufacture, warehousing and delivery of customized products are undertaken by the same resources as for catalogue products. Once the customized product is designed and made available to the customer it is assigned a product number and that customer can place orders for it just as for any other product from the catalogue. Sales of specials are forecast and purchasing and production planned accordingly, as for catalogue models. Order sizes are often in single units delivered to a specific outlet. Week on week variation in the demand for a product, whether a special or a catalogue item, tends to be high.

2.2.2   Computer assembler

Market and business environment: A 'sell then produce' business model has evolved in this sector therefore build to order is an important competency. A make-to-stock policy would run the risk of product obsolescence due to short component lifecycles. The market is very competitive and manufacturers are vying to offer the best specified machine at a set of common price points. Being quick to introduce latest components is important.

Strategy: This company is targeting specific selected sectors where the supply of IT services and software is as great or of greater importance than the supply of hardware, but where the ability to customize the computer offers advantages. For example, a significant proportion of customers appreciate receiving computers that are configured for their network with software pre-installed. Customers fall into several segments, which differ in terms of the services and service performance offered. In respect of the computer manufacturing division, this translates to different lead times per segment. There is strong seasonality in the sales with two high activity periods. The throughput difference from highest to lowest month can of the order of nine times. The assembly process has been designed to provide the necessary level of throughput flexibility by taking on seasonal staff as necessary. Assembly tasks are not complex and can be learnt quickly, with high quality achievable under vigilant supervision.

Product: A desktop computer is constructed from, typically, 20 to 30 components. In most cases customers select configurations of memory, processor type and so on to suit their budget. The other routine type of customization is the selection of components to conform to the customers existing systems such as choice of network card. Although a significant proportion of the computers assembled have unique configurations (20%) there is high commonality of components, as configurations can be different in one or two respects. Consequently the variation in demand of components is low even if there is high variety in the manufactured specifications. Shortfalls in components can be overcome by recourse to wholesalers. Changes to components are frequent, requiring a systematic product change process to log changes and prompt revalidation.

Operations: Order taking is direct via a number of methods – web, telephone, sales person. Integration of information systems enables availability of components to be checked at the time of ordering. Order size is from one to tens and upwards. Customers can change the specification up until the point of manufacture and can check on progress. The manufacturing schedule is extremely flexible. Orders are prioritized and the highest priority are skimmed from the system several times an hour. Routines within the planning software revise the priorities automatically with the aim of fulfilling each order within the promised delivery time. Assembly is organized in parallel mini-lines which feed a burn-in and software loading cell, and onto final testing, packaging and dispatch. Each product is given a unique job number and its progress is controlled by a central database, which uses the customer's specification to determine the BOM, software loading and testing regime. Components are picked for each machine in turn and handheld scanners are used to log each component picked, enabling the manufacturing control system to alarm when errors are made. In the packing and dispatch areas all other items such as keyboard, monitor, manuals etc are picked for the order. Staff are linked to the central database which navigates them to each item.

2.2.3 Communications component manufacturer

Market and business environment: This manufacturer is involved in a relatively new high technology market with implications for the rate of change of customer requirements, in particular, of product performance and quality as the product technology advances. The product range of the business consists of two main products with 70 products still in development. The market could be described as engineer to order because of the high variety of product.

Strategy: The aim of the business is to, where possible, increase the use of standard components by imposing its product standards on the customer. This policy of commonality extends to the engineering of products to suit application across product families. All products have complex process routings. Customization is offered based on process capabilities rather than a pre-defined product range. Often a customer approaches the supplier with a new design to suit their application and places an order for the manufacture of small batches of prototypes. Once the product is engineered the customer usually reorders to the same specification.

Product: The housing and interface is the feature most often customized due to there being few or no industry standards for these products. The housing makes up a high percentage of the product cost. Customers understand the product technology and collaborate in drawing up the specification. Design effort per product is high.

Operations: Manufacturing is organized as a main production line and a prototype line, both of which are a combination of automated and skilled manual assembly. All finished products are function tested. Because it is the housing that is customized, the internal components are common to a number of products, hence there is scope for postponing allocation and for reworking products if there is a large discrepancy between forecast and actual orders. Forecasting error is an issue as some component lead times are in the order of six months.

2.2.4   Mobile phone manufacturer

Market and business environment: Mobile phone handset manufacturing facilities serve global markets, with international supply chains and dispatch to customers worldwide. Consolidation in the mobile telecommunications market means network operators are also global and powerful. However, manufacturing competition is very strong with competitive factors being product customization and not least, production costs.

Strategy: This company has several manufacturing facilities, supplying network operators around the world. Each product is sold to many customers, and is manufactured on a make-to-order basis as far as possible, but any slack time is used to replenish a central warehouse with high demand models. It is imperative the manufacturing function is responsive to customers. At times this conflicts with the company's desire to have its products retain their brand identity.

Product: The product is the mobile handset, the box and its contents which includes manuals and, occasionally, promotional materials. The customizable attributes are: the body of the handset, which can be in single or dual colors; the flip down front on which a customer's logo can be printed (8 variants of color and logo combinations are in production); software of the handset (over 600 variants); the antenna (no customer has chosen to make this a distinct color from the body); labeling on the body (in accordance with product safety regulations); and the packaging (size of box and contents).

Operations: Orders are collated centrally and allocated to the manufacturing facility. Order size varies from a few (less than 10) to many thousands. Throughput at this facility is in the order of 6 million handsets per year. The body color is determined at the first stage of manufacture, as the body provides structural support to the internal electronics. The handset is assembled by a sequence of robots. A buffer of different colored semi-finished phones is held at the decoupling point. Up until this point the product is tracked by batch. To fulfill an order the semi-complete handsets are taken from the buffer to have the appropriate flip cover and antenna attached, software loaded and labels put on. At this point the handset is given its own identity. Finally the box and contents are assembled in readiness for the handset. Apart from software loading all of these tasks are manual. To protect against a new variant being problematic to manufacture its bill of materials is checked on paper and a test product may be sent through the facility. This tests the control systems as well as the processing of components. A request from a customer for a new software variant can be implemented in a couple of days. Other customizations depend on supplier responsiveness.

2.2.5   Commercial vehicle manufacturer

Market and business environment: Since the introduction of commercial vehicles the number of manufacturers and choice of products has grown greatly and competition is significant. It is the norm for a manufacturer to offer extensive choice over length, height and layout as well as over other factors such as engine capacity, decor, etc. For many customers this level of customization is sufficient. However, there are large numbers of body shops that customize vehicles in the aftermarket, with the capability to add a company's livery through to radical structural alterations. Some of these customizations can be extensive and involve installing special equipment, removing unwanted components, fitting internal racking, or putting a different body on a chassis. Examples include vehicles for emergency / recovery services, vehicles with articulated loading systems and vehicles for carrying livestock.

Strategy: As well as offering a wide range of standard choices, this company sees benefit from gaining a share of the special customizations being performed in the aftermarket. The fleet market is significant and customization capabilities can influence a fleet manager's buying decisions. Their requirements can be diverse and a pre-engineered range of options may be insufficient, hence the company has chosen to offer a specials customization service. It has formed close relationships with several body shops who collaborate in the design process as well as the production of specials. An advantage to a customer of having the OEM lead the customization is greater quality control and consistency in the product. Previously a fleet operator may have had to co-ordinate and manage several body shops, giving rise to different solutions of varying compatibility. The company prefers not to take on a 'special' unless there is an expectation of the customer purchasing a sufficient number of them to cover the engineering and mobilization costs. However it has no hard and fast rules and each request for a special customization is reviewed on its own merits and in light of the relationship with the customer.

Product: The standard range of the vehicle has a number of wheelbase lengths, heights, and drive-train options (e.g. fuel type, front or rear wheel drive). All special customizations start with these standard options. Customization can involve external livery, seating arrangements, modification of electrics, addition of aerials, special racking systems, addition of anchoring points, or the fitting of a unit onto the chassis. Expertise from the customer may be necessary, such as for fitting electronics and aerials. In some cases end-user needs are not straightforward to interpret and the first product is treated as a prototype. The engineering process involves cross checking for clashes between a modification and other options and the assessment is recorded in the product's master configurator. One of the difficulties of managing specials is that they can be affected by modifications or updates to standard features and hence can require revalidation.

Operations: Special customizations make up approximately 15% of production. The production is divided between a traditional automotive assembly line using synchronized track followed by finishing work at a body shop. A minority of specials involve additional tasks on the assembly line only, but the majority have additional tasks on the line as well as body shop work. To initiate a special product customers can approach dealers but in most cases they go straight to a dedicated liaison team at the OEM. Once a special has been engineered and added to the product catalogue a customer can order it in the same manner as ordering a standard option.

2.3   Customizable product attributes

The cases illustrate that products are customized in a variety of ways. There is no defined set of customized attributes in common usage, so to compare the case studies we have developed a taxonomy of ten attributes which view customization from the customer's perspective:

- Dimensional fit/size – adjusting or scaling (part of) the product to fit the customer's requirements;

- Hardware function – customizing functions by changing, adding or subtracting hardware features;

- Software function – altering the programming of the product;

- Property of the whole product – changing a product to be, for example, corrosion resistant, vibration tolerant or a low noise emitter;

- Grade – components are upgraded or downgraded for reasons such as cost, without (intentionally) altering the function or quality level of the product;

- Quality level – components are selected for performance reasons such as reliability requirements;

- Aesthetics and style – changing the shape or look of the product, such as by selection of décor for a vehicle;

- Personalization – for example, adding a corporate logo or altering the color (livery) of the product;

- Literature – documentation with the product is modified e.g. changing or adding manuals;

- Packaging – changing a product's packaging appearance or design, or inserting other items into the package.

The number of attributes that are customizable gives an immediate but approximate picture of the extent of customization offered by each case study company (Table 1).

2.3.1   Implications of customizable attribute

Often the customization of a product involves customizing a mix of attributes, e.g. tailoring a suit is a combination of dimensional, style and grade customization. Furthermore, one component may affect multiple attributes. For example, cashmere is perceived to be a high grade material for suits but is less hard wearing – a whole product property – than lower grade materials. Consequently, the mapping between customizable attributes and components is an issue that needs to be considered in the design of a customizable product.

Table 1:   Product attributes customized by each manufacturer

Manufacturer	Bicycle	Computer assembler	Communi-cation component	Mobile phone	Commer-cial vehicle
dimensional fit/size			✓		✓
hardware function		✓	✓		✓
software function		✓		✓	
property of the whole product			✓		✓
grade	✓	✓			✓
quality level					
style					✓
personalization	✓			✓	✓
literature			✓	✓	
packaging		✓		✓	

Not only is the link between customizable attributes and product architecture of interest, so to are the relationships between attributes and operations, in particular with inventory management and process control. For example, it is likely that grade customization involves a choice of raw materials or components, hence it is a form of customization likely to have impacts on inventory management. Function customization could dictate the production steps of a product, hence rarely selected functions would bring down the utilization of some production resources. The relationships between customizable attributes and operations is likely to be specific to a product, hence there may be no benefit in attempting to hypothesize universal rules. For a specific product the links between attribute and operations is fundamental in evaluating different customization strategies and in understanding trade-offs between the amount of customization offered and costs of fulfillment. Such an analysis can lead to a tempering or expansion of the customization envelope.

An appreciation of the implications of customizing attributes is given in Table 2 which presents the experience of the commercial vehicle manufacturer from its specials customization team. This shows the three most common motivations for requesting a special vehicle are hardware function, style and personalization. The first of these – hardware function – occurs even though there are many choices in the standard range, which demonstrates that customers have specific needs that are difficult to satisfy with generic options. Although a less frequent motivation, dimensional fit follows the same pattern, with special dimensional requirements being requested even though there are very many choices in the standard range. Unfortunately for the company, increasing the envelope of dimensional customization is a challenge in terms of engineering and production difficulties, with them rated as difficult and very difficult respectively.

Table 2:   Assessment of 'special' customization by attribute type (commercial vehicle manufacturer)

Attribute type	Customer motivation for requiring a 'special' customization[1]	Amount of choice standard range[2]	Amount of choice specials range[2]	Engineering difficulty in increasing customiz.[3]	Production difficulty in increasing customization[3]
Dimensional fit	occasionally	very many	some	difficult	very difficult
Hardware Function	often	many	many	average	average
Property of the whole product	occasionally	many	many	difficult	difficult
Grade	occasionally	low	some	average	average
Style	often	low	many	relatively easy	difficult
Personalization	often	none	some	relatively easy	average

1 Never, Rarely, Occasionally, Often, Always
2 None, Low, Some, Many, Very many
3 Relatively easy, Average, Difficult, Very difficult, Impossible

The operational feasibility of customizing an attribute is not the sole consideration when setting boundaries on the customization envelope. For example, how low a grade of product would an enterprise be prepared to produce that is compatible with its brand and reputation? In the mobile phone case study above, the company would have preferred not to have changed the packaging due to concerns over diluting the brand. Consequently companies may have self-imposed customization limits on all attributes.

2.4   Modes of operation

Comparison of the case studies highlights differences in how they fulfill customized orders. The computer assembler pre-engineers its products by testing component configurations before offering them for sale. Its order fulfillment resources are ready to source, assemble and deliver any configuration it offers.

The commercial vehicle and mobile phone manufacturers also pre-engineer a catalogue of options, but it is normal for them to engineer new configurations for customers. The mobile phone manufacturer will expand the color range of its handsets, alter the programming and change the packaging, including inserting promotional material that a customer has had produced elsewhere. Although reluctant to do so, they will alter their manufacturing resources if needed. Changing the manufacturing process and finding new suppliers is the norm for the commercial vehicle manufacturer in order to deliver the new options it engineers for its customers. The bicycle manufacturer is willing to engineer a customized product but only using components and manufacturing resources it has available.

Three modes of operations are apparent from contrasting these cases. Firstly there is the offering and manufacture of a large range of pre-engineered options (exemplified by the computer assembler). Second there is the mode of engineering one or more new features for a customer and adapting the order fulfillment resources accordingly (e.g. the commercial vehicle manufacturer). Thirdly there is the mode of engineering a product for a customer that can be manufactured and delivered using existing order fulfillment resources (as practiced by the bicycle manufacturer). Further reflection of the case studies uncovers another factor that differentiates customizing enterprises, which is whether an enterprise requires (an expectation of) repeat orders for it to be economic to customize, or whether it can customize a product that is produced once only. This is the case with both the commercial vehicle manufacturer and the bicycle company. The mobile phone manufacturer claims its systems could accept one-off customizations but with their customers being national or international networks this situation has not arisen.

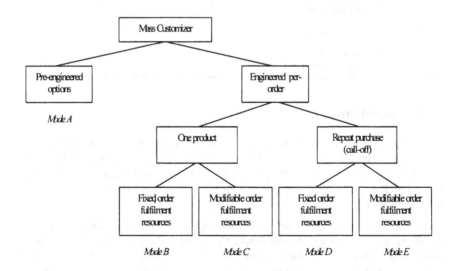

Figure 1:   Mass customization modes of operation

The three distinguishing characteristics lead to a set of five modes, A to E, as shown in Figure 1 (discussed in detail in [5]). Mode A is the operations model most often associated with mass customization, but it should not be thought of as the only legitimate model. Modes B and C can be thought of as the models that companies from the engineer-to-order sector move towards, and Modes D and E are the models that mass producers might shift into.

Two of the case study companies operate in more than one mode (Table 3). Both the commercial vehicle manufacturer and the mobile phone manufacturer operate in Mode A, with the former also engineering products for customers and adapting the order fulfillment processes when sufficient repeat orders are expected (Mode E), and the latter preferring to operate in Mode D but having to follow (occasionally) Mode E due to customer power. Only for Modes B and C is it a requirement that customers can order a single product. For the other modes the enterprise may set a minimum order size. None of the case study companies has a minimum order size. It is in terms of the eventual number of orders for a product that the bicycle company and commercial vehicle manufacturer evaluate each request for a special customization.

Table 3: Manufacturer by mode

Manufacturer / Mode	A	B	C	D	E
European bicycle company				✓	
Computer assembler	✓				
Communications component manufacturer		✓		✓	
Mobile phone manufacturer	✓			✓	(✓)
Commercial vehicle manufacturer	✓				✓

2.4.1 Modal drivers

Given that a company has chosen to pursue a mass customization strategy and given that its customers value the benefits of customization, which mode should the company follow? From observation and analysis of the cases we speculate there are five factors that influence the choice of mode, two external and three internal factors:

(i) The strength of customer desire for differentiation from other customers. Resellers are an example of customers that value differentiation. They seek to offer unique advantages to their customers and one approach is to offer distinctive and unique versions of products, or to promote the product in a unique manner. Consequently, to get unique products they can push customizers to go beyond the boundaries of their 'standard' customization options.

(ii) Customer power. Clearly the more powerful the customer the greater the pressure they can exert to reduce their compromise in a purchase.

(iii) Cost of engineering and mobilizing an additional option. Low costs are achieved either by efficiencies in engineering and mobilization of order fulfillment resources (including taking on new suppliers or procuring additional equipment), or due to re-usability of designs and processes such that the development cost can be shared across customizations.

(iv) Implications of a new option / variant on components. If further variants of a product are, in effect, unique permutations of a 'standard' set of components, a new variant may have little impact on sourcing arrangements and costs. If new variants require many new materials or parts that need to be sourced from new suppliers, the impact (and cost) can be high.

(v) Implications of a new option / variant on the process. As with the impact on ingredients, if a new variant requires new or modified manufacturing or delivery processes, the implications are higher costs.

These qualitative ratings are an estimation of the present customization offered by the companies and their present customers and circumstances. A rating of low against all five factors makes Mode A feasible and this is the case for the computer assembler which follows Mode A only. It can be expected that a company with high ratings for factors (i) or (ii) above will have to engineer products for customers rather than dictate options to them. If the ratings for factors (iii), (iv) or (v) are also high it can be expected that the company requires repeat orders and will be in Modes D or E. This is true of the bicycle producer, communications company and commercial vehicle manufacturer. The bicycle company is high for factor (iv) because of the low responsiveness of suppliers, resulting in inventory increasing to cope with a new customization.

The circumstances of the commercial vehicle manufacturer are such that it is unwise for it to restrict itself to Mode A. The customization of vehicles for large fleet customers represents a significant proportion of its business, but the customizations are not the sort that can be foreseen nor are many of them reusable for other customers. Fleets buyers are powerful and their needs are uncertain to the extent that it is common for the first special vehicle to be treated as prototype with improvements to the specification established through user trials. To attempt to predict the need of these customers would risk considerable engineering and process development costs. Even though the mobile phone manufacturer rates low for factors (iii), (iv) and (v), they at no time operate in modes B and C as might be expected. They themselves say their systems could handle one-off orders, but due to the size of their customers this situation does not arise. The communications company has to manufacture one-offs of customized products (Mode B) due to the power of its customers but the cost of engineering these products is high, hence this analysis suggests the company is facing a dilemma.

These modal drivers are not static and a change in one or more could force a customizer to adopt a different mode or change strategy. A shift in driver can be enforced on a customizer or self determined, as illustrated by the commercial vehicle manufacturer who chose to take on the aftermarket customizers and as a consequence could not follow Mode A alone but had to adopt Mode C or E and chose the latter.

Table 4:   Ratings of mode drivers for each manufacturer

Factor / Manufacturer	Bicycle	Computer assembler	Communi-cations components	Mobile phone	Com-mercial vehicle
i) Customers need differentiation from other customers	High	Low	Low	High	Low
ii) Customer power	High	Low	High	High	High
iii) Cost of engineering / mobilizing an additional option / variant	Low	Low	High	Low	High
iv) Implications of a new option / variant on ingredients (raw materials or components)	High	Low	Low	Low	High
v) Implications of a new option / variant on the process	Low	Low	High	Low	High

2.5   Conclusions

Case studies of manufacturers that exhibit mass customization characteristics and tendencies are useful sources for insights into mass customization. These five companies have allowed a set of operations modes to be constructed and have permitted speculation over the drivers of these modes. If the speculations have foundation, an emerging conclusion from this work is that no single operational template can be held up as the ideal mass customization model. Market and business environment factors and internal operational factors play a part in influencing the mode a mass customizer can (and should) follow. Difficulties are likely where there is a conflict between the appropriate mode and internal factors.

References

[1]   Pine II, B.J.: Mass Customization: The new frontier in business competition, Boston 1993.

[2]   Hart, C.W.L.: Mass customization: conceptual underpinnings, opportunities and limits, in: International Journal of Service Operations, 6 (1995), pp. 36-46.

[3]   Da Silveira, G.; Borenstein, D.; Fogliatto, F.S.: Mass customization: Literature review and research directions, in: International Journal of Production Economics, 72 (2001), pp. 1-13.

[4]   Yin R.: Case study research: design and methods, New York, 1994.

[5]   MacCarthy, B.; Brabazon P.G.; Bramham J.: Fundamental modes of operation for mass customization, in: Paper submitted International Journal of Production Economics, 2002.

Acknowledgements: We would like to acknowledge EPSRC (project GR / N11742 / 01) for their support of this work. In addition we would like to record our thanks to our consortium members and our colleagues at Oxford University.

Contact:

Professor Bart MacCarthy
Mass Customization Research Centre, School of Mechanical, Materials, Manufacturing Engineering & Management, Nottingham University, UK
E-mail: bart.maccarthy@nottingham.ac.uk

3 The Many Faces of Personalization

An integrative economic overview of mass customization and personalization

Kai Riemer and Carsten Totz
Institute of Information Systems, Muenster University, Germany

The emergence of internet technology results in manifold opportunities of cost-effective one-to-one relationships with customers. It is intended to provide customer oriented information and products etc. in an individualized one-to-one manner. This chapter will give a conceptual overview of the personalization concept and will discuss how mass customization (product personalization) can be useful accompanied by other personalization activities, e.g. personalization of communication and customer interaction. Therefore the concept of personalization is integrated into the online marketing mix. The marketing mix discussion leads to a personalization performance system which shows the potential objects of web personalization activities from a customer's point of view giving a guideline for planning personalization activities. The model consists of the three main layers product & services, website and communication. This chapter will also provide a definition of concepts and an economic motivation of personalization and mass customization. Doing so, we want to integrate the marketing view of personalization and mass customization with the visualization of the personalization performance system.

3.1 Introduction

The emergence of internet technology results in manifold opportunities of cost-effective one-to-one relationships with customers. E-commerce managers increasingly realize, that the standard distribution of products in online-shops leads to a strong competition among e-commerce businesses due to the internet's price transparency. If the price is the only differentiating aspect of an e-commerce business, customer loyalty and margins will be low. As one reaction to this dilemma, it is intended to better meet the customers' needs and expectations to increase satisfaction and thus increase customer loyalty [1]. In this context personalization is mentioned to be one valuable solution, as Riecken states:

"Personalization is about building customer loyalty by building meaningful one-to-one relationship; by understanding the needs of each individual and helping satisfy a goal that efficiently and knowledgeably addresses each individual need in a given context" [2].

Furthermore the individualization of products and services decreases product comparability and therefore leads to an increased differentiation from competitors facilitating a unique positioning within the market space. Personalization (and individualization, which we use synonymously) in general refers to matching one object's nature with one subject's needs. More precisely it means to customize products, services, content, communication etc. to the needs of single customers or customer groups. The special economic potential derives from the use of modern information technology concepts (esp. the internet) to realize personalization for masses of customers at the cost of a standardized approach. In this case, mass customization may be seen as the individualization of products (and services [Piller, Meier]), at the cost of one-size-fits-all [3]. In the following we will concentrate on web-based personalization and mass customization.

Personalization is based on the knowledge of customer needs. And different ways of gathering this knowledge exist. For a differentiated discussion of the personalization concept it is important to distinguish between the objects of personalization (products, services etc.) and the methods of gathering data and applying personalization. Personalization thus may be realized in different ways. The relevant data can be explicitly gathered from the customer in a configuration process or implicitly and more or less transparent to the customer by a software system. In the second case the system gathers and extracts data and adapts itself to it's users [4]. In this case the personalization is based on so called customer profiles. Algorithms learn how to automatically customize certain objects (content, look & feel, product selections etc.) to the customer's personal characteristics. Hirsh et al. call these systems "self-customizing software" [5]. Keeping in mind the differentiation between explicit configuration by the customer and the implicit self-customizing software systems, we will in the following concentrate on different personalization potentials and objects.

First of all, an economic motivation gives the introduction to the following integrative treatment of mass customization, personalization and online marketing which finally leads to an integrated personalization performance system, which embeds mass customization into the centre of a greater context of various other individualization opportunities.

3.2   Economic motivation of personalization

With the following considerations we will try to give an answer to the economic key questions of personalization: Which economic goals motivate the application of personalization concepts? What are the drivers of personalization success? Where and how approaches personalization economic success?

Beneath traditional ratios of economical success as for instance turnover, profit or gross margin, the defection ratio, the length of customer relationships or even the customer-lifetime-value ascended as effective marketing and management indices [6]. We do not intend to mirror the discussion on the influence of extended customer relationships on the economic situation of businesses (e.g. increased cross- and up-selling ratios, reduced acquisition costs on average etc.), but try to point out the effect of personalization on customer retention. Customer retention can be the result of technological (e.g. incompatibility of technological systems: Windows/Intel vs. Apple/PowerPC), contractual (e.g. contract period: 12 or 24 month cell phones contracts) or psychological obligations (e.g. brand preference caused by satisfaction etc.), which are lock-in effects caused by switching costs. Switching costs can be defined from the customer perspective as any kind of costs associated with the migration to a new supplier, vendor or service provider [7]. Although the establishment of contractual and especially technological lock-in situations can be a major source of competitive advantage, we will focus on the personalization-driven creation of psychological obligations. *Plinke* divides customer-relationship-specific switching costs into direct switching costs, opportunity and sunk costs [8].

3.2.1 Direct switching costs

Costs for searching new suppliers, vendors or service providers, for initiation, negotiation and arrangement of new relationships and potentially required investments are referred to as direct switching costs. Migration implies the identification and comparison of products or services regarding their suitability to serve customer needs. The personalization or individualization of products or services to specific customer requirements lowers their competitive comparability. The assessment of product or service attributes and their effects on sales price becomes complicated and time consuming. Furthermore, addressing the customer in a personal manner increases the familiarity within a customer relationship [9]. As a result of personalization initiatives, not only the comparability of products or services is decreased, but also emotional obligations are increased. The search for new suppliers, vendors or service providers with the promise to outperform a current customer relationship is made extremely difficult – challenging, time intensive, uncertain and therefore costly.

3.2.2 Opportunity costs

The uncertainty of potential searching activities leads to a second category of switching costs. Opportunity costs might be characterized in this context as the unrealized benefit of an alternative dedication of resources and the potential loss of net-benefit of the current relationship in case of migration. The personalization of products or services to customer-specific requirements increases their customer-specific net-benefit. This boosts the probability to lose the advantageous effects of a current relationship from the customer's perspective. The advantages of a well-trained self customizing system for example, which is based on detailed profile data gathered throughout the customer relationship with the company, will be lost for the customer by switching to another business. Furthermore, direct switching costs in terms of uncertainty and information seeking rise additionally.

3.2.3   Sunk costs

Although irrelevant to rational investment decisions, sunk costs are considered valid from the customer point-of-view. We refer to sunk costs in the context of switching costs and personalization as irreversible predetermined (customer) investments, intended to ensure and back up the success of the customer relationship. Not accepting a certain loss of value or benefit, these investments cannot be applied to a different deployment. Focusing on emotional or psychological lock-in situations, the adoption of product- or service-specific skills, as for instance the configuration of professional unified messaging services, mobile phones or favored websites, can be considered to be customer investments into a relationship. As personalization and configuration tasks require detailed information on customer needs, preferences and utilization purposes fundamentally, the customer often himself can carry them out best (explicit configuration, perhaps supported by decision support systems etc.). The time, intellectual and monetary effort a customer spends on the adoption of relationship-specific skills can be referred to as investments. Because in case of switching the customer would lose his investments and have to go again through the procedure of supplying information for product individualization [10]. And with an increasing complexity of product- or service-specific configuration tasks, the comparability of configuration processes decreases and customer investments turn out to be more specific and less reversible. It seems comprehensible, that the customer willingness towards investments into a new relationship decreases with an increasing level of sunk costs [11].

In the same manner the customer loses investments into a well-trained self-customizing web-system based on automatically gathered customer profiles. An Amazon.com customer for example, who uses and benefits from the recommendations of a recommender system based on a long history of transactions and click-streams, loses his transaction and click-stream investments by switching to another online bookstore (sunk costs). The example also shows that for this customer in this situation no equivalent other service in the market exists. If he switches, he loses the valuable recommender feature (opportunity costs).

3.2.4   Recapitulation of personalization economics

To summarize the influence of personalization on switching costs (see Figure 1):
- The individualization of product or service attributes and parameters to customer-specific needs and requirements holds the potential to increase customer satisfaction.

- Direct switching costs rise due to established trust towards the supplier, vendor or service provider and its capabilities to meet promised quality levels. Finding alternatives is made difficult by specificity of products or services.

- If customer satisfaction is positively influenced by personalization, then uncertainty and opportunity costs rise as a defecting customer risks to lose the net-benefits of the current relationship.

- Customer driven configuration or personalization tasks (like configuration of mass customized products) require relationship-specific investments of the customers. Can customers be persuaded to invest significantly into a specific relationship sunk costs are increased.

- Thus, personalization and mass customization might drive customer satisfaction, trust and investment specificity and increases switching costs – preventing customers from defecting.

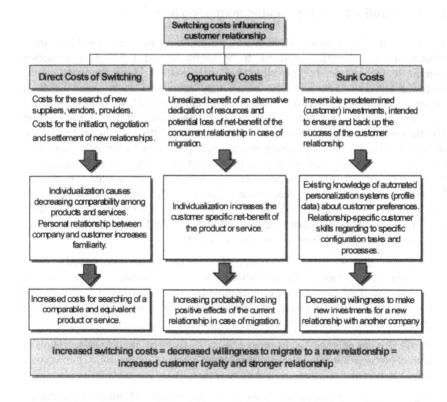

Figure 1: The influence of different types of switching costs on customer retention

3.3   Mass customization and the online marketing mix

The primary goal of this chapter is to give a conceptual overview of the personalization concept and to show why and in which way mass customization (product individualization) has to or may be usefully accompanied by other personalization activities. Therefore, we introduce the online marketing mix with its five fields of action, as an entire concept for planning e-commerce actions. Based on the online marketing mix we will point out significant implications of mass customization for the five mix areas and show further personalization potentials and opportunities to complement mass customization to an integrated and widespread personalization approach. A consolidation of the therewith identified personalization opportunities leads to the already mentioned personalization performance system, which will be introduced in the next paragraph.

3.3.1   Introduction of the online marketing mix

Starting from the classic 4-P-marketing mix with the areas product, price, placement and promotion, it seems to be useful to add a fifth field of action specific to the internet context which we call 'process'. In dependence on the extended services marketing mix, which may consist of up to seven fields of action, adding process, personnel and physical facilities [12], we have derived the five field marketing mix, because personnel and physical facilities are not relevant for the virtual sales channel internet. Process as the fifth segment addresses the online specific issues of planning and modeling customer interaction and web navigation processes. Because of a lack of direct customer contact the design of web interaction processes and interfaces is crucial for the success on internet businesses. Due to this, the field process subsumes all interaction relevant tasks and concepts to ensure frictionless customer web transactions. As a short introduction we will introduces the main concepts in the five marketing mix areas (for contents of the first four mix segments cp. [13]):

- *Product:* The core task is to identify the (internet specific) core products and services, complemented by additional services. In the web context, website content and features are part of the companies' online performance system and have ergo also to be taken into consideration. The mass customization of products is originated in this field of action.

- *Price:* Pricing of goods and services, price differentiation and the selection of web specific pricing mechanisms (auctions, demand aggregation, name-your-price etc.) are the core tasks of the price area.

- *Placement:* In the placement domain we can subsume all measures to ensure that products and services reach intended customer groups. Consequently, the placement of products and services in the online channel (selection of intermediaries), the coordination of various distribution channels (multi-channel-management, click-and-mortar approaches) and the design of fulfillment and logistics are part of this field of action.

- *Promotion:* We can divide this segment into mass communication concepts for advertising products and services and into dialogue communication with single customers. The first part is characterized by the selection of suitable online advertising instruments like banners, spots, sponsoring etc. The second part deals with (one-to-one) communication with customers within the scope of continuous information (newsletters), after sales services and consultancy, complaint management and other specialized interaction.

- *Process:* To ensure frictionless functioning of costumer web interaction processes a conscientious web design has to be realized. Two types of processes have to be taken into account: back-end processes, like order processing, stock and transport processes, billing, communication, e.g. in call-centers etc. and front-end processes like website navigation, special configuration or the web order processes.

First, we will identify mass customization implications for other measures in the five mix segments and then comment upon additional personalization opportunities in the second step.

3.3.2   Implications of the mass customization concept

Implications for the product and service portfolio

Mass customization has its origins in the marketing mix area product. The creation of mass customization ideas and the identification of suitable mass customizable products is part of the configuration of the (online) product and services portfolio. Within these actions, the application of mass customization implies the identification of product components, standardized modules, permissible product configurations and efficient production processes [10]. Especially the standardization of product components is seen as a critical task in designing cost efficient mass customization actions [14]. Additionally, the type of mass customization approach and the extent to which a product is customizable have to be defined. Gilmore and Pine distinguish four different approaches which differ in the extend of product and product representation changes (for further details see [15]). Besides, supporting service processes have to be adapted to the needs of mass customized products and service employees have to be instructed in the handling of modular and configurable products.

Implications for pricing

Regarding the pricing, mass customization is an appropriate concept to reduce or avoid the frequently discussed transparency of prices on the web (see exemplary [16]). With the individualization of products and services their specifity increases. This decreasing comparability offers new opportunities for one-to-one price differentiation, so that mass customizers may set prices for customized without risking a perception of this discrimination by other customers. Valued customers may be rewarded without rejecting others by creating an apparent inequality among customers.

Beneath the set of positive effects, mass customization may also have negative implications for pricing and price transparency. Mass customization may lead to an external transparency of component prices and beyond it to a certain pricing complexity. In general, two possibilities for the pricing of configurable products exist: first, the individualized products may not differ in many aspects, which means that the resulting products are more or less homogeneous with some cosmetic changes in colors and forms (e.g. the NikeID shoe [17]). In this case one ore few standard prices exist and pricing complexity is low. But if the configuration allows a high heterogeneity among the resulting products (e.g. computers at Dell [18]) the resulting price has to be calculated online from component prices. With the need to put prices online, a mass customizing firm has to decide to which extent it wants to make component and single feature prices transparent to customers and competitors. The risk is to become comparable in component prices to other businesses especially in standard components where only a comparability in product prices existed before implementing mass customization actions. In addition to this, single component prices may be irritating and therefore not understood and accepted by customers. From the customer's point of view, some component prices may seem to be too high compared to their benefits, which might result from complex production processes, that are not transparent to the customer. While they are not visible to customers in the case of non configurable products, their visibility in mass customization actions may lead to a lack of acceptance and avoid customers from selecting these components or from configuring and buying the product. In this case a careful calculation and communication of prices is necessary and tests with customers should help to calibrate them.

Implications for placement and channel management

The placement of mass customization actions in different channels leads to a much higher coordination complexity than the placement of standard products. This applies for example to a producer who wants to place mass customization in the market using sellers as intermediaries. Especially if a physical component is necessary to provide the solution, the producer depends on the distribution system of a seller for example to collect the relevant customer data. An example is the mass customization of shoes, where a 3D scanning solution is used to gather the necessary customer data. With the gathered data, the customer may in a second step configure several pairs of shoes via internet or by visiting the seller. This multi-channel approach provides best benefit of both worlds to the customer but requires a brief coordination of measures in the internet and the physical channel with the seller regarding pricing, assortment and communication. The complexity will further rise if an integration of the internet solution with the distributor's website has to be done. An example for a 3D scanning solution is the new mass customizing service for sports shoes by Adidas in Germany (mi adidas [19]) which requires a physical scan of the customer's feet as an initial action. Adidas has to coordinate a cooperation with several distributors, where a special 3D-scanning solution will be established to collect the relevant customer data for the individualization of the product – a novel sports shoe, which will be well-fitting to the individual form of the customer's feet. Another aspect of channel management in mass customization actions is the coordination of online configuration and communication and the offline after sales communication and services especially in the case of a decentralized sales structure with several outlets. The offline repair and maintenance service has to be trained for being able to deal with each possible product configuration.

Implications for promotion and communication

The promotion of mass customization activities differs from advertising standard products regarding the product complexity and the special role the customer plays by self-configuring the product. This leads to an increased need for information rich campaigns explaining the new prosumer (to *pro*duce and con*sume* [20]) role to the customer and giving him the right support to carry out the necessary configurations. Furthermore, dialogue communication with customers has to be adjusted to the one-to-one relationship which the company sets up with the customer by providing individualized products. A customer calling back with a question about his individualized product has to be known by the service person (e.g. in the call center), who has to be familiar with the details of the individualized product. Mass customization raises the customer expectations of being treated as an individual by the company. An anonymous and hasty follow-up interaction would cause irritation and a perception of inconsistency. The company has to meet this individual expectations as good as possible in all areas of interaction with the customer. Based on this, one-to-one communication and product individualization may be seen as two facets of the same comprehensive customer relationship approach.

Implications for process design and web interaction

Process and web front-end design is crucial for mass customization success, as the web interface is the initial point-of-contact in the mass customization service value chain and therefore a critical success factor for mass customization activities. From the customer's perspective the web-based configuration sequence is a critical part of the overall mass customization service process and it shapes its overall quality perception dramatically. Customer satisfaction is not only related to the quality of the customized product or service itself but as well to the quality of the web-based configuration process and interface, which essentially determine the customer's motivation and capability to adopt required configuration tasks and to finally purchase the customized product.

We have shown implications of mass customization and identified special needs for coordination and further personalization activities (e.g. individualized communication). Beyond this, additional opportunities for individualization exist to complete the above mentioned comprehensive one-to-one relationship approach. They will be illustrated in the following paragraph, once more based on the structuring concept of the online marketing mix. Common to all individualization instruments is the acknowledgement of the customer as an individual with individual preferences, interests and individual (buying) behavior etc.

3.3.3 Additional individualization opportunities

Product, service and content individualization: Apart from the individualization of (core) products, other product and related service objects may be individualized. Nearly all components of the web performance system mentioned before may be object of personalization activities. First of all services may be individualized as well as (physical) products. To identify additional services, it is appropriate to distinguish primary and secondary services, regarding their role in the perform-ance system. Primary or core services are the main deliverables of a service company, whereas secondary services are additional support services like maintenance, consultancy etc. [21]. The online flower-shop of 1-800-flowers.com for example provides a configurable online birthday reminder as an additional service [22]. Furthermore, another product related individualization can be identified: individual product recommendations and the individualization of product bundles for single customers. These measures are often based on profile-driven automated systems which analyze shopping history of the related or even other customers to provide suggestions on products or product bundling (for further details see [23] and Amazon.com example [24]). Another personalization object is content. Based on automated systems or explicit customer configuration websites may provide individual content modules for customers. In this case the separation from the individualization of interaction and navigation is difficult. In most cases content individualization on websites means to simply change the order of presentation, e.g. to configure an individual starting page with favorite content blocks (see the case of Yahoo! in [25]). In fact only the navigation structure is changed, which means for the website provider that he does not have to really change content but to rearrange it.

Price individualization: Cortese and Stepanek have used 'Good bye to fixed pricing' as headline for their Business Week Special on Electronic Commerce [26]. Their message is that consumers will increasingly be confronted with models for flexible, individualized pricing and product differentiation on the web. The individualization of prices for single customers usually is based on a customer value approach and has been discussed under the label of Weblining [27]. Weblining, a combination of the Web and redlining, (i.e., the differentiation of offers based on customer profiles) denotes a pricing strategy, which automatically differentiates Web-based offerings. It encompasses two elements: Based on detailed information about the customers – mostly transaction profiles –, companies evaluate and forecast the individual customer's (present and/ or future) value (micro-segmentation). The customer's value is then used on the web to automatically differentiate a company's offerings regarding to prices. As soon as a customer identifies him- or herself, the presentation of the Web offerings will be adapted according to the company's policy. Another measure for an indirect price individualization is the implementation of loyalty programs like Lufthansa's "miles-and-more" program [28] or the Beenz initiative [29]: customers get individual reimbursements or incentives for an extensive usage of a service or for creating high individual turnovers.

Individualization of logistics options (placement): The placement area of the marketing mix contains only little individualization potential. Whereas the placement of products in the sales channel and the coordination of channels have little individualization potential, the design of the logistics and fulfillment system offers some more individualization opportunities. The intention to best meet the customer's wishes regarding to products and services culminates in the individualization of logistics and transport parameters. The customer may choose from several logistics options like payment, packaging and transport. Real individualization is about individual packaging, e.g. gift wrapping and individual greetings to enclose to the package, or the determination of individual delivery times. Especially the last aspect provides a real benefit for customers by respecting their individual time schedules.

Individualized one-to-one communication: The individualization of promotion and communication has been widely discussed in the context of one-to-one marketing and customer relationship management. Because of our intention to give an integrative overview, we won't intensify the discussion of single one-to-one measures (for further details exemplarily see the work of Peppers and Rogers [30]). Nevertheless, some fundamental argumentation is of interest. Because of the distance between company and customer and the computer mediated nature of interaction, internet based interaction is to a certain extent defected in social and personal matters: website interaction normally seems to be anonymous and distanced [31]. One-to-one communication tries to identify each customer as a single individual, to set-up a dialogue communication and to personally address each customer. In fact this individualized communication is often machine based and pursued automatically. Yet, it helps to overcome the perceived anonymity and to give the customer a feeling of being personally attended to. Furthermore, one-to-one communication implies interactions with the customer over various communication channels. Therefore, it is necessary to coordinate the interaction at several customer touch points, which may be supported by suitable customer databases and special customer relationship management systems.

Individualization of website interaction: The individualization of navigation processes and interaction allows the customer to adjust website appearance and interaction to his preferences and needs. Many websites provide special configurable website sections like Yahoo!'s myYahoo!. MyYahoo! is a customized personal copy of Yahoo!, where the user may select content modules from a list. It provides users with the latest information on every subject, but with only the specific items they want to know about [25]. The configuration ergo allows the customization of content, layout and navigational order.

Using the online marketing mix for structuring, we have shown several implications of mass customization for online marketing and derived further individualization needs. In a second step we have identified additional personalization opportunities to usefully support or accompany (core) product or service individualization. Before we consolidate these findings and present the integrated web performance system, Table 1 gives a short summary.

Table 1: Summary of individualization aspects within the online marketing mix

Marketing mix element	*Implications of mass customization*	*Additional individualization opportunities*
1. Product	Identification of components, standardized modules, permissible product configurations and efficient production processes; select mass customization approach; set-up specific supporting processes	service individualization content individualization individualization of offers and product bundles
2. Price	decreasing product comparability leads to greater scope for price differentiation (new) transparency of single component prices may irritate customers	customer specific (value based) pricing (weblining) loyalty programs
3. Placement	coordination of multi-channel effects coordination of intermediaries when placing mass customization actions in the sales channel	individualization of logistics options (packaging, delivery time etc.)
4. Promotion	communication of new prosumer role to the customer; customer demand for one-to-one communication	one-to-one communication addressing customers personally
5. Process	web design and interaction quality crucial for web-based configuration processes	individualization of website look & feel individualized navigation and interaction processes

3.4 Personalization performance system

The personalization performance system integrates the above mentioned findings on individualization from the customer's point of view, shows the potential objects of web personalization activities and gives a guideline for planning personalization activities. From an external customer point of view three main individualization related performance layers may be distinguished: (1) products and services as the centre of the performance system, (2) the website as the relevant mediating channel and (3) the communication with the company. The model is based on the performance system by Belz and has been adapted with respect to individualization activities [32]. Belz places the product in the centre of his performance system and completes it with (added) services, consultancy and a relationship layer representing interaction and communication with the customer. We adjusted this system considering the important role of the website and the special nature of web-based mass customization and individualization and reassessed interaction and communication aspects by establishing two additional layers: a website and a communication layer.

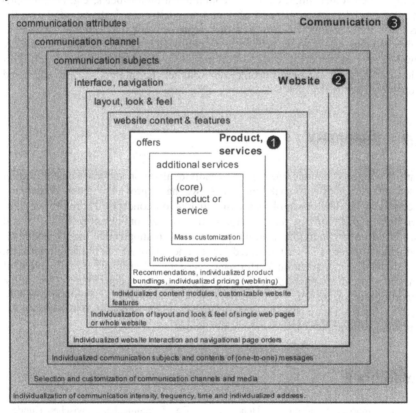

Figure 2: The personalization performance system

In a second step we have sub-divided each layer into three sublayers to structure the several aspects of individualization within the layers. Figure 2 shows the resulting nine layer model. The product layer contains the individualization of core products and services, of supporting services and product offers, bundles and recommendations. Price individualization has been subsumed under the individualization of offers. Website individualization starts with individualized content and information. Whereas content belongs to the product segment of the marketing mix from the company's point of view, it is regarded as an integral part of the website from the customer perspective. Besides this, the individualization of website layout and look & feel as well as of navigation and interaction processes completes the website layer.

The individualization of communication starts with the individualization of communication subjects, for example the selection of newsletter subjects and contents. The selection of preferred communication media or channels (website, e-mail, telephone, pager, short messages (SMS) etc.) belongs to the second part. Finally the customer might configure communication attributes, e.g. an appropriate intensity. The personalization of e-mail newsletters for instance, may give the customer the opportunity to select the frequency and the time of newsletter deliveries (e.g. only one newsletter a week delivered at the weekend versus a daily newsletter). Furthermore, the way of addressing the customer might be configurable: Letting the customer decide on whether he prefers to be addressed anonymously or personally.

3.5   Summary

Starting with an economic motivation of individualization and mass customization based on the theory of switching costs, we have provided a conceptual and integrative overview of web-based individualization concepts. Therefore we have identified implications of mass customization on the online marketing mix and introduced further individualization opportunities to supplement mass customization actions in terms of an integrated one-to-one customer relationship approach. The derived personalization performance system consists of several layers of possible individualization actions from the customer perspective. The online marketing mix discussion – summarized in Table 1 above – and the performance system (Figure 2) may provide useful support for e-commerce related businesses in planning and implementing one-to-one actions.

References

[1]   Homburg, C.; Giering, A.; Hentschel, F.: Der Zusammenhang zwischen Kunden-
        zufriedenheit und Kundenbindung, in: Bruhn, M.; Homburg, C. (Eds.): Handbuch
        Kundenbindungsmanagement, Wiesbaden 1999, pp. 81-112.

[2]   Riecken, D.: Personalized Views of Personalization, in: Communications of the
        ACM, 43 (2000) 8, pp. 27-28.

[3]   Gilmore, J.H.; Pine II, B.J.: Introduction: Costumization That Counts, in: Gilmore,
        J.H.; Pine II, B.J. (Eds.): Markets of One: Creating Customer-Unique Value
        through Mass Customization, Boston 2000, pp.vii-xxv.

[4]   Göker, M.H.; Smyth, B.: Delivering Personalized Information: What you get is
        what you want, in: Künstliche Intelligenz, 9 (2001) 1, pp. 17-21.

[5]   Hirsh, H.; Baus, C.; Davidson, B.: Learning to Personalize, in: Communications of the ACM, 43 (2000) 8, pp. 102-106.

[6]   Reichheld, F.F.; Sasser, W.E.: Zero Defections. Quality comes to Services, in: Harvard Business Review, 68 (1990) 5, pp. 105-111.

[7]   Jackson, B.B.: Build Customer Relationships that last, in: Harvard Business Review, 63 (1985) 6, pp. 120-128.

[8]   Plinke, W.: Grundlagen des Geschäftsbeziehungsmanagements, in: Kleinalten-kamp, M.; Plinke, W. (Eds.): Geschäftsbeziehungsmanagement, Berlin 1997, pp. 1-62.

[9]   Plötner, O.: Das Vertrauen des Kunden: Relevanz, Aufbau und Steuerung auf industriellen Märkten, Wiesbaden 1995.

[10]   Reichwald, R.; Piller, F.T.: Mass Customization, in: Possel-Doelken, F.; Zheng, L. (Eds.): Strategic Production Networks, Berlin / New York 2002, pp. 389-421.

[12]   Meffert, H.; Bruhn, M.: Dienstleistungsmarketing, 2nd Ed., Wiesbaden 1997.

[13]   Meffert, H.: Marketing – Grundlagen marktorientierter Unternehmensführung, 8th Ed., Wiesbaden 1998.

[14]   Svensson, C.: A discussion of future challenges to "built to order" SME's, in: Mass customization: A threat or a challenge?, 4th SMESME International Conference 2001.

[15]   Gilmore, J.H.; Pine II, B.J.: The Four Faces of Mass Customization, in: Gilmore, J.H.; Pine II, B.J. (Eds.): Markets of One – Creating Customer-Unique Value through Mass Customization, 2000, pp.115-132.

[16]   Stanley, M.; Witter, D.: The Internet and Financial Services, on www.msdw.com/ techresearch/financeser/internet_financial_servicespart1.pdf

[17]   Nike Inc.: on www.nikeid.com.

[18]   Dell Inc.: on www.dell.com.

[19]   My Adidas: on www.miadidas.com.

[20]   Toffler, A.: The Third Wave, New York et al. 1990.

[21]   Piller, F.T.; Meier, R.: Strategien zur effizienten Individualisierung von Dienstleis-tungen, in: Industrie Management, 17 (2001) 2, pp. 13-17.

[22]   1-800-Flowers: on www.1-800-flowers.com.

[23]   Resnick, P.; Varian, H. R.: Recommender System, in: Communications of the ACM, 40 (1997) 3, pp. 56-58.

[24]   Amazon.com: on www.amazon.com.

[25]   Manber, U.; Patel, A.; Robinson, J.: Experience with Personalization on Yahoo!, in: Communications of the ACM, 43 (2000) 8, pp. 35-39.

[26]   Cortese, A., Stepanek, M.: Goodbye to Fixed Pricing, on www. businessweek.com/ 1998/18/b3576023.htm.

[27]   Stepanek, M.: Weblining, in: Business Week, April 3rd 2000, pp. 14-20.

[28]   Miles & More Program of Lufthansa AG: on www.lufthansa.com.

[29]   Beenz.com Inc.: on www.beenz.com.

[30]   A set of selected books by Pepper & Rogers: on www.1to1.com/Building/ CustomerRelationships/content/books.jsp.

[31]   Cassell, J., Bickmore, T.: External Manifestations of trustworthiness in the interface, in: Communications of the ACM, 43 (2000) 12, pp. 50-56.

[32]   Belz, C. et al.: Erfolgreiche Leistungssysteme, Stuttgart 1991.

Contact:

Dipl.-Wirt.-Inform. Kai Riemer
Institute of Information Systems, University Muenster, Germany
E-mail: kai.riemer@wi.uni-muenster.de

4 Economic Evaluation of Mini-Plants for Mass Customization

A decentralized setting of customer-centric production units

Ralf Reichwald, Frank T. Piller, Stephan Jaeger and Stefan Zanner
TUM Business School, Department of General and Industrial
Management, Technische Universitaet Muenchen, Germany

In this chapter we will present a new setting of mass customization value creation. Main elements of our approach are scaleable, geographically distributed and networked facilities – so-called mini-plants – each of them covering the majority of all value chain activities and located in close proximity to a particular local market. In addition, customization will be not only limited to physical goods, but extended to customized product-service bundles. The objective of this chapter is to examine whether such a decentralized scenario of value creation could provide a suitable framework for the efficient production of individualized goods. Our evaluation criteria are economical ones, i.e. the financial effects arising from such a setting. We will discuss whether the additional costs and hurdles of mass customization in mini-plants can be counterbalanced by the advantages of such a decentralized fulfillment situation (compared to both mass production and to centralized mass customization). Advantages could arise from both (1) new cost saving potentials as a result of a decentralized mass customization system and (2) a higher consumers' willingness to pay for a customized solution coming out of such a mini-plant.

4.1 A vision: a look into the future

Let us take you on a journey to the year 2010! Suppose you need a complex surveillance and control solution for your house. The solution should integrate seamlessly with your wire bound and mobile communication system. You head for the next sales office of SURCON, a leading company for sophisticated surveillance solutions. However, when you enter the sales room, you realize that it is in fact a whole factory. They call it a mini-plant where not only sales processes take place but also manufacturing and product development. A specially trained sales person welcomes you. In an informal conversation you provide details of your surveillance and control requirements and outline your current communication situation. One important aspect for you is that you need numerous remote control functions such as closing windows, watering plants, turning on and off lights and heating and answering the door. Prior to your visit prior to your visit you have sent to SURCON electronically details of your house such as a sketch and wiring plan. Using an dedicated interaction tool kit, the sales consultant guides you through a process of transforming your wishes into a complete solution, based on your information and the details of your house. Your customized solution is a bundle of standard and individualized product and service components, consisting of the core product – the mobile device and the operating controls in your house – and accompanying services such as the planning and implementation of all surveillance elements, the programming, and taking over the surveillance while you are on a holiday.

To deliver the customized solution, the sales consultant checks the company's knowledge base whether the requested solution (or a similar one) has already been created in a different mini-plant of the SURCON company. She finds a dataset containing a solution that has been created for a customer in the Miami mini-plant. It contains a concept for the in-house implementation and the product data for the mobile device. Along comes a list of required components for remote services and design suggestions for the in-house elements of the surveillance system. Using this concept as a basis your sales consultant works on the details of your solution like the properties of the mobile device. You also decide on the operating controls such as shape, size, color, pattern etc. However, as you realize that your wife would also like to specify the "look and feel" of your device, the configuration is placed on an online system. This system provides you with access to the product data and allows to fine tune the device at home. After working with your wife on your customized solution for another day or two, you decide that your configuration is ready and place the order.

After a short period of time you receive an e-mail from SURCON saying that your mobile device will be assembled in two days in their mini-plant. Some components are built in the mini-plant, others were ordered from a central pre-fabrication site of the company. You are invited to follow the final assembly of your personal product. As you have not seen anything like that before, you pay a visit to the mini-plant where your device is built in front of your eyes. You even have the change to make some small adjustments in direct contact with the factory workers. Meanwhile all the installation work in your house has been taken care of. SURCON offers you additional services such as reprogramming your surveillance system, taking care of your house during your vacation, and basic maintenance services such as fixing broken elements or firmware updates for the devices. After delivering your product, all information from your order is fed into the knowledge system where it will become the basis for other orders of future customers.

4.2   From centralized mass production towards mini-plants for mass customization

While this vision might sound rather simple on a first view, it is a future scenario breaking with many elements of both traditional systems of industrial value creation and traditional views of a mass customization system. In the business-to-consumer market, but also in large parts of the business-to-business market value creation is (still) dominated by the mass production paradigm: "Mass production has become a paradigm not only of production but of management, its precepts encompass the entire firm and all its many functions across the value chain" [1]. Based on just a few simple principles like breaking up the production process in single tasks and assigning these to specialized staff members and machines, mass production focuses on standardizing and stabilizing all products, product parts and processes [2, 3]. This is the foundation of huge cost saving and rationalization potentials, namely economies of scale. Their condition is that the mass production system can work in a stable and controllable market- and process-related environment. Thus, management's focus in such a firm is on optimizing flow and operational efficiency and increasing manufacturing and sales volumes. In order to not disturb the stable conditions, incremental product innovation, separated product and process development and a clear separation between production and sales is strongly preferred. If these conditions are given, mass production is – without doubt – the cost-minimizing production system. This is valid today like in former times.

However, today's markets are changing faster and consumers are more demanding than ever [4, 5, 6]. Thus, mass customization was emerging in the last decade as a solution to address the new market realities while still enabling firms to capture the efficiency advantages of mass production. Like mass production, mass customized manufacturing is seen in most cases as taking place in a central production facility in order to capture economies of scale when manufacturing modular component (economies of modularity) and to guarantee stable processes [7, 8]. Recently, some papers also discussed a network structure for mass customization [9]. Network arrangements are characterized by the fact that different stages of the production process are performed by different participants of the network. Suppliers are integrated into the manufacturing of customized parts for a specific order. The distributed value activities come together in a centralized assembly and distribution of the customized products.

In this chapter we will present a new setting extending the network approach. Main elements of our approach are scaleable, geographically distributed and networked facilities – so-called mini-plants – each of them covering the majority of all value chain activities. The mini-plants are located in close proximity to the customers and include next to production also sales and customer service activities. Further more, product and process development activities are performed in each mini-plant. The objective of this chapter is to examine whether such a decentralized scenario of value creation could provide a suitable framework for the efficient production of individualized product/service bundles. Our evaluation criteria are economical ones, i.e. the financial effects arising from such a setting. We will discuss whether the additional costs and hurdles of mass customization in mini-plants can be counterbalanced by the advantages of such a decentralized fulfillment situation (compared to mass production and to traditional centralized mass customization). Advantages could arise from both (1) new cost saving potentials as a result of a decentralized mass customization system and (2) a higher consumers' willingness to pay for a customized solution coming out of such a mini-plant.

The remaining of this paper is organized as follows: In the next section we will describe in larger detail the constituting factors of a mini-plan and how these structures can capture some of the challenges of mass customization. Afterwards we will present an economical analysis of such a value creation setting. We will present a framework distinguishing between factors increasing consumers' willingness to pay (in order to represent to differentiation benefit of mass customization), additional costs by such a decentralized solution and new cost saving potentials resulting exactly from the same situation. Please note that the influence between the specific setting of mini-plant supplying customized product-service-bundles and mass customization performance as indicated in this chapter is of theoretical nature only and not supported by empirical research yet. Thus, we will just formulate hypothesis here, no answers.

4.3   Mass customization of product-service bundles in decentralized mini-plants

The increasing heterogeneity of demand in many markets leads to a situation in which a single product is less and less able to satisfy today's customer require-ments. Consumers' interest and demand shift from sole products towards integrated problem solutions. Our concept is focused on delivering rather complex mechatronical products containing both mechanical parts and constructions as well as electronics and control software. We define a *problem solution* as a bundle of one or more physical products accompanied by a number of services. This solution is supposed to be tailor-made to the specific requirements of each customer. For an optimum solution it is therefore necessary to conduct an integrative approach in developing and providing fitting services and the physical product(s). Each product-service bundle should be produced truly on-demand (following a made-to-order or even engineer-to-order approach [10, 11, 12] according to the specifications and desires of each customer. The degree of customizability should be rather high. Functionality, fit (measurements) and (aesthetic) design can be adopted to each customer's wishes. To match the customization demands, both the physical core product and the supporting services can be used.

Examples for such an individualized product/service-bundle could include

- a home surveillance and control solution as presented in the scenario in the first section;

- a mobility solution as provided by US mass customization pioneer "Model E", i.e. a subscription model consisting of a build-to-order vehicle and an all-inclusive-service package (from registration and insurance to maintenance, connectivity and road side assistance);

- a modern home cleaning robot being adapted to a specific cleaning situation (floor material, geographical conditions, allergies of owners, pets etc.) and fitting aesthetic design needs. In this example customized services could include homecare and cleaning personnel for unusual, rare cleaning situations which cannot be performed by the robot.

Individualized product-service bundles demand a high degree of interaction between each customer and the company. As already discussed in the introduction of this book (and in many other chapters), mass customization implies the integration of the customer into value creation. One can look at the relation between the customer and supplier as a cooperation providing benefits for both sides, but demanding inputs of both participants, too. In mass customization systems, integration of the customers is required primarily during configuration. Every order implies coordination about the customer specific product design [13]. The individual wishes and needs of each customer have to be transformed to a unique product specification. The costs arising from customization consist largely of information costs. They are accounted for by the investigation and specification of the customers' wishes, the configuration of individual products, the transfer of the specifications to manufacturing, an increased complexity in production planning and control, the coordination with the suppliers involved in the individual prefabrication, and the direct distribution of the goods. All theses activities are characterized by a high information intense compared to traditional mass production. Thus, customer-related value added is produced on the information level [14].

While for simple products customer integration and product configuration often take place on the internet, a closer relationship between consumer and manufacturer is necessary to deliver complex product-service bundles. The relationship does not end with the last part of the order process but continues during the whole life cycle of the customized solution. Thus, interaction and production facilities may be required to be located in proximity to the users. However, implementing many production facilities worldwide leads to an enormous amount of fixed costs as these facilities represent redundant plants. The logical consequence is, that the size of the plants has to be reduced. Thus, the idea of mini-plants came into the focus of our research.

Mini-plants for mass customization are scaleable, geographically distributed and networked facilities. A mini-plants covers the majority of all required activities of the value chain of mass customization [6, 14]. The mini-plants are located in close proximity to the customers and include next to production also sales and customer service activities. Further more, product and process development activities are performed in each mini-plant (see Figure 1). Being close to the target markets offers a number of advantages such as better adoption of local customer preferences, minimizing or eliminating transport and customs costs as well as the opportunity for an intensified relationship to the customer. The structure of each single mini-plant as well as the interactions between the particular plants have to be designed in such a manner that a profitable delivery for smaller, local markets could be realized. Therefore a mini-plant is the smallest possible production unit which could be composed of configurable and scaleable (process) modules.

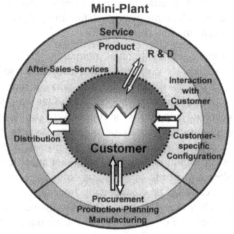

Figure 1:   Customer integration in the creation process of product/service-bundles

However, mini-plants are no traditional craft workshops which meet most of these characteristics, too. All mini-plants belong to one company and are strongly connected. They are networked among each other and exchange knowledge gathered during the configuration process about customized product/service concepts, ways and methods of production as well as marketing relevant information. Exchanged information may include feasibilities of solutions, cost, duration, technical elucidations and experience as well as customer profiles. Finally, mini-plants can be supported by a centralized unit providing pre-fabrication of components, basic development activities as well as training and support activities.

The German mass customizer "Kueche Direkt" (kitchen direct) has established such a mini-plant network for their operations. The company offers fully customized high quality kitchen furniture including electrical appliances to prices similar to large discount chains like IKEA. Each unit consists of a highly automated manufacturing system, the process control software, a configuration and sales unit and the construction software translating a customer's order into a parametrical product design. The whole factory is operated by just two workmen and one sales assistance. Basic furniture designs, customization options and machine control programs are provided by a central support center (operated by the machinery company INA distributing the whole system on a franchise-alike basis).

Another mini-plant example provides Reflect.com, a sister company of Procter&Gamble. Reflect is offering customized cosmetics on the internet and in dedicated stores. Using interactive software, visitors create their own cosmetic line, mix and match various options like colors, scents, and skin-care preferences to create a unique product. While the internet orders are manufactured in a &G facility in upstate New York, orders that are configured in a Reflect store are customized on the spot in a specialized mini-plant just in the store. The ratio behind this approach is to give customers an ultimate experience without any delivery time and to get better access to customer knowledge (we will describe this case example in more detail below). Examples for mini-plants in regard to our understanding could be also the famous infamous portable manufacturing units footwear companies are said to operate on ships moving from one cheap labor market to another. Finally, "Model E", mentioned above, plans also to produce its cars in local mini-plants, called regional micro factories.

4.4   Economic Evaluation

A mini-plant is distinguished from a large-sized plant by lower entry barriers into local markets, a rather low investment risk and its reduced process complexity due to its scaleable and decentralized character. These characteristics of the mini-plant – small, flexible, decentralized and in proximity to the customer – frame the main competitive advantages of a decentralized, customer-centric production of customized product/service-bundles. We will evaluate these advantages in more detail in the following. Our framework of evaluation is taken from Piller/Möslein [12]. It will be described briefly in the first paragraph of this section. Afterwards, we will evaluate how the supply of customized product-service-bundles in a mini-plant effect the influencing factors and variable of this framework. As noted above, our conclusions of the influence between the specific setting of mini-plant and mass customization performance are of hypothetical nature only and not proofed by empirical research yet.

4.4.1   Framework for evaluation

Our framework for evaluation can be summarized as shown in Figure 2. As more and more customers express a desire for products or services which fit their needs exactly, suppliers can charge a price premium for a customized product which reflects the customers' increasing willingness to pay. As we will discuss in more detail below, the willingness to pay is influenced by the user's need for customization and the perception of a supplier's mass customization program. However, competitive pressure and additional risks faced by customers during the buying process may reduce possible rent surpluses to a larger extent. The possibility of gaining higher margins is also challenged by additional costs of mass customization which will be described in more detail below. Therefore a firm has to seek mechanisms to counterbalance the additional costs. Traditional approaches to reduce costs described in the mass customization literature either try to reduce the complexity of products, processes and information resulting from customized manufacturing and individual customer interaction or try to handle this complexity better [6].

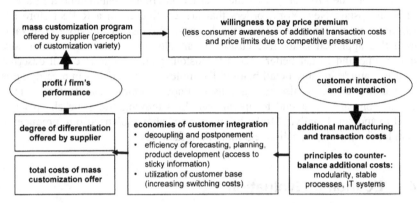

Figure 2:   A model of value creation in mass customization [12]

However, interacting with the customers during the course of configuration is not only a driver of costs, but also the source of new cost saving potentials. The integration of the customer can help a firm to fulfill specific processes more efficiently. We will call these cost saving potentials 'economies of integration'. They are based on three different sources to reduce costs and efficiency, which will be discussed in the following sections. The firm's capability to manage and counterbalance the additional costs of customization result in the total costs of offering a customized good. Mass customizers studied in earlier research [6, 15] have chosen a wide variety of instruments and mechanisms to deal with the additional costs and to realize benefits from customer integration. Together with influencing factors not considered here like brand value, channel access, communication policies, etc., a firm's differentiation position is set. The profits gained have to be partly reinvested in order to keep the interaction with a customer active and to gather feedback in-between sales. The profits have also to be used to continuously update the basic product architectures and to improve the elicitation (configuration) mechanisms. In the following, we will analyze the different parts of the framework in more detail. We will especially comment on how a decentralized setting for mass customization influences the different cost and value drivers within this framework.

4.4.2   Differentiation potential and consumers' willingness to pay

Traditionally, customization is connected with the possibility of charging premium prices derived from the added value of a solution meeting the specific needs of a customer [16]. The individualization of product and service attributes to customer-specific needs and requirements holds the potential to increase customer satisfaction. The firm enters a quasi monopolistic status as its offer is unique and, at least to a certain degree, not comparable with the other products in the market segment. A firm could be able to charge prices representing the willingness to pay of each single customer, capturing the whole consumer rent [17].

The possible extent of a price premium depends on the heterogeneity of customer demands in regard to specific attributes of a product and, thus, the willingness to pay for a customization solution compared to a standard product. Customer demands are reflected in the mass customization program of a supplier. Products that require matching different physical dimensions allow often a higher premium than products that are customized just by the possibility of changing colors or design patterns. In the sport shoe market, Adidas can charge higher premiums (up to 50 percent) for its customized sport shoes brand "mi adidas" compared to the personalized sneakers of Nike (NikeID brand, 5-10 per cent premium). Adidas makes it possible not only to choose between some colors and to place a name on the shoe, but also customizes the shoes with regard to comfort, fit and functionality for each buyer.

The proposed production scenario of distributed mini-plants emphasizes this differentiation potential in a number of ways. Main driver is the massive impact of customer proximity. Being close to a customer leads to a stronger identification with the manufacturer, which is mainly due to the positive association towards a local brand. Thus, the neighborhood of a mini-plant may create an emotional relationship between the customer and product as well as between the customer and the mini-plant [18]. Customer proximity of a mini-plant supports a high service level during the whole life cycle of a solution as well as short delivery times and high delivery certainty [19]. Customer proximity has also an impact on the quality perception as in case of problems or complications the company is nearby and can be easily reached to handle the problem.

This is a very important aspect especially when delivering product-service bundles as the quality of a service can be evaluated only marginally before its purchase or, respectively, delivery. Customers therefore rely on substitutes such as company reputation or brand names to judge service quality. Having the manufacturer nearby may have a positive influence on the quality perception. Another important aspect increasing customer attraction may be the strong personal involvement of each customer into the development and production processes. Although customer integration is a distinctive factor of every mass customization setting, being close to the customer offers the advantage of face-to-face communication. Thus, the perceived amount of control over the whole process by a customer may be significantly enhanced [20]. All these factors may increase the perceived utility of the customized solution from the customers' point of view and, thus, may increase their willingness to pay.

4.4.3   Additional costs of mass customization

These additional premiums of mass customization (compared to traditional mass production) are challenged by additional costs connected with this system. Due to high competitive pressures in many industries, even high levels of differentiation do often not justify much higher prices (as it is the rule within a traditional differentiation strategy [16]). The cost-benefit relation alters because buyers demand relatively high standards of quality, service, variety or functionality even when the sales price is favorable or, vice versa, suppliers have to meet additional requirements in pricing when a product is markedly differentiated [21, 22]. Thus, management of the additional costs becomes strategically crucial in every mass customization system. In the context of the discussion in this chapter, we have to distinguish between additional costs compared to (1) a mass production system and (2) compared to a traditional mass customization system with centralized and customer interaction and manufacturing.

Ad 1: Compared to a made-to-stock (mass production) system, higher costs derive in a mass customization system in all activities of the value chain. In sales, the configuration and interaction with the customer and the distribution of smaller lot sizes lead to additional costs. As we have stated already above, a major cost driver are (transaction) costs accounted for the investigation and specification of the customers' wishes and the configuration, the transfer of the specifications to fulfillment, an increased complexity in production planning and control, coordination with the suppliers, and so on. This includes not only investments in configuration systems and other information handling equipment. A firm has to establish also mechanisms to minimize the burdens of customization from the customers' point of view. The direct interaction between manufacturer and consumer during the whole life-cycle of the product-service bundle entails a higher information and communication intensity. Corresponding measures demand investments in customer service centers, highly qualified staff, or trust building promotion activities – leading to additional costs, often seen as the "natural" costs of differentiation [23]. Further cost surpluses result from a loss of economies of scale (specialization and standardization) in comparison to mass manufacturing. Higher set up costs, more complex and detailed quality control, costs for better qualified labor and the complex manufacturing planning increase the cost level. Additionally, inventory of components may rise, and higher capital investments in advanced flexible production units and appropriate information systems result in additional machinery costs (see [24, 25] for a more detailed discussion of the additional variable and fixed costs of mass customization).

Ad 2: Our concept of mini-plants may offer new opportunities, however, it is challenged, first of all, by additional costs compared to a centralized manufacturing. One important aspect is that a number of scaleable, largely independent production units represent redundant facilities. The proposed scenario also leads to additional costs for organizing the knowledge exchange throughout the mini-plant network. Effort has to be taken to collect, prepare and provide the knowledge and to turn it into an useful and usable knowledge base. In general, mini-plants are characterized by the loss of economies of size & scale. Larger centralized manufacturing units allow the concentration and centralization of know-how, realize centralized learning effects, need lower total investment costs and imply often less coordination needs. Finally, large plants can often make use of strong market power and "easier" relationship towards the suppliers. A mini-plant for mass customization has to proof to provide advantages that can counterbalance this loss of economies of size & scale.

4.4.4   Economies of integration to counterbalance the additional costs

To deal with these costs of customization, the literature on mass customization has developed various approaches building important principles of mass customization: They include firstly the use of modular product architectures (concept of reusability and product family design approach) and related organizational processes. Modularization can be seen as one main enabler of mass customization (for a more detailed discussion see [6, 10, 26, 27, 28, 29]. Secondly, mass customization is characterized by limiting the customization possibilities. This reflects the need for stable product architectures and stable processes (on the contrary, a craft customizer reinvents not only the products but also the fulfillment processes [7, 10, 29, 30, 31, 32, 33]). Thirdly, specialized information systems are used for configuration, manufacturing planning, order tracking, and relationship management [13, 27, 28, 32, 34, 35, 36].

However, as we have indicated above in our framework, the additional costs of mass customization can be reduced not only by specific information or manufacturing technologies and modular product families but could be also counterbalanced by new saving potentials resulting from the direct interaction of a firm with each single customer during the configuration process. The integration of the customer can help a firm to fulfill specific processes more efficiently. We will call this new class of cost saving potentials 'economies of integration' [6, 12]. They go beyond the differentiation advantages of customized manufacturing which are expressed in the price premium. Economies of integration represent the efficiency when a firm gains deeper knowledge about its environment and establishes value processes that eliminate waste on all levels. From our experience with firms performing mass customization we know that many mass customization managers do not explicitly realize the benefits of customer integration in this regard – even if they make use of the effects implicitly. Many firms still focus their attention on managing additional costs by flexible manufacturing and information systems. Customer interaction and integration are seen rather as a necessity than an asset to increase efficiency of business processes.

Integrating customers into value creation and interacting intensively with them creates saving potentials that are based in different fields. Economies of integration are formed by a bundle of cost saving potentials consisting of three major sources:

(1) By decoupling the value chain into an order specific and a customer neutral part, cost savings arise from the postponement of activities until an order is placed. By doing so, a firm wins certainty and prevents costs of misplacement of activities due to non precise planning information.

(2) By integrating the customer into value creation, a firm gets access to so called "sticky information" [37]. The aggregation of this customer information to better and more perceive market knowledge increases the efficiency of market research and product development activities.

(3) By using the course of customization to increase switching costs for the customer, a firm builds stable relationships with its clients, allowing a better utilization of its customer base ("re-use" of existing customers for additional sales). Thus, costs for marketing activities and customer acquisition may decrease.

In the following, we will describe these sources of economies of integration in more detail and analyze them in particular before the background of supplying customized product-service bundles in mini-plants.

(1) Decoupling and postponement

Firstly, economies of integration are the result of the build-to-order approach connected with mass customization. Build-to-order means postponing some stages of fulfillment until the order has been placed (manufacturing on demand). This implies splitting the fulfillment system into a standardized and a customer specific part at the decoupling or postponement point [1, 6, 10, 27, 29, 38, 39]. However, in contrast to mass production, decoupling in the sense of mass customization implies that the supplier has to integrate each single customer into its fulfillment processes. Information on the output of the configuration process is the planning input for the following manufacturing, assembly, and delivery steps. While these processes are cost drivers as discussed above, they are also the basis for economies of integration in regard to decoupling and postponement:

- *inventory: reduction / elimination of inventory* in distribution chain; reduction of safety stock;

- *planning: reduction of planning complexity, adaptation costs* (of planning decisions), fashion risk, and development costs (product flops);

- *capacity utilization and stability:* no bull-whip-effect, stable processes, reduction of the over capacity required in made-to-stock systems to adopt to short-term changes of trends;

- *sales:* avoidance of lost sales in retail due to out-of-stock items, prevention of discounts at the end of a season; opportunities for better channel management, reduction of error costs.

The savings from these effects can be huge. For the clothing industry, cost savings up to 30 percent are reported as a result of the prevention of discounts and overstocks and the reduction of the forecasting (fashion) risk if mass customization based manufacturing on demand can substitute the existing production on stock following seasonal fashion cycles [38]. Sanders [40] calculates that today almost $300 billion are wasted annually due to wrong forecasting, heavy inventory, fashion risks, and lost profits as a result of necessary discounts. This amount provides an enormous opportunity to counterbalance additional costs mass customization.

One can argue that decentralized production settings are better suited to capture these saving due to postponement and decoupling compared to a centralized setting. The advantage of centralization – economies of size (see above) – can be captured in relatively stable and homogeneous mass markets, but not in heterogeneous, flexible and fast changing environments [41]. In highly dynamic markets, sales figures for specific selections of a basic product-service bundle are rarely predictable. Decentralized mini-plants may be able to react and adjust production and enterprise processes to market requirements more flexibly and faster. This implies not only forecasting on a material and component level but also setting the shop layout. Building up large-sized plants implies very risky investment decisions. The concept of scaleable mini-plants could weaken this limiting factor as it allows an easy gradual adjustment to new market realities. Further advantages of a mini-plant compared to a large-sized plant include reducing the risk of overcapacity. In summary, mini-plants can postpone more activities economically and, thus, capture more savings in inventory, planning, capacity utilization and stability, and sales.

Other cost saving effects result from economies of learning. Learning effects are traditionally interpreted as one of the main reasons for economies of scale [42]. However, as literature has shown, they can also take place in a mass customization setting [6]. The decentralized production scenario implies a distributed creation of knowledge, experience and skills. To realize learning effects it is necessary to manage the exchange of this knowledge and to enforce its re-usage. Knowledge for exchange can enclose all kind of technical know-how on the development and production regarding technical solutions for certain tasks, feasibility information, and manufacturing information of specific solution bundle. Marketing related information include pricing strategies, customer preferences and customer profiles, impact of certain advertising channels, etc.

Most important of all, however, is to save and transfer knowledge regarding the linkages between all the know-how aspects mentioned above in conjunction with the ability to draw conclusions for future orders. Being able to retrieve such kind of information represents a huge potential to shorten learning cycles and therefore contributes to reducing costs.

(2) Access to sticky information and generation of customer knowledge

A second source of economies of integration is based on the aggregation of customer information in order to get access to more perceive market information.

The self configuration by the buyer allows access to „sticky local information" [37, 43]. Sticky information increases when the costs of information exchange between two different actors are higher than processing this information within one unit. They originate in location specific costs like technological and organizational activities of decoding, transmitting and diffusing the information. We can argue that often customer specific information is sticky in that sense. Tastes, design patterns, and even functionalities are rather subjective and difficult to describe objectively. Many customers are not able to describe their needs precisely and therefore to transfer their wishes into a concrete product specification that allows the company to build a customized product or to deliver a customized service. By integrating the customer into the design of a product or service, economies of integration represent the saved costs of getting easier access to the sticky information. Note that this argument is only true as long as the information the customer needs from the supplier to execute the co-design has a low level of stickiness. Modern internet technologies are a major enabler of this condition. Design tools and configurators should, if developed thoroughly, be able not only to construct a customized product from a functional point of view but also have mechanisms to explore a customer's demand set. They have to provide an open and easy to understand platform for interaction and co-design. Using these tools, customers can often specify their product implicitly. This is especially true in the case of complex customized solutions consisting of physical products and service components which are in the center of our analysis.

By transferring customer needs and wishes into customized products, a company gains access to the sticky information and can transfer it to explicit knowledge. By aggregating this knowledge, the company can generate better market research information and more accurate forecasting concerning customer needs. This is especially true when the firm (still) operates a mass production for anonymous markets alongside the customized business [6,31]. For the portion of business that is (still) manufactured on stock, the customized segment provides panel-like market research information without the common panel effects biasing the results. The information gained here can be used to improve forecasting and sales planning of products made to inventory. In regard to this source, economies of integration are represented by reducing (traditional) market research costs, the prevention of market research biases (like panel effects), and the gain in information quality due to the access to implicit information. Additionally, new product development and continuous improvement can benefit substantially from such user information.

One firm that has installed mechanisms to explicitly harvest this information on a regularly base is Reflect.com. We have described this company already above. Reflect allows customers to configure a product as many times as they want. The site acts as a "life panel" for all P&G cosmetics operations. Its customer base contains more than one million buyers creating their own cosmetics and thereby formulating dermatological needs, evaluating new scents, bundling variants, choosing packages, and, thus, developing new products. The customized order specifications are matched with the socio-demographic profiling information of each customer and the feedback or change of specifications after the sale. For P&G, reflect.com was reported in our research to be a very efficient market research tool – saving the mother company a large sum of investments in traditional market research.

Decentralized mini-plants in close customer proximity support this potential in different ways. Firstly, face-to-face interaction with the customer ensures an easy transfer of customer preferences and know how to the company. Complaint management workshops could, for example, supplement knowledge capturing during configuration if customers are queried about existing solutions and what extensions they would appreciate. Especially for products where customer preferences vary with location, customer proximity of the mini-plants is supportive. Secondly, the knowledge sharing mechanism within the mini-plant network supports a time-pacing strategy known from innovation management literature [44]. Having access to solutions that have been created in other mini-plants opens up the opportunity to approach customers whose profiles are already known with suggestions for new or extensions to existing solutions.

To stress the scenario in the first section of this chapter again, suppose a client in Malaysia requesting a surveillance and control solution with a remote humidity control for certain rooms of his house. Once this special feature is developed, implemented and then saved into the SURCON knowledge database, it may become an option for all further sales. A customer consultant in, e.g., Miami, may offer this solution to some of his existing customers. One of them may be a customer cultivating orchids, another may be a wine fanatic. Both could find an additional humidity control an useful extra feature. Applying this time-pacing-strategy helps to create additional revenue by making use of product (technology) based economies of scope. In both cases, the direct involvement of the customer in the design and configuration process intensifies the identification and emotional relationship of customers with their local manufacturers.

(3) Efficient utilization of customer base

Economies of integration can further be based on cost savings connected with increasing customer loyalty resulting from customization. If one interprets the number of different customers of a firm as a cost driver, then does high customer loyalty not only decrease transaction costs during configuration, but can also reduce marketing efforts and eliminate inefficiency in advertising. From a transaction cost perspective, the expenditures and efforts resulting from interacting and communicating with a customer during the first sale (configuration) can be used for further sales as well [45]. Economies of integration are, in this sense, a special form of economies of scope [46], resulting from a better "utilization" of the customer base (versus a better utilization of manufacturing resources in the case of traditional economies of scope). Once lock-on has been achieved, the enterprise can stretch its brand into other revenue-generating opportunities at relatively low cost. The company has no acquisition cost and low marginal cost, because information, knowledge and relationships have already been established [44, 47].

However, the prerequisite for this potential is that customer integration as a principle of mass customization is able to increase the intense of relationships between a supplier and its customers and, thus, to increase customer loyalty. As Riemer and Totz show in Chapter 3 of this book, customer loyalty can be seen as a result of switching costs, opportunity costs, and sunk costs based on technological, contractual, and psychological obligations faced by a customer [48, 49]. All sources of switching, sunk, and opportunity costs within a mass customization system are positively influenced by customer integration as prerequisite of mass customization. Here, the relationship level of mass customization is based [49].

Customized product-service bundles offer many potentials in this regard. While the core physical system will be bought only once, additional services may be demanded regularly (or will often be even subscribed). Customer proximity and face-to-face communication in the mini-plants lead to an intensified customer relationship management, too. According to the media richness theory [41, 50] complex information such as the life situation and the derived requirements are communicated best via rich media with face-to-face representing the optimum solution. This in turn creates a strong tie when customer preference profiles are supplemented and completed with every repurchase – increasing the switching and sunk costs for the customer.

4.5   Conclusion

Within this chapter we presented a decentralized approach for mass customization. We propose to supply rather complex, mechatronical customized solutions consisting of physical products and additional services by a network of small scaleable units – so-called mini-plants – in which all or at least most activities of value creation take place. These mini-plants are placed in close customer proximity and are networked among each other to exchange information regarding feasibility, cost, duration, technical solutions and experience as well as customer profiles. While there are still many questions unanswered like the design of this network, the set up of the knowledge sharing processes, the split of value creation between a mini-plant and a central pre-fabrication, the objective of this chapter was to present an evaluation of such a setting from an economical perspective.

Using a framework by Piller/Möslein [12], economical effects were examined in three categories: the differentiation potential (willingness to pay), the additional costs and new cost saving potentials. While these effects exist for every mass customization system in comparison to a made-to-stock system, we discussed how customer proximity, face-to-face communication and flexible adaptation to local requirements influence these effects positively. However, it has also to be stated that additional costs arise (in comparison to a centralized production scenario). These are both production (in terms of redundant productive factors such as machines, staff, buildings) and information (in terms of information handling) related. However, these additional costs may be outweighed by a number of cost saving effects, referred to economies of integration in this chapter. Further (empirical) research has to provide insight into the variables influencing the decision to implement a centralized or decentralized setting in a mass customization scenario, The approach presented in this chapter was developed to provide customized solutions consisting of a rather complex, mechatronical physical core product and accompanying services. Future research has to prove for which other product categories a decentralized setting of mini-plants is applicable, too.

References

[1]   Pine II, B. J.: Mass Customization, Boston 1993.

[2]   Milgrom, P.; Roberts, J.: Complementarities and fit – strategy, structure and organizational change in manufacturing, in: Journal of Accounting and Economics, 19 (1995) 2, pp. 179-208.

[3]   Wigand, R.; Picot, A.; Reichwald, R: Information, organization and management, Chichester and New York 1997.

[4]   Alfnes, E., Strandhagen, J.: Enterprise Design for Mass Customization: The Control Model Methodology, in: International Journal of Logistics: Research and Applications, 3 (2000) 2, pp.111-125.

[5]   Cox, W. M.; Alm, R.: The Right Stuff: America's Move to Mass Customization, in: Annual Report of Federal Reserve Bank of Dallas 1998.

[6]   Piller, F.T.: Mass Customization, 3rd Ed., Wiesbaden 2003.

[7]   Pine II, B. J.: Mass Customization – the new imperative in business, in: Anderson, D.: Agile product development for mass customization, Chicago 1997, pp. 3-24.

[8]   Victor, B.; Pine II, B. J.; Boynton, A.: Aligning IT with new competitive strategies, in: Luftman, J. (Ed.): Competing in the information age, New York and Oxford 1996, pp. 73-95.

[9]   Piller, F.T.; Reichwald, R.; Lohse, Ch.: Broker Models for Mass Customization Based Electronic Commerce, in: Goul, M.; Gray, P.; Chung, M. (Eds.): Proceedings of the Americas Conference on Information Systems – AMCIS 2000, 2 (2000), Long Beach 2000, pp. 750-756.

[10]   Duray, R. et al.: Approaches to mass customization: configurations and empirical validation, in: Journal of Operations Managements, 18 (2000), pp. 605-625.

[11]   Lampel, J.; Mintzberg, H.: Customizing customization, in: Sloan Management Review, 37 (1996) 1, pp. 21-30.

[12]   Piller, F.T.; Möslein, K: From economies of scale towards economies of customer integration: value creation in mass customization based electronic commerce, in: Working Paper No. 31 of the Department of General and Industrial Management, Technische Universität München, 8/ 2002.

[13]   Hibbard, J.: Assembly online: the web is changing mass production into mass customization, in: Information Week, 4/1999, pp. 85-86.

[14]   Reichwald, R.; Piller, F.T.; Moeslein, K.: Information as a critical success factor for mass customization, in: Proceedings of the ASAC-IFSAM 2000 Conference, Montreal 2000.

[15]   Piller, F.T.; Stotko, C.: Four approaches to deliver customized products and services with mass production efficiency, in: Durrani, T. S. (Ed.): Proceedings of the IEEE International Engineering Management Conference IEMC-2002: Managing Technology for the New Economy, Cambridge University 2002, pp. 773-778.

[16]   Porter, M.E.: Competitive Strategy, New York 1980.

[17]   Chamberlin, E.H.: The Theory of Monopolistic Competition, 8th Ed., Cambridge 1962.

[18]   Porter, M.E.: Clusters and the New Economics of Competition, in: Harvard Business Review, 76 (1998) 6, pp. 77-90.

[19]   Zahn, E.: Fuehrungskonzepte im Wandel, in: Bullinger, H.-J.; Warnecke, H.-J. (Eds.): Neue Organisationsformen im Unternehmen: Ein Handbuch fuer das moderne Management, Berlin 1996.

[20]   Franke, N.; Piller, F.T.: Key research issues in user interaction with configuration toolkits, forthcoming in: International Journal of Technology Management, 2003.

[21]   Davis, S.: Future Perfect, Reading 1987.

[22]   Ramirez, R.: Value co-production: intellectual origins and implications for practice and research, in: Strategic Management Journal, 20 (1999) 1, pp. 49-65.

[23]   Porter, M. E.: Strategy and the Internet, in: Harvard Business Review, 79 (2001) 2, pp. 62-78.

[24]   Agrawal, M.; Kumaresh, T. V.; Mercer, G. A.: The false promise of mass customization, in: The McKinsey Quarterly, 38 (2001) 3, pp. 62-71.

[25]   Zipkin, P.: The limits of mass customization, in: Sloan Management Review, 42 (2001) 3, pp. 81-87.

[26]   Jiao, J.; Tseng, M. M.: Design for Mass Customization, in: Annals of the CIRP, 45 (1996) 1, pp. 153-156.

[27]   Sahin, F.: Manufacturing competitiveness: Different systems to achieve the same results, in: Production and Inventory Management Journal, 42 (2000) 1, pp. 56-65.

[28]   Tseng, M.M.; Jiao, J.: Mass customization, in: Salvendy, G. (Ed.): Handbook of Industrial Engineering, 3rd Ed., New York 2001, pp. 684-709.

[29]   Victor, B.; Boynton, A. C.: Invented Here, Boston 1998.

[30]   Ahlström, P.; Westbrook, R.: Implications of mass customization for operations management, in: International Journal of Operations & Production Management, 19 (1999) 3, pp. 262-274.

[31]   Kotha, S.: Mass customization: Implementing the emerging paradigm for competitive advantage, in: Strategic Management Journal, 16 (special issue 1995), pp. 21-42.

[32]   Lee, C.-H.; Barua, A.; Whinston, A. B.: The Complementarity of Mass Customization and Electronic Commerce, in: Economics of Innovation & New Technology, 9 (2000) 2, pp. 81-109.

[33]   Rautenstrauch, C.; Tangermann, H.; Turowski, K.: Manufacturing Planing and Control Content Management in Virtual Enterprises Pursuing Mass Customization, in: Rautenstrauch, C. et al. (Eds.): Moving towards mass customization, Heidelberg 2002, pp. 103-118.

[34]   Fulkerson, B.; Shank, M.: The new economy electronic commerce and the rise of mass customization, in: Shaw, M. et al. (Eds.): Handbook on electronic commerce, Berlin 2000, pp. 411-430.

[35]   Piller, F.T.: Customer interaction and digitizability, in: Rautenstrauch, C. et al. (Eds.): Moving into Mass Customization – Information Systems and Management Principles, Berlin et al. 2002, pp. 119-138.

[36] Wind, J.; Rangaswamy, A.: Customerization. Journal of Interactive Marketing, 15 (2001) 1, pp. 13-32.

[37] von Hippel, E.: Sticky information and the locus of problem solving, in: Management Science, 40 (1994) 3, pp. 429-439.

[38] Feitzinger, E.; Lee, H.: Mass customization at Hewlett-Packard: the power of postponement, in: Havard Business Review, 75 (1997) 1, pp. 116-121.

[39] Toffler, A.: Future Shock, New York 1970.

[40] Sanders, F.-H.: Financial rewards of mass customization, in: Opening keynote of the 2001 World Conference on Mass Customization and Personalization, Hong Kong University of Science and Technology, Hong Kong 2001.

[41] Wigand, R.; Picot, A.; Reichwald, R.: Information, organization and management. Chichester and New York 1997.

[42] Milgrom, P.; Roberts, J.: The Economics of Modern Manufacturing: Technology, Strategy, and Organization, in: The American Economic Review, 80 (1990) 6, pp. 511-528.

[43] von Hippel, E.: Economics of product development by users, in: Management Science, 44 (1998) 5, pp. 629-644.

[44] Brown, S. L.; Eisenhardt, K. M.: Competing on the Edge: Strategy as Structured Chaos, Boston 1998.

[45] Vandermerwe, S.: How increasing value to customers improves business results, in: Sloan Management Review, 42 (2000) 1, pp. 27-37.

[46] Peters, L., Saidin, H.: IT and the mass customization of services, in: International Journal of Information Management, 20 (2000) 1, pp. 103-119.

[47] Pine II, B.J.; Peppers, D.; Rogers, M.: Do you want to keep your customers forever?, in: Harvard Business Review, 73 (1995) 2, pp. 103-114.

[48] Jackson, B.B.: Build customer relationships, in: Harvard Business Review, 63 (1985) 6, pp. 120-128.

[49] Riemer, K.; Totz, C.: The many faces of personalization, in: Tseng, M. M.; Piller, F. T. (Eds.): Proceedings of the 2001 world conference on mass customization and personalization, Hong Kong University of Science and Technology, Hong Kong 2001 *(see also chapter 3 of this book)*.

[50] Daft, R.L.; Lengel, R.H.: Organizational Information Requirements, Media Richness and Structural Design, in: Management Science, 32 (!986) 5, pp.191-233.

Acknowledgments: This research is part of a National Research Cluster on Mass Customization (SFB 582, see www.sfb582.de), funded by the Deutsche Forschungsgemeinschaft (DFG) (German Academy of Science) at the Technische Universität München. The authors thank Christoph Ihl and Christof Stotko for their valuable input and contribution to this article.

Contact:

Prof. Dr. Dr.h.c. Ralf Reichwald
TUM Business School, Department of General and Industrial Management,
Technische Universitaet Muenchen, Germany
E-mail: reichwald@wi.tum.de

5   Customer Driven Manufacturing Versus Mass Customization

Comparing system design principles for mass customization and (traditional) customer driven manufacturing

Klaus-Dieter Thoben
University of Bremen and Bremen Institute of Industrial Technology and Applied Work Science (BIBA), Bremen, Germany

For the last years the concept of mass customization has gained broad attention within various branches of industry. Mass customization has been identified as a competitive strategy by an increasing number of companies. Accordingly theoretical, technical as well as managerial aspects have been studied aiming at a better understanding of this new paradigm. However, especially in Europe there is a long tradition of designing and manufacturing customer specific products such as machinery, ships and even cars. For this chapter we have analyzed various industrial cases, consultancy projects as well as research work in the broader field of customer driven manufacturing. We will identify a number of design principles for the appropriate design of customer driven manufacturing systems. Doing so, we will discuss concepts and principles for the design of manufacturing systems delivering a wide range of products and services that meet specific needs of individual customers. Synergies, similarities as well as limitations and potentials of both mass customization and (traditional) customer driven manufacturing will be evaluated.

5.1   Nothing new?

Mass customization (MC) relates to the ability to provide customized products or services through flexible processes in high volumes and at reasonably low costs. The concept has emerged in the late 1980s and early 1990s and may be viewed as a natural follow up to processes that have become increasingly flexible and optimized regarding quality and costs. In addition mass customization appears as an alternative to differentiate companies in a highly competitive and segmented market [1]. However, at the same time there is a long tradition especially in Europe in designing and manufacturing customer specific products such as machinery, ships and even cars [2, 3]. Whereas the concept of mass customization mainly addresses the customer in terms of a consumer the concept of customer driven manufacturing addresses mainly the customer in terms of a manufacturer and/or service provider, i.e. the capital goods industry. However, starting from different perspectives both approaches are aiming at a cost efficient customization (and personalization) of products and services. The objective of this chapter is to present a rough systematic overview of both approaches. The main aim is to support a better understanding of the potentials and the synergies. Synergies, similarities as well as limitations and potentials of both approaches will be discussed.

5.2   Mass customization

Despite the increasing attention Mass customization has been receiving in practice as well as in literature, there is no common understanding about the term. Literature still provides a broad spectrum of definitions. Whereas some authors give a very broad and sometimes very general definition others provide more narrow and practical definitions. In a practical sense mass customization can be seen as an approach that uses information technology, flexible processes, and organizational structures to deliver a wide range of products and services that meet specific needs of individual customers (often defined by a series of options), at a cost near that of mass-produced items [4].

From an economic point of view, implementing mass customization concepts requests a careful determination of the range in which a product or service can be customized efficiently. The level or the levels of individualization of the offered items seem to be critical for an appropriate definition. Pine [5, 6] e.g. suggests five stages of modular production: customized services (standard products are tailored by people in marketing and delivery before they reach customers), embedded customization (standard products can be altered by customers during use), point-of-delivery customization (additional custom work can be done at the point of sale), providing quick response (short time delivery of products), and modular production (standard components can be configured in a wide variety of products and services).

Based on an analysis of existing frameworks to categorize the various levels of customization of a product, Da Silveira et al. [1] propose a set of eight generic levels of mass customization, ranging from pure customization (individually designed products) to pure standardization (see Table 1). Design as level 8 refers to collaborative product development, manufacturing and delivery of products according to individual customer preferences. The next level (Level 7: fabrication) refers to manufacturing of customer-tailored products based on predefined designs. Assembly as level 6 deals with the arranging of modular components into different configurations according to customer orders. On levels 5 and 4, mass customization is achieved by simply adding custom work (e.g. Ikea furniture) or services to standard products (often at the point of delivery). In level 3, MC is provided by alternative approaches for distributing or packaging of products (e.g. different labels and/or box sizes according to specific market segments). In level two, mass customization occurs only after delivery, through products that can be adapted to different functions or situations. Level 1 refers to pure standardization, a strategy that according to Da Silveira et al. can still be useful in many industrial segments.

Table 1: Generic levels of mass customization [6]

Generic Levels	Approaches by Gilmore/Pine [7]	Strategies by Lampel et al. [8]	Stages by Pine [5]	Types by Spira [9]
(1) Standardization		pure standardization		
(2) Usage	adaptive		embedded customization	
(3) Package & Distribution	cosmetic	segmented standardization		customized packaging
(4) Additional Services			customized services; quick response	providing additional services
(5) Additional Custom Work			point of delivery customiz.	performing add. custom work
(6) Assembly		customized standardization	modular production	unique config. out of standard components
(7) Fabrication		tailored customization		
(8) Design	collaborative, transparent	pure customization		

5.3 Customer driven manufacturing

Apart from mass customization there is a long tradition especially in Europe in designing and manufacturing customer specific products such as machinery, ships, etc. In the following, these approaches of customer driven manufacturing will be evaluated and compared with mass customization in the next section.

5.3.1 Customer order decoupling point

In stable and evolving dynamic markets the product has been classified based on the way its demand, from the manufacturer's perspective, is generated. In other words, the point up to which a customer is involved in the final specification of the product. Based on the ratio between those parts of the manufacturing process which are driven by customer orders and those which are driven by forecasts according to Wortmann [10] a manufacturing organization's operation can be classified by introducing the notion of the Customer Order Decoupling Point (CODP). The CODP refers to the point in the material flow from where customer-order-driven activities take place (see Figure 1). The customer's influence on the product can range from the definition of some delivery-related product specifications in advanced phases in the product life-cycle (e.g. packaging, transportation) to a modification of the ultimate functions of the product in the very early phases (e.g. customer-related product specifications).

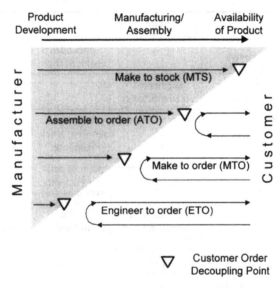

Figure 1: Typology of customer order decoupling point (CODP)

Make-to-stock (MTS): MTS typifies the manufacture of products based on a familiar and relatively predictable demand mix, where product life cycles are reasonably long and stable. In an MTS organization, a stock of finished products is maintained, from which customer orders are filled. Interaction with the customer is minimal and the production volume of each sales unit is high. Whilst an MTS system offers quick product delivery times, disadvantages of this system include high inventory costs and minimal customer interaction.

Assemble-to-order (ATO): ATO organizations maintain a stock of semi-finished products, so that following receipt of an order for a particular configuration, the relevant sub-assemblies can be assembled. In the ATO environment, the same core assemblies are generally used for the majority of products. Furthermore, whilst product routing is fixed, product delivery time is of relatively short length and is based on the availability of major subassemblies. A typical example of this type of manufacturing is that of personal computers.

Make-to-order (MTO): MTO organizations maintain a stock of standard components, so that following the receipt of an order for a particular design; the product is manufactured from these components. Whilst the product is not specified until a customer order is received, finished products from this system are partially one of a kind, but not pure one of a kind because the final product is not usually designed from a particular specification. Interaction with the customer is extensive and is based on sales and engineering. Product delivery times range from medium to long, whilst promises for completion of orders are based on the available capacity in manufacturing and engineering. The manufacture of machine tools and many capital goods are examples of this type of MTO manufacturing.

Engineer-to-order (ETO): ETO represents an extension of the MTO system, with the major difference being that the engineering design of the product is almost totally based on customer specifications. Customer interaction is greater and true one of a kind products are engineered to order. Figure 2 indicates potential customer driven activities in order processing in the capital goods industry.

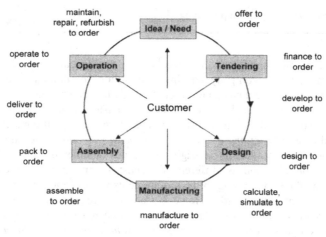

Figure 2: Potential customer driven activities in engineer-to-order (ETO) environments

5.3.2 Typology of production situations

Apart from the CODP, Wortmann proposes a second concept which is important in understanding customer- driven manufacturing. The second concept is concerned with the amount of investment made in developing products, production processes or specific resources independently of the customer order (see Table 2). According to this concept a company can be called *resource-oriented* or *capability-oriented* if it has invested substantially in resources (human, machinery, etc.) but not in specific processes or even products (independently of a particular customer order). A company is *product-oriented* if it has made substantial investments in product development independently of any specific customer order. Quite often, a product-oriented company has also invested in resources.

In addition a company is *process-oriented* or *work-flow-oriented* if it has made substantial investments in production process development independently of customer orders. Quite often, a process-oriented company has also invested in resources. As there are also companies which are both product-oriented and process-oriented, the three concepts used in this dimension are not mutually exclusive.

Table 2:   Typology of production situations [10]

	Engineer-to-order	*Make-to-order*	*Assemble-to-order*
Product-oriented	Packaging machines	Machine tools	Medical systems
Process- oriented	Printing	Fine paper	Service industries
Resource- oriented	Ship building	Repair shop	Construction company

A few examples are given in the following to illustrate the concepts. An aircraft-manufacturing company has invested billions of dollars in developing products without having a customer order. Nevertheless companies manufacturing aircrafts are able and usually have to add some customer driven engineering to each individual aircraft sold to a customer. Therefore, an aircraft company can be categorized as a *product-oriented engineer-to-order company*. The same is often true for companies producing packaging machines or other machinery systems. The production situation is different for most of the aerospace companies producing satellites. According to the requirements most of the satellites are built completely to a customer's order. However, satellites are built by companies being able to develop, manufacture and assemble customized satellites right the first time. Such companies can be categorized as *resource-oriented engineer-to-order* companies. The same holds as well for shipyards building ferries or cruise-liners.

Finally, consider a printing company specializing in weeklies. Suppose that this company subcontracts all the work to other companies, but that it standardizes and organizes the flow of work. Thus, the company has invested in quality control, logistics and in blanket contracts with other specialized companies. At least such a company might be able to subcontract not only the detailed layout and all related preparatory work, but also the printing and finishing and even the distribution of these weeklies. This company can be categorized as a process-oriented or work-flow-oriented engineer-to-order company.

5.4   Comparing mass customization and customer driven manufacturing

5.4.1   The origins

The concept of mass customization (MC) as well as the concept of customer driven manufacturing (CdM) originates from traditional manufacturing paradigms. Whereas the concept of mass customization originates from the mass production paradigm (MPP), the concept of customer driven manufacturing originates from the one-of-a-kind production (OKP) paradigm [2, 3]. However, today both underlying approaches i.e. MP as well as OKP are facing tremendous pressures: Customers are no more interested neither in buying standardized products nor in paying premium prices for customized products or even customized product features. Table 3 gives a first overview of the differences between MC and CdM.

Table 3:   Mass customization vs. customer driven manufacturing

	Mass Customization	*Customer driven Manufacturing*
Complexity of product	low (usually commodities)	high (usually capital goods)
Level of customization of product	low to medium (often restricted to limited variety of configurations / product types)	high to very high (what the customer wants is what he get)
Integration of customer into order processing	usually low	usually high (often fully integrated)
Type of customer addressed	usually consumer	usually suppliers, service providers, etc.

Figure 3 depicts a diagram considering product variety on the one axis and process stability on the other axis can be used to classify the various approaches. Whereas MP can be characterized by a high process stability (i.e. no changes, no modifications are needed during order processing) and a low product variety (in extreme cases there are no variants), the potential of OKP can be characterized by high product variety and low process stability (product specifications as well as process specifications are changing from order to order), (see Figure 3a & b).

As the stability and the smooth operation of a production process is a prerequisite for cost efficiency the MC approach tries to keep the stability of the process but to increase the variety of products (see Figure 3c). To pursue these seemingly excluding goals has become possible, since modern manufacturing and information technologies offer the ability to companies to deliver a certain range of products and services that meet even the needs of individual customers at a cost near that of mass-produced items. Whereas MC mainly addresses the customer in terms of a consumer the concept of customer driven manufacturing addresses mainly the customer in terms of a manufacturer and/or service provider, i.e. the capital goods industry. Originating from customized production approaches CdM is aiming at an increase in process stability without a decrease in the variety of the products to be offered to the customers.

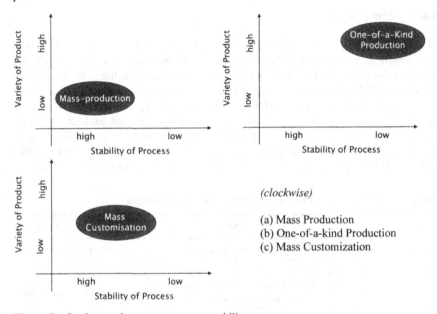

Figure 3: Product variety versus process stability

5.4.2 Common objectives

Starting from different perspectives both approaches do have a common objectives in cost efficient customization (and personalization) of products and services. Using the concept of the customer order decoupling point (CODP) the different starting points of both approaches become obvious (see Figure 4). As discussed above in Table 1 the main focus of MC so far is on ATO (Assemble-to-order) and MTO (Make-to-order). The Engineer-to-order (ETO) approach so far is not realized in practice or - if implemented - restricted to the customer driven variation of single and predefined product features.

Figure 5 depicts a set steps or even changes to undergo for a mass-producer as well as for a one-of-a-kind manufacturer to become a mass-customizing company. By increasing the customer focus through the implementation of the proposed changes a mass producer has to take care of his cost-advantages. Losing the cost-advantages and failing in achieving an increased customer focus leads to a critical situation for a company. By improving the cost focus a one-of-a-kind producer has to take care of his competitive advantages based on his customer focus. To lose the competitive advantage based on the customized offerings and to fail in achieving cost-advantages leads to a critical situation for this company as well.

However in both cases a careful change management process is crucial for a success development of the company.

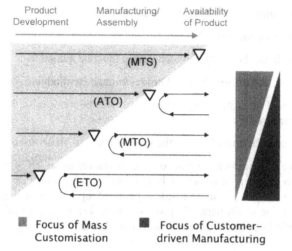

Figure 4: Classifying MC and CdM by using the customer order decoupling point (CODP)

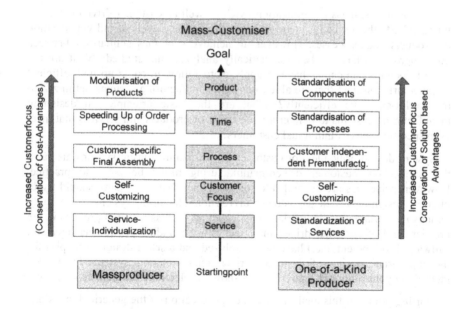

Figure 5: Two starting points, one ultimate goal: from mass production and one-of-a kind production to mass customization (based on [11])

Da Silveira et al. have identified six success factors most commonly emphasized in literature when mass customization is applicable [1]:

- Customer demand for variety and customization must exist.
- Market conditions must be appropriate.
- Value chain should be ready.
- Manufacturing and information technology must be available.
- Products should be customizable (Products must be modularized, versatile and constantly renewed).
- Knowledge must be shared across the value chain.

As the successful implementation of MC relies on these success factors MC cannot be seen as every company's best strategy. To summarize, a successful implementation of MC involves major aspects of operations including product configuration, value chain network, process and information technology, and the development of a knowledge-based organizational structure.

5.4.3   System design principles

Both approaches, mass customization (MC) as well as customer driven manufacturing (CdM) theoretically intend to cover the various levels of individualization of products and processes. However, to achieve synergies similarities between both approaches have to be systematically analyzed and studied: What are the basic principles companies have to apply or to follow to come up with efficient manufacturing system that are able to design manufacturing systems optimized to realize lot-size one efficiently? What are the basic principles in designing manufacturing systems to deliver a wide range of products and services that meet even specific needs of individual customers?

System design principles for customer driven manufacturing systems are rules that show how customer driven-manufacturing might be put into practice. However, principles are general rules or guidelines that a company should try to obey in designing efficient customer driven manufacturing systems. In a study various industrial cases, industrial, consultancy as well as research projects in the broader field of customer driven manufacturing (including project management, software development, etc.) have been analyzed and a set of design principles for the appropriate design of customer driven manufacturing systems have been identified by the author [12, 16]. Table 4 shows a selection of principles identified.

Starting point for this analysis was the specification of the generic dimensions of a customer specific order. According to decision theory various strategies are applicable in dealing with the dimensions of such problems. Problems can be anticipated, avoided, ignored, outsourced, minimized, controlled, etc. From a problem oriented perspective order processing in customer driven manufacturing can be seen as a problem solving process conducted by a company. Analyzing the various facets of this type of "problems" a set of so-called problem dimensions can be defined. Accordingly a customer order can be characterized by a number of dimensions such as its cost-criticality, time-criticality, novelty, uniqueness, complexity, in-transparency, momentum or uncertainty (about goal).

Table 4:   Design principles in customer driven manufacturing

Design principles in customer driven manufacturing		
▪ Small Interfaces	▪ Repetition	▪ Co-operation
▪ Self organization	▪ Feedback	▪ Simulation
▪ Decentralization	▪ Rough vs. detailed	▪ Late Commitment
▪ Pragmatism	▪ Modularization	▪ Push vs. pull
▪ Prototyping	▪ Synchronization	▪ Integration
▪ Parallelism		

A template to characterize the concept as well as the potentials of the design principles in a standardized way has been developed. This template includes sections like "relationships with other principles", "potential to solve a problem", "potential for application" (i.e. application domain), etc. Figure 6 shows the template to systematically characterize the principle "late commitment". Some of the principles identified in the study are of generic nature, whereas others are very specific. Generic principles might be applied to the different organizational levels (strategic, tactical and operational) within a company, whereas the field of other principles is limited.

First steps in analyzing industrial cases, literature and research projects related to MC have shown, that design principles are applied here as well. However, as MC and CdM originates from different manufacturing paradigms (see above), there are many differences as well. "Late differentiation" e.g. is used as one major principle in MC. Applying this principle allows companies to have standard processes until the point of differentiation, very often a prerequisite for cost efficient process design in MC. Apart from the principle "Late differentiation" (synonym in CdM is "Late individualization") in CdM the principle "Late Commitment" is used quiet frequently.

This principle again is well known in decision theory: „Make safe decisions first / dangerous decisions very late". Even in product data modeling a similar principle is being applied: „Early Binding" vs. „Late Binding". De Vin proposed to apply the strategy called "Design by least commitment" in such cases where designers have to take decisions in very unstable environments [13]. Similar proposals have been made as well by Marri et al. and by Knackfuß for decision making in process planning [14, 15].

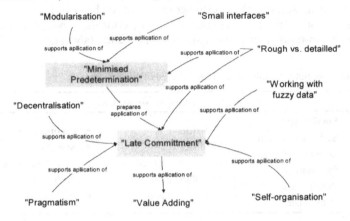

Relationships with other design principles

Potential to solve problem

Strategy	Cost-criticality	Time-criticality	Novelty	Uniqueness	Complexity	In-transparency	Momentum	Uncertainty about goal
To anticipate	–	–	–	X	–	–	–	–
To eliminate	–	–	X	–	–	–	–	–
To controll	–	X	X	–	X	X	X	–

Potential for Application

Type of system to be considered

Resource-related	Process-related	Object-related
–	X	–

Figure 6: Characterization of design principles in CdM (example) [12]

5.5 Conclusion

Both approaches, mass customization as well as customer driven manufacturing theoretically intend to cover the various levels of individualization of products. Whereas the mass customization concept originates from the mass production paradigm (MPP), the customer driven manufacturing concept originates from the one-of-a-kind production (OKP) paradigm. Accordingly the mass customization concept mainly addresses the customer in terms of a consumer and the customer driven manufacturing concept addresses mainly the customer in terms of a manufacturer and/or service provider, i.e. the capital goods industry. However, it can be stated that the more complex the products, the higher the level of customization of the product and the more intensive the integration of the customer into the order processing the applications of mass customization seems to be limited. At the same time the CdM concept seems to be weak when the objective is to deliver customized products at a cost near that of mass-produced items. Both approaches, mass customization and customer driven manufacturing, do have strong as well as weak points. Identifying and applying generic design principles for manufacturing systems providing customized products seems to be a promising approach.

References

[1]   Da Silveira, G.; Borenstein, D.; Fogliatto, F. S.: Mass Customization: Literature review and research directions, in: International Journal of Production Economics, 72 (2001), pp. 1-13.

[2]   Wortmann, J.C.; Muntslag, D. R.; Timmermanns, P. J. M.: Customer driven Manufacturing, London et al. 1997.

[3]   Hirsch, B.E., Thoben, K.D.: One of a Kind Production: New Approaches, Amsterdam 1992.

[4]   Hart, C.: Mass customization: Conceptual underpinnings, opportunities and limits, in: International Journal of Service Industry Management, 6 (1995) 2, pp. 36-45.

[5]   Pine II, B.J.: Mass customizing products and services, in: Planning Review, 21 (1993) 4, pp. 6-13.

[6]   Pine II, B.J.; Victor, B.; Boyton, A.: Making mass customization work, in: Harvard Business Review, 71 (1993) 5, pp. 108-111.

[7]   Gilmore, J.; Pine II, B.J.: The four faces of mass customization, in: Harvard Business Review, 75 (1997) 1, pp. 91-101.

[8]   Lampel, J.; Mintzberg, H.: Customizing Customization, in: Sloan Management Review, 38 (1996) 1, pp. 21-30.

[9]   Spira, J.: Mass customization through training at Lutron Electronics, in: Computers in Industry, 30 (1996) 3, pp. 171-174.

[10]   Wortmann, J. C.: A typology of customer driven manufacturing, in: International Journal of Service Industry Management, 6 (1997) 2, pp. 59-73.

[11]   Reiß, M.; Beck, T. C.: Fertigung jenseits des Kosten-Flexibilitäts-Dilemmas – Mass Customization als Strategiekonzept für Massenfertiger und Einzelfertiger, in: VDI-Z, 136 (1994) 11/12, pp. 28-30.

[12]   Thoben, K.D.: Kundenspezifische Produktion – Anforderungen, Prinzipien, Methoden und Werkzeuge, Habilitation, University of Bremen, Bremen 2001.

[13]   De Vin, L. J.: Co-operation between Manufacturing Functions in Sheet Metal Part Manufacturing, in: Martensson, N; Mackay, R.; Björgvinsson, S. (Eds.): Changing the Ways We Work – Advances in Design and Manufacturing, 8 (1998), pp. 425-431.

[14]   Marri, H. B.; Gunasekaran, A.; Grieve, R. J.: Computer Aided Process Planning: A State of the Art, in: International Journal of Advanced Manufacturing Technology, 14 (1998), pp 261-268.

[15]   Knackfuß, P.: 1992, Ein Konzept zur merkmalsgestützten Konstruktion –
        Arbeitsvorbereitung und Fertigungssteuerung bei der Herstellung von Blechteilen,
        Dissertation, Universität Bremen, Bremen 1992.

[16]   Hirsch, B.E.; Thoben, K.D.: Why Customer driven Manufacturing, in: International
        Journal of Service Industry Management, 6 (1997) 2, pp. 33-44.

Contact:

Prof. Dr. Klaus-Dieter Thoben
University of Bremen, Bremen Institute of Industrial Technology
and Applied Work Science (BIBA), Bremen, Germany
E-mail: tho@biba.uni-bremen.de

6   User Modeling and Personalization

Experiences in German industry and public administration

Thomas Franke and Peter Mertens
Bavarian Information Systems Research Network (FORWIN),
University of Erlangen-Nuremberg, Germany

Whereas the theoretical foundations of user modeling and personalization techniques have been the subject matter of many works for several years, their practical implementation in IS has been neglected for a long time. In order to change this, we developed a couple of computer-assisted information, consulting, decision support and offering systems in our research institute in cooperation with German firms and public administration. A guideline in all of our works is to pragmatically individualize the dialogue between man and machine by user modeling. The operational area of the experiments that will be outlined below ranges from personalized management information systems (MIS) to training and advising systems. One particular experiment will be considered in some detail. This is an online tourism and spare time advising system; it stores information about the user's cultural and leisure interests and generates individualized city tours or activity programs for longer stays. In order to do so, it uses content based as well as social filtering.

6.1   Introduction

One of the long term goals of developing information systems (IS) is a reasonable fully-automated information processing. In this context, 'reasonable' means that the computer should handle any task at which it is better or at least on the same level as a human being [1]. During long evolutionary processes, the human brain has become a very capable information processor. Therefore, one could define as a sub-goal on the way to full-automation that computers should become more manlike. This is illustrated in Figure 1. It is also shown that an IS has to be easily accessible as a first step towards full-automation, so people will be inclined to use it for solving their problems rather than doing it manually on their own. Since IS in general are utilized by a very heterogeneous audience, they have to be able to adapt to the respective individual in order to achieve the highest conceivable accessibility for as many people as possible. We might put our main question as follows: "How does the computer get to know the user?" One technique to reach this end is to employ user models that store interests, preferences and other data of the individual.

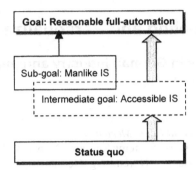

Figure 1:   Paths to fully-automated IS

6.2   Experiments with user models

Although the theoretical background of user modeling and personalization techniques has been investigated into some depth for several years, their practical implementation in IS is still in its initial stage [2]. Therefore, we decided to carry out some experiments with applications that strongly rely on those elements in order to explore possible hindrances and incentives of their utilization. A number of examples from the domains of management information systems (MIS), training and advising systems will be described in the following paragraphs.

6.2.1   Personalized MIS

A lot of executives suffer an 'information overload' from the vast amount of data they have at their disposal, either from the internet, from other external sources or from internal IS. In this situation, user modeling techniques can contribute to a solution in different ways.

MINT - management information from the internet

The MINT system was developed in cooperation with the SAP AG, a German developer of standard software, in order to enhance SAP's R/3 Business Information Warehouse product. The idea behind it was that traditional executive information systems focus on quantitative, internal data, which is provided by the company's ERP system (e. g. costs of production, contribution margins of different products or productivity indices of staff and machines) [3].

In a fast changing environment with strong competition, however, it becomes more important to also consider qualitative and external data, like competitors' market shares, press releases or duration and expirations of their patents, before making operative, let alone strategic decisions. In order to provide the decision-makers with this vital information, the MINT-system contains an 'editor's workbench', with the help of which the user searches, records and categorizes news items from the internet [4]. After that, individual reports for the managers are assembled, either triggered by certain dates or specific patterns of key performance indicators. The personalization of these reports is achieved by means of a hierarchical user model, which is defined in a module called 'Profile Manager'. The model differentiates between the company, role and user levels of customization (see Figure 2).

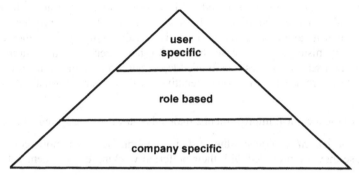

Figure 2: Layers of the MINT user model

The pyramidal form indicates that the stored information becomes more and more specific ascending through the layers. On the company level, the user model comprises e. g. industry and enterprise-type specific factors whereas the role level includes the objective information requirements of the user with regard to his position in the company. Finally, the user layer contains an individual's personal preferences concerning the representation of data and the previous knowledge the user has. By taking into account all these pieces of information, the MINT-system assembles reports that match both the addressees' objective and subjective information needs as well as their reading habits. Meanwhile, innovative elements of the MINT concept were implemented in SAP's SEM (Strategic Enterprise Management) initiative.

Adaptive briefing book

This prototype is the product of a collaboration with Siemens A&D, a division of the Siemens Corporation. Its aim is to provide executives with the relevant data in an environment of fast changing information requirements. In order to do this, it analyzes internal information stored in a huge Data Warehouse and generates a briefing-book [5]. As the generation of ad hoc reports takes rather a long time (up to 15 minutes) due to the size and structure of the Data Warehouse, the briefing-book system tries to accelerate the process by means of pre-generated statements. For this purpose, it has to anticipate the user's future data queries. This is where the user modeling aspect sets in. The system stores the manager's query history in order to extrapolate future demands. Individual settings of the query form, like data dimensions to visualize (customer, region, period etc.), preferred performance indicators or sorting criteria, complete the user's profile.

The content of the briefing-book is determined using the method of single exponential smoothing for predicting the probability of a future request for a certain report from the past frequency. Subsequently, the system performs a kind of ABC analysis on the calculated figures and chooses those reports that have the highest probabilities and cover 80 % of the potential queries when combined. These can be generated during periods of low system load (e. g. during the night) and are ready for use straight away the next morning. The main problem of this method is that it cannot be used with recently recruited employees as they do not have a query history. In this case the briefing-book generator uses stereotypes based on the user's attributes 'position', 'department' and 'education' in order to present the addressee reports from executives with similar information requirements.

Besides the technical improvements, main advantages of the system are:

- For *experienced managers*, the system handling becomes more convenient because they are provided with their preferred working environment automatically.

- *Unpracticed users* get suggestions for reasonable analyzes.

- The system can now be employed in *day-to-day business* due to a reduction of the response time from 15 minutes to approximately one second.

Although we could demonstrate that the idea works, the total project was not successful because of difficulties with the size and the structure of the data warehouse.

UNTERNEHMENSREPORT II

The cooperation partner for this project was the DATEV eG, Nuremberg, a large German cooperative association. This association supports German tax consultants with information technologies. The typical clients of the tax consultants are small to medium-sized enterprises. The system UNTERNEHMENS-REPORT II generates textual reports (expertise) concerning the economic performance of these enterprises by analyzing their balance sheets, profit and loss statements and by intercompany comparisons. Beyond that, it calculates their financial standing with regard to future loans from banks.

After having analyzed the data and generated the tables and charts, the challenge is the 'configuration' of text passages. For this, templates for sentences that consist of fixed letters and text variables are stored. Depending on the result of the analysis, these variables are substituted by suitable text passages in order to get complete phrases. This is especially difficult since German grammar is rather complicated.

The addressees of those reports might be tax consultants, entrepreneurs or even banks who want to decide whether they will grant another credit or not. Depending on the recipient, the user can select whether the report should contain tables or not, whether it will include diagrams and if so, which kind of graphical presentation would be preferred (see Figure 3, left). Another option is the possibility to choose the level of detail for the expertise. For this reason, a hierarchical design that consists of different layers was implemented for the expert system. The different types of reports that can be generated depending on the recipient can be seen in Figure 3 (right).

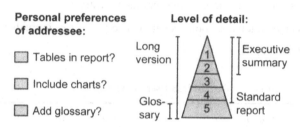

Figure 3: Addressee Orientation in UNTERNEHMENSREPORT II

Especially if the report is intended for a top manager of the client's organization, recommendations as to how to react to a specific situation may be useful. However, it is not advisable to use evaluations and critiquing sentences for all addressees (especially external ones). The recommendation and judgment problem was solved by providing the respective pieces of text with a special attribute. The default choice for a report is to compile expertise without these marked passages [6].

In the case of banks as recipients, one cannot speak of user modeling but rather of addressee modeling. The peculiarity in this context is that the banks' guidelines in carrying out credit investigations partly vary with regard to performance indicators and business terms. UNTERNEHMENSREPORT II takes this into account by storing the banks' specific vocabularies separated from the rest of the knowledge base. UNTERNEHMENSREPORT II is a running system. About 30 reports per day are produced in quite different tax consultants' offices.

6.2.2   Personalization in training and advising systems

Training and advising systems are a classical domain of user modeling. The first, because they can explicitly request knowledge about the users, hence the user models can be generated more efficiently than in any other domain. The latter, because they need to know details about the user in order to make deliberate recommendations.

AMPreS – adaptive multimedia presentation system

The goal of this project, which was implemented in cooperation with the study group of the German Ceramic Society, is to present rapidly increasing knowledge in a complex domain to a heterogeneous audience using multimedia techniques. In the developed prototype (an industry catalogue of technical ceramics) the main purpose of user modeling is information filtering and the compiling of individual guided tours through the domain of knowledge. Potential target groups for this prototype are [7]:

- engineers and businessmen in the ceramic industry;
- teachers at vocational schools, vocational counselors or journalists;
- students and pupils.

Table 1:   Dimensions of the AMPreS User Model [7]

Dimension	Parameter values			
intention	overview		details	
time budget	low	medium	long	
knowledge (domain)	layman	specialist	expert	
knowledge (system)	beginner		experienced	
cognitive style	holist		serial	
presentation type	intellectual	photographic	cineastic	audible

In order to address the different motivations, knowledge and skills of those people, the system classifies them by the dimensions shown in Table 1. In addition to these stereotypes, the system also stores individual values like bookmarks or search items. AMPreS acquires the information necessary for the assignment of the parameter values in the different dimensions with the help of a special starting routine. This consists of a few explanatory pages that contain, for example, terms a specialist or expert in the ceramic domain should be familiar with, but which a layman is unlikely to know. In this case if the user requires additional information about a specific term (via a hyperlink), the system assumes he is not an expert. Similar mechanisms are used to determine the values in the other dimensions. Concluding from the attained data, AMPreS selects the relevant data pages and media from the catalogue and arranges them into a sensible tour. Such a tour would contain more tutorial-like passages and explanatory charts if the addressee was a pupil with little domain knowledge and more technical facts and figures as well as engineering drawings in the case of an expert (see Figure 4).

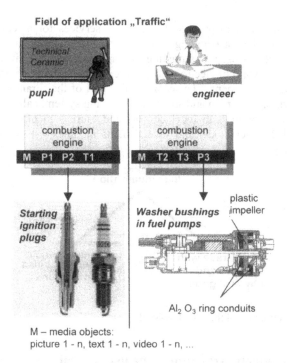

Figure 4: Individualized guided tours in AMPreS

6.3 Personalized online regional tourism and spare time advising system

6.3.1 Project background

Another example for personalization in the field of training and advising systems will be discussed in greater detail in the next paragraphs. This is an online tourism and spare time advising system for the region around Nuremberg in Germany. In March 1999 the RegioSignCard concept, which was introduced by the region of Nuremberg together with some surrounding cities, succeeded in becoming one of three candidates to win the Media@Komm competition that was initiated by the German Ministry of Science, Research and Technology. The main goal of this initiative is to create a regional platform for both municipal and private services on the internet using multimedia technologies and the digital signature. To achieve this, a multi-functional chip card is to be introduced, which will serve as a central means of communication between all applications.

The 'Franken-Mall' is one of the so called private-public-subprojects within the RegioSignCard concept. Its fundamental idea is individualization (see Figure 5).

Firstly, it is intended to provide added value services for a regional electronic market-place for the participating cities and their surrounding areas. This market-place is primarily focused on attracting the small and medium sized enterprises of the region to the world of electronic commerce and helping them get started in this new business environment. The second major part of the 'Franken-Mall' consists of the said regional tourism and spare time advising system, called TourBO, which is intended to support both local citizens and tourists in planning their leisure activities [8]. Its user model and several of its modules will be the topic of the following chapters.

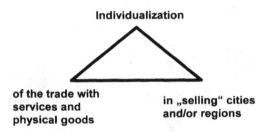

Figure 5:   Central idea of the 'Franken-Mall'

6.3.2   Content and structure of the user model

TourBO needs individual data from its users for most of the services it makes available for them; only very few modules, such as a search machine for events or attractions, do not use this information [8]. The main part of this user model consists of an interest profile that stores the users' leisure preferences in the following way: the domain of leisure activities is structured in six categories, namely 'sightseeing', 'music', 'sports', 'cultural activities', 'nightlife', and 'shopping', which are subdivided into a total of 31 smaller rubrics (e. g. 'theatre', 'opera', 'art and exhibitions', and 'children's theatre' in the case of 'cultural activities'). In each of these subdivisions the user may specify his interests in six degrees from 'no interest at all' to 'very high interest'.

These verbal statements are converted into decimal numbers, internally, for better handling in later calculations and stored in a database. Additionally, the customer can tell the system his age and gender as well as whether he prefers rather unhurried or active leisure programs. Table 2 shows the classification of the profile at hand within the framework for user models Mertens et al. proposed [1]; the appropriate attribute values are shaded in a darker gray than the other ones. In detail, this means that the used profile aims at selecting those pieces of information from a large amount of data which is most relevant for an addressee rather than rehash and present data on the screen according to his preferences. The said addressee may be a single customer of the system or a whole group (e.g., a tourist party), and in most cases the addressee of the gathered information is also the operator of the system.

Table 2: Classification of TourBO's user model

Attribute	*Attribute values*		
purpose	selection	*presentation*	
		domain	system
object	*addressee*		operator
	customer	role	organization	group	
accuracy	individual	stereotypical	
character of information	soft information	hard facts	
variability	static	dynamic	
duration	long dated	short dated	
visibility	transparent	nontransparent	
extraction	implicit	*explicit*	
		ex ante	ex post
acquisition of knowledge	manually	automatically	

On the one hand, each customer may specify his own individual profile rather than being assigned to a stereotypical group. However, an elaborate stereotypical approach is also considered and developed, in order to reduce the user's effort of entering his data. We try to overcome the two major drawbacks of stereotypes (the limited variety of the starting profiles and the ambiguity of the assignment process) by introducing fuzzy logic methods [9]. Furthermore, a weighing mechanism mixes two or more stereotypes to generate an initial profile, if appropriate.

The profile stores only hard facts, which the user has explicitly keyed in before entering the system (ex ante), and does not make any implicit assumptions about other 'softer' factors. By-and-by the customer may give feedback about recommendations he was given and thereby adapt his profile (explicit, ex post extraction). As the entered data is stored in a database between two sessions, it is a long term profile, which can be viewed by the customer at any given time (transparent). The dynamic user model can be changed manually during advising sessions as well as in between them but never does so automatically [10].

6.3.3 Functionality of TourBO

The following sections will introduce the most important of TourBO's modules. Their functions range from generating individual sightseeing tours or event newsletters to the recommendation of suitable leisure partners and eventually to a construction kit for package tours.

Recommendations for suitable leisure partners

The target group for this module are mainly local citizens and in particular those who have moved in recently and do not have many acquaintances yet. If such a person wants to play tennis, badminton, squash or the like, there is always the problem of finding a suitable partner. Another typical situation is that somebody wants to go out clubbing, to the theatre or other cultural activities but has no idea where to go and with whom. This is where the partner recommendation module can be of use.

The system may be utilized in two different ways. Firstly, a customer who would like to undertake some leisure activities with another person, but has no concrete plans, can enroll in a pool of potential partners, from which the recommendations are taken later. Secondly, a person who seeks a partner for a certain activity (e. g. playing squash) on a certain date may initiate an active search [11]. For both modes of utilization the respective user model, as described above, is augmented with several KO-criteria regarding the recommended partners, e. g. desired age, gender and number of partners (group or single). With this information, the module calculates the best suited partners in a three level process. In the first step it verifies the KO-criteria and singles our all candidates that do not meet them. Subsequently, it tries to find out those candidates whose profiles closely resemble the seeker's. In order to do so, it takes into account two things: the absolute distance between the profiles and their degree of synchronism.

The former figure is calculated using an adapted version of the square Euclidean distance or second order Minkowski metric [12]:

$$(1) \quad d_{A,B} = \frac{\sum\limits_{i=1}^{N} w_i \cdot (r_{Ai} - r_{Bi})^2}{(r_{max} - r_{min})^2 \cdot \sum\limits_{i=1}^{N} w_i}$$

with:

$d_{A,B}$	:=	distance of profiles A and B, $d_{A,B} \in [0;1]$
r_{Ai}	:=	interest value of profile A in rubric i
r_{Bi}	:=	interest value of profile B in rubric i
w_i	:=	weight of i-th rubric, $w_i \in \{1, 2, 3\}$
r_{min}	:=	lowest possible interest value
r_{max}	:=	highest possible interest value
N	:=	number of rubrics in a profile

As it is a square distance, one large divergence has a higher influence than several minor ones with the same absolute value. The changes that were made to the conventional Euclidean distance have the following effects: Firstly, the different rubrics of the profile are given different weights according to their relevance for the desired activities. Those activities the user seeks a partner for are weighted to a value of three, rubrics in the same superior category are given a weight of two and the rest get one. As the sum of weights can differ between two searches (due to a different number of rubrics within a category), the result is finally normalized to a range of values between zero and one by dividing it by the largest possible distance. All candidates whose profiles exceed a marginal distance value (e. g. 0.3) are sorted out in this step (see Figure 6, the arrows indicate the distance).

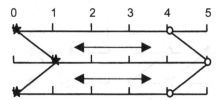

Figure 6: Absolute distance of two profiles

Now the remaining potential partners pass through the next level of the matching process. Here, the similarity of the profiles regarding their synchronism is checked. This means, two profiles that are exactly parallel to each other are considered to be most similar regardless of their absolute distance. This is because we assume that if the absolute distance between two profiles falls below an acceptable limit their synchronism becomes more important than a more exact matching. The q-correlation coefficient, which calculates and compares the divergences of the single values in a profile from its average value, is able to perform the task of determining this kind of similarity [12]:

$$(2) \quad S_{A,B} = \frac{\sum_{i=1}^{N}(r_{Ai} - \bar{r}_A) \cdot (r_{Bi} - \bar{r}_B)}{\sqrt{\sum_{i=1}^{N}(r_{Ai} - \bar{r}_A)^2 \cdot \sum_{i=1}^{N}(r_{Bi} - \bar{r}_B)^2}}$$

with:

$S_{A,B}$	:=	similarity of profiles A and B, $S_{A,B} \in [-1;1]$
r_{Ai}	:=	interest value of profile A in rubric i
$\bar{r}_A$	:=	average of all interest values in profile A
r_{Bi}	:=	interest value of profile B in rubric i

$\bar{r}_B$:=   average of all interest values in profile A

N :=   number of rubrics in a profile

Figures 7 (a) and (b) show two examples of profiles with the same absolute distance. However, in (a) they run exactly parallel to each other, which would lead to a q-correlation coefficient of 1, whereas in (b) they are inverted resulting in a value of -1. According to their performance in the final matching step, the system sorts the partners and recommends the best three to the seeking customer.

As it is more likely that a person who actively seeks a leisure partner agrees to his e-mail address being disclosed to a third party, the recommended partners are informed as well and provided with this address. Then they may contact each other and arrange a date and place for a meeting.

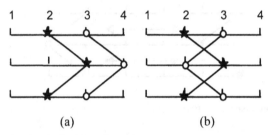

(a)   (b)

Figure 7:   Examples for similarities of profiles

Personalized sightseeing tours via evolutionary algorithms

One of the focal points of our research work is the integration of soft computing techniques in tourism consulting. This module generates sightseeing tours that take into account individual interests as well as the location of the attractions in order to find as good a round trip as possible using an evolutionary algorithm. We use these evolutionary algorithms to improve an initial solution in analogy to a Darwinian process in nature. Since only the best solutions of a problem are used to derive the next generation, the generated solutions tend to become increasingly better with each reiteration of the procreation and selection process because the 'fitter ones' will dominate the others [13].

In our case the starting population consists of a number of timetables that contain the different sights which should be visited during the tour as well as the breaks that are necessary in order to get from one location to the following and the respective durations of both. Additionally, the population contains so called 'strategy parameters' that store the probabilities with which one of the following genetic operators is applied [14]:

- *Insertion:* Adds another activity to a timetable.

- *Deletion:* Removes a random activity from a timetable.

- *Mutation:* Exchanges an activity; that means one is deleted and another one is inserted.

- *Inversion:* Changes the sequence of the activities within a timetable.

- *Recombination:* Splits two timetables and recombines their parts crosswise.

The utilization probability of the operators is adjusted depending on their performance: The better the effects of an operator on the fitness of a solution are, the more often it will be used. In general, this leads to a faster approximation of the theoretical optimum than an algorithm without strategy parameters. Having generated the descendants of the initial population, the system has to determine the fitness values of the different timetables. To do this it makes use of a so called 'matrix of contentment' which has the structure depicted in Figure 8.

The first column (❶) includes the content of the timetable, that is to say the sequence of activities and breaks. In the following columns the ratings of different members of the sightseeing group are situated (❷). For example, member one gives 1 point to activity A whereas member two awards 7 points to the same activity. Breaks and idle times that are the result of going from one location to another are negatively rated, more so with the duration of a break increasing, less if a user desires an unhurried tour. The length of the pauses is calculated from the coordinates of the activities without considering means of transportation, traffic density etc. This method could be refined by calculating the duration with the help of weighted, directional graphs that represent different traffic networks for different means of transportation and/or daytimes.

Now the sums of the column values indicate a single member's individual approval of the timetable in question (❸). If this value is very low, the system could either dynamically increase the weight of this member's rating in order to lessen his disadvantage or recommend that he should rather take his tour together with another group that fits better to his interests.

On the other hand, the sums of the lines show the group's collective opinion of a single activity (❹): the higher this value, the bigger the contribution of the associated activity to a high overall rating of the time table. Furthermore, the matrix provides another indicator for the quality of an element: In the rightmost column the standard deviation of the group members' opinions from the average is calculated (❺). This gives a clue about the degree of consensus within the group. For example, activity A and C have the same overall rating whereas the standard deviation of A is way beyond that of C. This means there is a greater degree of consensus on activity C in the group hence it should be preferred by the system.

At this point however, the algorithm only takes into account the overall sum of the matrix (❻) in order to calculate the fitness value of the timetable, albeit the discussed refinements will be implemented in the next version. The 'fittest' solutions are now selected for another turn of applying the genetic operators to generate a population of descendants; the process starts over again until a sufficient solution is reached.

❶	grade 1	❷ grade 2		grade N	❹ Σ	❺ σ
activity A	1	7		2	10	2,6
break	-3	-2		-1	-6	0,8
activity B	6	9		10	25	1,7
break	-6	-3		-2	-11	1,7
activity C	4	3		3	10	0,5
⋮	⋮	⋮	.....	⋮		
activity N	10	2		5	17	3,3
Σ	12 ❸	16		17	45 ❻	2,2

Figure 8: Matrix of contentment

Construction kit for package tours

Another module that focuses on visitors from abroad is TourBO's configurator for package tours. Its intention is to make up complete journeys from larger building blocks rather than from single activities, as it is the case with the sightseeing tour generator. Thus, fewer interfaces between the separate parts exist, which on the one hand reduces the complexity of composing the travel and on the other hand makes it easier to offer the whole package from one vendor with 'one face to the customer'.

According to this, a travel generated by TourBO comprises several building blocks, each of which has a certain theme or motif, e. g. a basic module containing the accommodation, a culture module consisting of a theatre or opera visit and a sightseeing tour and finally a wellness module including a stay at a spa. In order to assist the operator in creating these building blocks the system provides an editor which allows to design templates that are filled out if required whenever a customer starts a configuration session. A building block consists of several components which again are comprised of one or more elements [15]. Therefore, the design process of the templates has three levels. Firstly, the user creates a new building block and enters its basic data, like its name, a short description and the overall duration of the block. Subsequently, the different components of the module have to be defined. Figure 9 depicts part of the component definition screen.

Besides the component's name and description, the following specifications, which are necessary for the configuration process later, are determined here:

- *the number of elements* (meaning single activities or attractions) the component is made of (e. g. 10);

- the *relative weight* the component is given within the building block (e. g. 4);

- the *rubrics of the interest profile* to which the component is related (e. g. 'churches/cemeteries', 'prominent buildings' and 'fountains').

Basic and component data together make up the framework for a building block, which is filled with content from the underlying database just-in-time when-ever a configuration process is initiated.

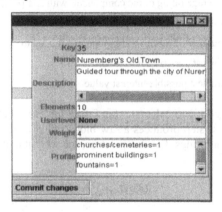

Figure 9: Component definition screen

After having discussed the back office system of the construction kit, we will have a closer look at the front-end and its functionality. The system offers four different search modes, the first of which is the so called 'convenience search'. This is what we call an 'active configurator', meaning that the interaction level with the user is very low [16]. In fact, he has to enter nothing more than the period of time he wants to plan for, the desired price category and the interest profile he wants to plan with. Based on this information, the system generates a whole travel automatically, selects the suitable building blocks and arranges them. As opposed to that, the 'building block search' could be called 'passive configurator'. Here, the customer chooses the modules he wants to have integrated in his journey from a list and the system merely puts them together properly. These two search modes have in common that they produce complete package tours, whereas the missing two ones are intended only for doing some fine tuning of already generated travels, either by adding single activities from specified categories with the 'component search' or by entering free text to describe the desired activity in the 'keyword search'.

As mentioned above, in the 'convenience search' the configurator has to select the most suitable building blocks for a given individual based on its interest profile. In the following paragraph we will introduce the used method in some detail. Figure 9 shows that a component may consist of several single activities (elements). Therefore, in the first step the system calculates the average component rating c:

$$(3) \quad c = \frac{\sum_{i=1}^{N_r} r_i}{N_r}$$

with:

c	:=	average rating of the component, with $c \in [r_{min}; r_{max}]$
r_i	:=	interest value of the profile in rubric i
N_r	:=	number of rubrics a component is related to
r_{min}	:=	lowest possible interest value
r_{max}	:=	highest possible interest value

After having used (3) to calculate this figure for every component of the block in question, the next step produces a raw block value as the weighted sum of the separate component ratings. For this purpose, the weight that can be seen in Figure 9 is used:

$$(4) \quad B_r = \sum_{i=1}^{N_c} c_i \cdot w_i$$

with:

B_r	:=	raw block value, $B_r \in [r_{min}; r_{max}]$
c_i	:=	average rating of component i
N_c	:=	number of components in the building block
w_i	:=	weight of component i in the block, $\Sigma\, w_i = 1$

Up to this point, the calculation takes into account only the absolute interest values in the profile and neglects the varieties of interest in the different components. However, we want to select building blocks that have high absolute interest values and at the same time maintain a balanced interest throughout all components. The variance of interest is an indicator for this:

$$(5) \quad \sigma^2 = \sum_{i=1}^{N_c} (c_i - B_r)^2 \cdot w_i$$

As (in case of a finite and discrete scale) this figure reaches its maximum if one half of the single values is situated at the lower end of the scale and the other half at the upper end, the range of values for the variance can be calculated as:

$$(6)\ \sigma^2 \in \left[0; \left(\frac{r_{max} - r_{min}}{2} \right)^2 \right]$$

Using (6), we derive an additional factor that adjusts the raw block value within a symmetrical interval around B_r depending on the variance of interest in the components:

$$(7)\ B_f = B_r \cdot \left(1 + \lambda - \frac{8 \cdot \lambda \cdot \sigma^2}{\left(r_{max} - r_{min}\right)^2} \right)$$

with:

B_f := Final block value, $B_f \in \left[(1 - \lambda) \cdot B_r; (1 + \lambda) \cdot B_r \right]$

λ := adjustment parameter for range of final block value

In case of no variance at all the raw block value is increased by $(100 \cdot \lambda)\%$ (e. g. $\lambda = 0,1$ results in a 10% enhancement) whereas in the opposite case it is reduced by the same amount. All other variances lead to results between these two extremes. This means, the better balanced the components in a building block are, the bigger the upgrade of the absolute interest values and vice versa. When the system has calculated B_f for all available building blocks, it selects the best ones successively until either their cumulated duration reaches the available time in the travel period or the final block value falls below a predefined threshold. The customer may store the resulting journey in a kind of 'shopping basket' in order not to lose previously generated results when starting another configuration run, and to be able to compare the different results directly on screen.

Individual event newsletters via e-mail

This module is directed mainly to the local population of the region of Nuremberg. It is also the first module that presently makes the step from prototype status to running system. The idea behind it is to provide information about current events on a regular basis with a push medium in an individualized way. The first problem in this context is how to fill the underlying event database with high quality content. In order to achieve this, we identified two possible ways of bringing data into the system. The first one is an automatic import via an XML interface [17]. This is mainly intended for bigger event organizers with a corresponding number of events per period. Alternatively, we want to provide a web based input dialog for the smaller organizers, whose minor amount of event announcements does not justify the installation of the automatic import tool.

As on the one hand a high quality of data is essential in order not to displease the customers with events that do not take place at the published date or location, and on the other hand the editor would suffer from an immense work overload if he was to check all incoming events for completeness, consistency and plausibility, we devised a dynamic sampling algorithm which determines how closely an organizer has to be watched depending on the quality of data he supplied in the past. Figure 10 depicts the procedure of dynamic sampling. In our case the procedure has three main testing levels; so at least one in eight events is checked. As the current testing level represents primarily the short term quality of an organizer's data, the number of correct events that is necessary to advance to the next level could be varied dynamically according to the long term quality (e. g. expressed by the overall percentage of 'faulty' events).

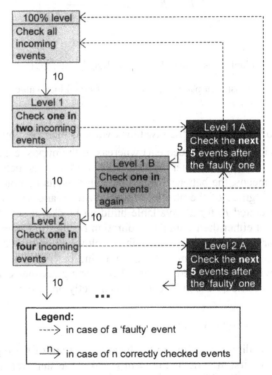

Figure 10: Dynamic sampling procedure [16]

After having filled the database with content, the next problem is to select the most suitable entries for a given individual and his newsletter. For this module, we made some changes to the underlying user profile. The categorization of leisure activities is transposed into a three-level hierarchy instead of the prior two-level one in order to achieve a higher accuracy of recommendations. In each of the resulting rubrics the customer merely specifies whether he is interested in it or not. Additionally he may enter a number of keywords which describe his special interests e. g. in a certain composer, artist or author. Correspondingly, the event organizers classify their activities in the same hierarchy. The generated message will consist of four major divisions, namely:

- 'Personal Highlights', based on exact profile matches on the third level of the hierarchy.
- 'The prospector's treasure trove', based on keyword hits.
- 'What else might interest you', containing less accurate profile matches.
- 'Soon to come', comprising advance notices.

In order to compose the personalized mail, the system passes through the following compilation procedure for each subscriber. First of all, it selects all potential recommendations, i.e. events that could be of interest in any way, from the event database. After that it rates all candidates as shown in Table 3.

Table 3: Rating of potential newsletter recommendations

Description of candidate	Awarded points
Keyword match	10
Exact match on level 3	7
Partial match on level 2	4
Partial match on level 3 (parallel category)	1
Insider's tip (by editor)	3
Editor's recommendation	1

Figure 11 depicts part of an exemplary user profile with the respective ratings. The darker shaded categories indicate that the customer is actually interested in 'operettas', which leads to an internally represented partial interest in the superior category of 'musical theatre' as a whole, and that he has entered the composer Franz Léhar as a keyword. The awarded points are cumulative; that means an operetta written by Franz Léhar would achieve 17 points, one from another composer 7 and Léhar's only opera 'Kukuschka' would get 11 points. In case of multiple profile matches, however, only the most detailed match counts (e. g. one does not add the 4 points for musical theatre into the result of the first example).

In the next step the system selects the N events with the highest ratings from all potential recommendations, with N being the maximum capacity of the newsletter. At this point, the process could already end if no complications are encountered. Unfortunately, there is a relatively high probability of having 'tied events' in the final stage of the selection process. This means there are e. g. five candidates with the same number of points for only two empty slots to be filled in the newsletter.

In this case the system calculates the distribution of the already accepted events on the rubrics of the user profile, on the divisions of the newsletter (keyword hits, advance notices etc.) and on the different days of the week. Then it upgrades those events from the most underrepresented interest category with a temporary bonus point. This is in order to achieve a balanced mix of equally rated events from different rubrics. If only one event is affected by this upgrade, the algorithm adds this one to the newsletter, deletes the temporary point(s) and starts over again with the calculation of said distributions until the capacity of the newsletter is reached. If not it subsequently repeats the upgrading process described above with events from the most underrepresented division of the newsletter respectively the most underrepresented day of the week, again with a well balanced newsletter in mind.

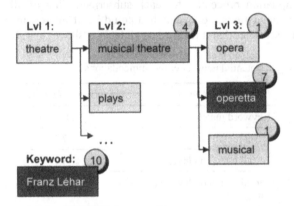

Figure 11: Rating of exemplary user profile

As the different upgrades are logically linked with a Boolean AND-operator, the probability of 'tied events' becomes smaller and smaller proceeding through the different steps of the process. Nevertheless, if there still are any after the final upgrade, the system chooses one of the remaining, equally rated candidates randomly. In order to have a better control over the generated newsletters, this compilation procedure is situated in a simulation environment; only if the editor has looked over and endorsed a summary of the generation process (including statistics etc.) the actual mailing takes place.

6.3.4   Privacy issues

The whole subject-matter of privacy in personalized web services intrinsically consists of two different problems: The first one is the legal situation, which is very complicated especially in Germany. Besides the general BDSG (federal law for the protection of privacy), there are two specific laws that deal with privacy in electronic services, namely the IuKDG (law for information and communication services) and the MDStV (federal treaty for media services). Their relevancy is determined by the distinction whether a service is directed at an individual or at the public and whether the editorial processing of data is a main part of the service. As can easily be seen, already the ascertainment of the relevant legal basis for a personalized online service poses a nontrivial problem.

As all the laws only regulate the handling of data that is linked to a person, in our case we tried to avoid this problem by working with anonymous or pseudonymous data (like a web-based e-mail address) only. In the near or mid-term future we will be able to store genuine personal data as well, because the city of Nuremberg wants to issue a multifunctional chip card including a digital signature function, which allows a user to give his approval to the storage and utilization of his personal data online, and thus safeguards the service provider against the privacy laws [18].

The second and more subtle problem in the privacy context is the difficulty to build up trust in the service providers on the part of the users. On the one hand, advising systems depend on detailed data about their customers in order to generate satisfying results. On the other hand, the users have a much more negative attitude towards the survey of personal data in the internet than in everyday life, as they traditionally regard online anonymity as a protected good. This phenomenon, coined 'the web trust gap' by Weitzner [19], is illustrated in Table 4.

Table 4: Effects of the 'web trust gap' [19]

Activity	User's assessment	
	offline (e.g. local bookstore)	online (e.g. Amazon.com)
track individual buying habits	personal touch	intrusive
inform customer when new book arrives	thoughtful	spam
make suggestions based on other's preferences	good salesmanship	big brother inside

One of the international standards we studied during our survey of online user modeling is the Platform for Privacy Preferences Project (P3P), an initiative that could possibly solve this dilemma [20]. On the one hand, this project provides an XML-framework to store user data in several standardized categories, ranging from physical contact information to demographic and socio-economic data and on to preference data. On the other hand P3P obliges service providers to publish their privacy policies both in machine and human readable formats before collecting personal data. It also provides a means for the user to specify conditions that have to be met by a data collector in order to grant him access to specific data elements. The aspired negotiation process could help to 'close the web trust gap'.

For our purposes of online tourism consulting the applicability of such a standardization effort is high: most of the information categories needed in TourBO can be mapped on the P3P framework [18]. However, P3P in its current state also has some minor drawbacks: either the user cannot sue data collectors who violate their privacy policy because of a missing legal basis or the approaches of the P3P initiative do not reach far enough as the effective laws of a country include stricter regulations (like in Germany) [21].

6.4   Conclusion

It was shown that there is a wide spectrum of possibilities to enhance IS with personalization functionality. In three of the most important fields like MIS, training and advising systems, we presented some ideas and experiments which for the better part are picked up, adopted or adapted by private and public enterprises. Furthermore, a new Bavarian research initiative called FORSIP (Forschungsverbund Situierung/Personalisierung (Situation/Personalization)) has been founded recently. Its goal is to integrate user models in terms of profiles which are valid for a longer time with the actual situations of the persons. The latter can be divided into her location (which is important for M-Commerce and applications like TourBO (location based services, cf. [9])) and her emotional state as detected by voice, prosody, the dynamics of hand movements when writing etc. (see Figure 12).

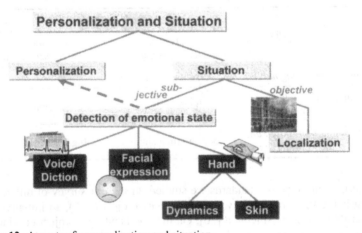

Figure 12: Aspects of personalization and situation

There is a feedback from the detection of the emotional state to user modeling. So if the system would find out that a user often gets angry immediately even after a small irritation, the system might add the descriptor 'choleric' to his or her profile. Therefore, we continue our research in this field, particularly concerning the two main hindrances of user modeling: privacy and acceptance. In case of the tourism consulting system, one way of building the trust necessary for acceptance could be the inclusion of an explanation component. This module explains to the user why the system gave certain recommendations to him. However, it should not simply rely on the same interest profile as the recommendation engine itself, but should also include psychological aspects of the explanation. Therefore, we are working to supplement our user model regarding the major domains of the user's personality (e. g. extraversion, neuroticism or openness), in order to make better and more fitting explanations to the customer. Another field of research will be the personalization of a stakeholder management system (SMS) on role-oriented basis.

References

[1] Mertens, P. et al.: Wie lernt der Computer den Menschen kennen? – Bestandsaufnahme und Experimente zur Benutzermodellierung in der Wirtschaftsinformatik, in: Scheffler, W.; Voigt K.-I. (Eds.): Entwicklungsperspektiven im Electronic Business – Grundlagen – Strukturen – Anwendungsfelder, 2000, pp. 21-52.

[2] Lindner, H.-G.: Sind benutzeradaptive Anwendungssysteme marktreif?, in: Schäfer, R.; Bauer, M (Eds.): ABIS-97 – 5. Workshop „Adaptivität und Benutzermodellierung in interaktiven Systemen", 1997, pp. 91-96.

[3] Mertens, P.; Griese, J.: Integrierte Informationsverarbeitung, in: Planungs- und Kontrollsysteme in der Industrie, Bd. 2, 9th Ed., 2002.

[4] Meier, M.; Mertens, P.: Linking Key Figures and Internet Business News for Personalized Management Information, in: Journal of Systems and Information Technology, 4 (2000) 2, pp. 13-32.

[5] Cas, K.; Bingler, D.: Adaptive Briefing Books mit Hilfe einer Client-Server-Architektur mit integriertem Abfragejournal, in: Timm, U. J.; Rössel, M. (Eds.): ABIS-98 – 6. Workshop „Adaptivität und Benutzermodellierung in interaktiven Systemen", 1998, pp. 1-6.

[6] Mertens, P. et al.: Towards Active Management Information Systems, in: Humphreys, P. et al. (Eds.): Implementing Systems for Supporting Management Decisions – Concepts, Methods and Experiences, 1996, pp. 305-325.

[7] Rössel, M.: Ein System zu individualisierten Informationsvermittlung – dargestellt am Beispiel eines multimedialen Branchenkatalogs der Technischen Keramik, 2000.

[8] Schuhbauer, H.: Advising Functions for Regional Online Tourism Information Systems, in: Maurer, H.; Olson, R. O. (Eds.): Proceedings of WebNet 98 – World Conference on the WWW, Internet & Intranet, 1998, pp. 1181-1182.

[9] Franke, T.: Extended Personalized Services in an Online Regional Tourism Consulting System, in: Wöber, K. W. et al. (eds.): Information and Communication Technologies in Tourism 2002, pp. 346-355.

[10] Schuhbauer, H.: Ein WWW-basiertes Stadtinformationssystem zur individuellen Freizeitberatung – Grundlagen und Prototyp TourBO, 1999.

[11] Franke, T.: Konzeption und prototypische Implementierung eines personalisierten Zugangs zu einem Freizeitberatungssystem, 1998.

[12] Backhaus, K. et al.: Multivariate Analysemethoden – Eine anwendungsorientierte Einführung, 9th Ed., 2000.

[13] Koza, J. R.: Genetic Programming – On the Programming of Computers by Means of Natural Selection, 1992.

[14]   Mihailov, K.: Einsatz genetischer Algorithmen zur Aktivitätenplanung im Freizeitbereich, 2000.

[15]   Depner, U. et al.: Konfiguration von Pauschalreisen aus Reisebausteinen im Rahmen eines Online-Freizeitberatungssystems, 2001.

[16]   Mertens, P.: Integrierte Informationsverarbeitung, in: Administrations- und Dispositionssysteme in der Industrie, Bd. 1, 13th Ed., 2001.

[17]   Bray, T. et al.: Extensible Markup Language (XML) 1.0 (Second Edition) – W3C Recommendation, on www.w3.org/TR/2000/REC-xml-20001006.

[18]   Franke, T.; Barbian, D.: Platform for Privacy Preferences Project (P3P) – Grundsätze, Struktur und Einsatzmöglichkeiten im Rahmen des Franken-Mall-Projekts, 2000.

[19]   Weitzner, D. J.: Building Trust on the Web: Platform for Privacy Preferences (P3P), on www.w3.org/Talks/1999/0511-www8p3p/.

[20]   Cranor, L. et al.: The Platform for Privacy Preferences 1.0 (P3P 1.0) Specification – W3C Candidate Recommendation, on www.w3.org/TR/2000/CR-P3P-20001215.

[21]   Franke, T.: P3P – Platform for Privacy Preferences Project, in: Wirtschaftsinformatik, 43 (2001) 2, pp. 197-199.

Contact:

Prof. Dr. Dr. h.c. mult. Peter Mertens and Dipl.-Wirtsch.-Inf. Thomas Franke
Bavarian Information Systems Research Network (FORWIN),
University of Erlangen-Nuremberg, Germany
E-mail: franke@forwin.de; mertens@wiso.uni-erlangen.de

7 Art Customization

Individualization and personalization are characteristics of art

Jochen Gros
C-LAB / Department of Design, Hochschule für Gestaltung
(Academy of Art and Design), Offenbach Germany

It appears that people have almost forgotten that applied art once was an important part of the industry. In 1907, it was still possible for the Viennese architect Adolf Loos to assert that "without ornamentation we would only have to work four hours a day." This sounds plausible if, for example, we consider the percentage of artistic output involved in building a cathedral or building the first, delicately chiseled brass telescopes. Today, the link between art and consumer goods has been broken as a result of the industrialization of processes and products. Applied art has been subsequently ousted from architecture and design. However, we will discuss in this chapter whether the new technologies and processes of mass customization could renew the association between art and consumer goods. In view of the ever-growing possibilities offered by computer-controlled tools and the general trend towards individualization and personalization, a renaissance in applied arts could become much likely a feasibility. Applied arts could embody the highest degree of expressing individualism and personality. This movement is supported by the capabilities of modern technology: In our digital age, the applications for art have no longer to be found only in the realms of handicrafts (or in industrial processes) but could derive out of the conditions of a mass customization system – resulting in, so to speak, art customization.

7.1 Introduction

William H. Davidow and Michael S. Mallone describe mass customization as a cultural revolution involving sweeping changes in all spheres of life, comparable to the way industry supplanted handicrafts.[1]. The question now is whether the digital technologies and processes of mass customization really do re-channel our cultural development with such explosive force, and, if so, what are the resulting opportunities and perspectives? The following article concentrates on one of these perspectives: the relationship between mass customization and applied art.

What mass customization and art have in common is the fact that they both aim at individualizing and personalizing artifacts. In the one case, that of mass customization, the search for individualization and personalization is just starting;

in the case of art we are reminded of the age-old traditions of individuality and personality, for example in the case of arts and crafts.

In order to gain a clearer picture of the esthetic potential of mass customization the first thing we need to focus on is a specific section of the history of applied art, starting with the technological and economic conditions of industrialization, the principal cause of the decline of applied art and the death of ornamentation. I will subsequently look at certain trends in contemporary art which, judging by appearances, are now embracing the new technologies and processes of applied art, resulting in a situation that could perhaps be dubbed the renaissance of applied art – or art customization.

These academic reflections are complemented by a feasibility study commissioned by the German Federal Ministry of Education and Research (BMBF) and conducted at the Academy of Art and Design's C-LAB in Offenbach [2]. The study not only investigated the theoretical aspects of the term art customization, it also linked the process to experiments in art and design, using for example the construction of several items of customized furniture.

7.2   Arts and industry

Industrialization was the decisive event in the decline of applied art. The mass production of art has proved impossible. While in former times, arts and crafts had been able to develop a symbiotic relationship, to date there has been no seriously attempt to create a term from the words art and industry.

Ornatus, 1450: Industrialization and the decline of ornatus began with the printed book. At the time, ornatus was not only the term for what we call ornament today. Instead, the word described the whole gamut of image-related illustrations and representations including fable, metaphor and rhetorical figures of speech. Consequently, it was not only a question of the pictorial arts when lead composition supplanted illustrations and calligraphy in book design. From this point onward generalized, unilateral abstract thought began to dominate the common consciousness. The success not only of science and technology, but also of abstract art and minimalist design are the result of abstraction, the fundamental intellectual direction pursued by the Moderns. This approach to design also abstracted from cultural diversity, regional peculiarity, individual influences and personal individualism.

Today, however, in the age of digital word and image processing, the tide has started to turn. Even conservative dailies include a high proportion of pictures, television has penetrated to the forefront of our consciousnesses and computer processes of great complexity are conducted using simple metaphors such as that of the desktop or a graphic user interface.

Arts and Crafts, 1860: The first resistance to the negative side-effects of industrial production occurred in England. With the support of philosophers such as John Ruskin and William Morris, the Arts and Crafts Movement criticized both the "inhumane" working conditions in industry and the esthetic devaluation of industrial products. Seeking however to reconcile a desire for social reform with esthetically conservative objectives, the only solution the Arts and Crafts Movement could come up with was to revert to traditional handicrafts, arts and crafts.

Historicism, 1870-1900: When industrialization permeated, the majority of everyday consumer goods, the ornament lost its value. This however did not immediately cause the death of ornamentation. Instead, for a short time ornamentalism even flourished. Historicism can largely be attributed to the experiences of a middle class which grew up with industrialization and suddenly found itself able to indulge in an wealth of ornamentation such as previously only the aristocracy could afford. In reality, however, there was nothing aristocratic about these conveyor belt ornaments. They were cheap goods, scarcely fit to be described as status symbols and deprived of their individuality as mass goods. This double devaluation only gradually penetrated the consciousness of the customers.

Art Nouveau, 1900: Art Nouveau, which also was critical to historicism, developed subsequent to the Arts and Crafts Movement. Aiming at esthetic innovation, the new movement ventured beyond the retrograde shapes of the Arts and Crafts Movement, even creating a new form of ornamentation with clear, flowing, organic lines, like those on the entrances of the Parisian metro. In the first decades of the 20th century, the reform spilt into two diametrically opposed schools of thought. On the one hand there was Art Deco with its unrestrainedly luxurious ornamentalism, choking on its baroque meaninglessness, and in the other, industrial design came into being. According to the latter movement in the industrial age neither handicrafts nor ornamentation could save applied art.

Industrial design, 1925: Initially, the transition from applied art to design was fascinating in its consistency. By means of polemics scarcely less vehement than religious zeal, ornamentation was declared taboo and applied art was banned from architecture and product design. It was clear to everyone involved from the outset that this would mean the end of cultural traditions, regional and individual differences. This however was not perceived as a loss, but was stylized into a new ideal. The ideology of classical industrial design, which is also known as Functionalism, expressly defined its esthetic ideals as international, impersonal and above individuality. The public never completely understood this. Nevertheless the thinking of architects and designers was governed by Functionalism right up to and into the 1960s.

Design criticism, 1965-75: Design criticism was first unleashed as a reaction to the general shock provoked by the uniform impersonality of satellite towns. This criticism questioned the functionalist ideals, seemingly radically. Psychologists called it inhumane, sociologists lamented the lack of social and cultural differentiation, and architects and designers called for more sense and sensibility. Above all, by now the market had well and truly fulfilled the demands of the post-War years and was becoming increasingly saturated. Consequently, it became necessary to start dividing mass markets into target groups in order to appeal to them rather more personally with apparently more individual designs.

New design, 1980-90: An avant-garde amongst designers drew radical conclusions from the criticism of this design, which was only functional in practical terms, and went almost to the opposite extreme. Italian groups such as Alchimia and Memphis provoked unusual interest at art exhibitions and in magazines worldwide with their anti-functional furniture. The Memphis Group, in particular, also attempted to go into industrial production with the new design to market it, however, the success of this initiative was rather moderate. One of the reasons for this was industrially produced ornamentation. For the first time since Art Nouveau, Alchimia and Memphis had once again developed a new form of decoration. But the more often the "Laminati" decoration was published, the less valuable it appeared as a sign of product differentiation and individualization. In Germany, the new design was largely hand-crafted. Operations were formed in backyards in all the larger towns and cities, focusing particularly on furniture and lamps. More than anything else, they were a media event. The topic of art and design was broached in a large number of symposia, articles and on television programs. The movement went by the name of Neues Deutsches Design (New German Design), thought in art categories and failed in as an arts and crafts enterprise.

Designer-Individual products, 1980-today: New Design has had a lasting effect. The conflict between art and industry flared up again acrimoniously, to be followed by a compromise – in theory a poor compromise. Product individualization has at least now been promoted to being a generally desirable aim. Since then, individuality, goods with character and product personality, have become fixtures in industrial design's catalog of objectives. Seemingly, the way to achieve these objectives was with designer personalities such as Philippe Starck. They formed what is known as author design, it has become possible to sell particularly high quantities of designer-individual products by cleverly differentiating them from mass goods. But what now appears to be happening is a transition from designer-individual to customer-individual products. For the first time, mass customization promises a principally new opportunity to further segment the markets and differentiate between products, linking the esthetics of individuality to an actual individualization of production. In this way, at least in theory, mass customization can overcome the basic conflict between art and industry.

7.3   Tools and content

Whether in future customer-individual production will open up new avenues in applied arts depends initially on the new technologies. If we consider the computer numerical control (CNC) manufacturing processes from this viewpoint, the first thing that strikes us is that these new CNC tools are already much more advanced than our ability to use them for design or artistic purposes. Let us take for example the process of translating graphic designs into line grids and using a CNC cutter to transfer the images onto laminates. A joinery near Frankfurt likes to advertise the process using the clichéd image of Marilyn Monroe (Figure 1).

Figure 1: Photo engraving by a joiner's CNC mill

The examples produced by the Eurolaser company would also be difficultly described as art. They simply demonstrate the new dimensions of feasibility (Figure 2). Another development allows burning CAD drawings under the surface of glass. The first use of it is, again still far away from art and ornament. It only shows the new potential of laser sculpturing inside of glass. Individualization has also reached the manufacturing of perforated metal sheet, which is largely used in architecture and design. Here the new technology is able to produce not only customized patterns but also picture grid perforations called "perfoart" (Figure 3).

The practice of CNC sculpturing is till now widely restricted to the copying of antique style furniture. But in fact 5-axis CNC milling systems are providing nearly universal possibilities for art customization, especially in combination with a 3D-scanner. The question remains only how to apply these possibilities in the design of actual consumer products (Figure 4).

Figure 2: Laser engravings on acrylic glass and wood

Figure 3: Customized perforation of metal sheets called "perfoart"

One of the advanced examples of new art technologies is emerging from the computer controlled embroidery. Despite the fact, that there is already a world wide web community developing and sharing digital embroidery patterns, one can still hardly find anything else as traditional ornaments and comic figures (Figure 5). An example for the wide range of new art technologies was given when two years ago CNC-equipment was introduced to manufacture mosaic mats by arranging the stones individually following a graphic pattern. Only one demo example was available – a smiley. What would the artists who furnished us with such wonderful mosaics as those in the Roman baths have had to say about that? As we can see, applying the new technologies by designers and artists is equivalent to somebody banging out nursery rhymes on a concert piano. We have access to highly sophisticated instruments, but what we lack are the requisite compositions and the artists to interpret them. In other words, art customization is less a problem of tools than it is of artistic substance.

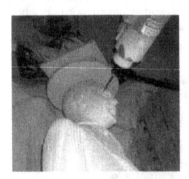

Figure 4: CNC sculpturing   **Figure 5:** Computer controlled embroidery

7.4   Art to become applied

Now that we have the technical means for a renaissance in applied art the question of the artistic contents becomes more and more relevant.

The post-moderns, 1980: The philosophy of the Post-Moderns encouraged the interest in history in the fields of art, architecture and design. Unlike in the case of historicism, historical shapes are not simply reproduced, but reconsidered and quoted with ironical distance. In theory, this represents a new means of justifying this revived interest in all the ornamentalism our cultural history has to offer. In practice however design once again found itself torn between ornamentation and the demands of the conveyor belt. As long as the Post-Moderns argued on a purely esthetic level it was impossible to come up with any new solutions to the problem. The Post-Moderns are at a definite disadvantage vis-à-vis the Moderns, whose doctrine was supported by the notion of a modern technology. By his time, however, the Post-Moderns may find a way of supporting their beliefs. So from their viewpoint, the new technologies are not only post-industrial, they are also post-modern. Individual production now makes it possible for us to quote cultural history in a completely new way.

The second moderns, 1990: After the Post-Moderns came the Second Moderns, whose central doctrine is the return to the abstract shape, but with the artist's individual signature. According to Heinrich Klotz: "The new abstraction is no longer the modern art of the geometric shape... Unlike avant-garde art, that finally ended up using calculated geometry and aimed at attaining the highest possible degree of objectivity transcending the individuality of the artist, the new abstraction celebrates the subjective gesture." [3]. In 1999 the concept of the Second Moderns was experimentally applied to design. Examples of the declared motto, "The product as a constructed sketch" [4] included a little cupboard by Tobias Cunz, where the front sections are cut using a laser beam following the pattern of a sketch that has been scanned on (Figure 6).

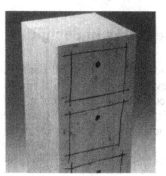

Figure 6: Drawers using laser technology by Tobias Cunz

Reornamentalization, 2001: A comprehensive exhibition by Foundation Beyeler entitled "Abstraktion und Ornament" (Abstraction and Ornament) [5] uses a concept by Markus Brüderlin to try to retrospectively reinterpret all of modern painting as ornamentation. As long ago as 1993 Brüderlin wrote "The ornamental nature of a large proportion of current (artistic) endeavors lends weight to the assertion that the ornament is one of the keys to an understanding of the 1980s and the 1990s. It would be possible to conclude that in general, contemporary German art has suddenly discovered the ornament." In conclusion, Brüderlin writes that

"As a style, art is beginning to come to terms with satisfying certain existing tastes, in the same way that ornamentalism is both the expression of and, simultaneously, a way of fulfilling the intellectual and visual needs of a particular time [6]."

Icon language, 2010 or later: The question of a truly new form of ornament remains open. One possible approach to this question was hinted at in an exhibition entitled "Elementarzeichen" (elementary signs). [7] The exhibition not only contrasted cave paintings and hieroglyphics with the pictographic art of Joan Miro and Keith Hearing, it also placed the latter works in a series of modern pictographs, starting with Otl Aicher. Finally, in terms of perspective, these elementary lines of art have something in common with the visions of a new "icon-language". Thimothy Leary for example says: "A new language is going to be a language of icons, it is going to be graphics". The principal background to these visions is the graphic user interface on a computer and the new possibilities of manipulating images electronically.

Of course, these processes are not always art-related. We also use pictograms and logograms (word pictures) to formulate representations of language, i.e. figures of thought, patterns of feeling, visual expressions. But at least this appears to be a way of solving the problem of a new and meaningful ornament. To get an idea of the icon-language or logo-language ornament we should have a closer look to the decoration of sports and the presentation of business and politics in some television studio backgrounds.

7.5   Revising the theory of design

If we now assume that not only the technical possibilities but also their esthetic substance allow for a revival of applied art, then the main impediments to art customization are the accepted ways of thinking and the ideals of the classic industrial design. One of the major reasons why ornamentation was rejected was a polemic essay entitled "Ornament and Crime" [8], written by Adolf Loos in 1907. However, two quotations are enough to demonstrate just how shaky the foundations of this classic modern design theory are from today's viewpoint:

"The people in the common herd used to have to use different colors to distinguish themselves. Modern man does not require clothing as a mask. His individuality is so unbelievably strong that it can no longer be expressed in items of dress." ... "If modern man tattoos himself, he is a criminal or a degenerate... If a tattooed person dies a free man, then he has simply died some years before committing his murder. The urge to ornament one's face and everything within reach is the primitive origin of pictorial art" [8].

Today, designer and journalist Walfried Pohl proclaims the contrary: "Lack of ornamentation is a crime". Initially, however, Pohl avoids the inflammatory word ornament, demanding instead the reintroduction of "structures composed of small parts". Taking as his starting point the distinction between macro and microstructures, he describes the Modern style as "an act of self-mutilation" which has banished structures composed of small parts and which the latter were only able to compensate for by overestimating macrostructures and "grand gestures". Consequently, our close-up view does not achieve anything and "looking at something once only exhausts its esthetic information." Additionally, with structures composed of small parts we lose "an almost endless wealth of means of differentiation. Since the disappearance of structures composed of small parts (ornaments), design and art have an abrupt, gruff relationship with each other" [6]. Another survey of new positions on the theory of design is provided by A.-Chr. Engels-Schwarzpaul in her dissertation: "Myth, Symbol, Ornament – The Loss of Meaning in Transition" [9].

We have to reconsider the whole theory of industrial design including the argumentation against ornaments because this theory is based only on the classical patterns of mass production. It is no question that as a consequence of the emerging mass customization common design concepts have to change at the core. Postindustrial production methods are a contradiction to industrial-design. Where the reference point is no longer called industry but customization, in theory, the style and the business plans of industrial-design are losing their traditional foundation. Instead we have to look for a new concept of customization-design including another style with pure surface differences, ornament and applied art. In other words: with all this powerful means of customization and personalization.

7.6   Vision and feasibility

After the vision of art customization, the first thing that we consequently did at our C-Lab was to conduct a feasibility study [2]. As well as investigating the scientific foundations of the vision in greater depth, our objective was to work with designers and artists, conducting design experiments on the "pre-competitive" practice of design. At this point however, it is only possible to address a certain number of the facets of this feasibility study.

CNC fitting fonts and icon-fonts: One of the most obvious ways of individualizing products is to use personal inscriptions. This method of design is often employed in mass customization. But if we wish to engrave a message on wood or cut it out of a panel or slab, for technical reasons we need new fonts especially designed for the purpose. From the point of view of feasibility, the first question that comes to mind is how to design this kind of fonts and, for example, to integrate them into computer-controlled cutting systems. This is not too much of a problem. After we however done this we can continue to speculate about how, in future, to create a new inscription typography with smooth transitions into ornamentation. The next step includes CNC-fitting pictographic systems and even larger icon-fonts. In this case there is nearly no more difference between expressing a non-verbal meaning and looking at it as a form of decoration. Best example are the old Egyptian hieroglyphs.

Non traditional ornaments: One experimental pattern design suitable for computer-controlled manufacture is called "Chairs of the 80s" (Figure 7). This pattern consists of pictograms of the seven most striking chairs of New Design, drawn freehand. It thus also quotes design history (however, a relatively recent one). The picture on the right demonstrates how this pattern is engraved onto a bench.

Figure 7: Chairs of the 1980s

Brush embossing: Another example of artistic applications for the new tools is the result of a collaboration between the C-Lab and sculptor Frank Reinecke. The process consisted of taking photographs or water-colors and transforming them into three-dimensional CAD drawings (using software that transposes halftones into contour lines) and then cutting them out to make a relief. Instead of involving lengthy carving work, the artistic process is thus reduced to light brushstrokes (Figure 8). This is not only interesting in esthetic terms, it also illustrates a new economy in applied art.

Digital pattern books: All designs in which art customization proves feasible – fonts suited to CNC, engravable patterns, CAD/CAM-based reliefs etc. – have one great advantage. We can transmit them via the internet, change them on the computer, rearrange them and produce them individually. When looking for such designs however, it now turns out that it was of course much easier to ban ornamentation than to develop a new culture of applied art. As in the past, the academies of art seem to play a particular role in this, as does the establishment of pattern books, in this case, of course, digital "pattern books". In a renaissance of applied art electronic pattern books could even play a much more significant role than their historical predecessors. For at least two reasons, digital patterns no longer provide an example for manual work, they replace the latter and, like the sale of virtual products they are of course published worldwide.

Figure 8: Chest, CNC engraving by Frank Reinecke

AC modules and AC studios: But who produces these digital patterns? Who pays for them to be developed? Who varies them and adapts them to the individual wishes of the customer? First, the software, which allows product configurations for mass customization will ever more contain modules with structures composed of small parts and ornaments – or will be linked up to digital pattern books, i.e. databases, online. Customization per art will then pay a greater role in mass customization – particularly in the more sophisticated and value added forms of individualization and personalization.

But when the focus of the mass customization of furniture will shift to applied art we will probably find start-up studios of artists or designers. This kind of art customization studios requires staff and investment in work that is something between handicraft and industry. This means that what will come into being is probably medium-sized companies with a high output, comparable with the manufacturers of the century before last – we call them *technofacory*.

7.7 Prophecy

To summarize: The term art customization reflects the new technological conditions and the economic changes in artistic production. It considers substantive trends in today's art which could be brought into line with mass customization. It re-examines the positions held by the theory of design and finally investigates the feasibility of the vision through experimental designs.

In view of the above, let everybody judge the following "prophecy" for them-selves: "Sophisticated CAD (computer-assisted design) programs and new composite materials are fueling a kind of re-revival of art nouveau. What Adolphe Retté, a writer of the original art-nouveau period, described as the condition of his time seems hauntingly similar to our own: 'spiritual anxiety, debris from the past, scraps of the present, seeds of the future, swirling, combining, separating under the imperious wind of destiny'. In another five years or so we just might be neck deep in a period of, well, call it nouveau art nouveau" (NEWSWEEK April 24, 2000).

References

[1] Davidow, W.H.; Malone, M.S.: The Virtual Corporation, Structuring and Revitalizing the Corporation for the 21st Century, Boston 1993.

[2] Gros, J.: Art Customization – feasibility study commissioned by the German Federal Ministry of Education and Research, Offenbach, 2000.

[3] Klotz, H.: Kunst im 20. Jahrhundert, Moderne, Postmoderne, Zweite Moderne, 1994.

[4] Gros, J.: Das Produkt als gebaute Skizze, in: Das Jahrhundert des Design, 2000, pp. 258-261

[5] Exhibition catalog: Ornament und Abstraktion, Fondation Beyeler (Switzerland), 2001.

[6] Brüderlin, M.: Ornamentalisierung der Moderne, in: Kunstforum International, 1993, p. 132.

[7] Exhibition catalog: Elementarzeichen, Neuer Berliner Kunstverein e.V., 1985.

[8] Loos, A.: Ornament und Verbrechen, 1907.

[9] Engels-Schwarzpaul, A.-Chr.: Myth, Symbol, Ornament – The Loss of Meaning in Transition, Auckland 2000.

Contact:

Professor Jochen Gros
C-LAB / Department of Design, Hochschule für Gestaltung
(Academy of Art and Design), Offenbach Germany
E-mail: gros@em.uni-frankfurt.de

Part III: Customer Centric Design and Development

Developing product families for customization and efficient manufacturing

Mass Customization aims at satisfying individual customers' needs with near to mass production efficiency. The implications of this new paradigm have positive as well as negative impacts for customers and manufacturers. On the one hand customers benefit from the availability of wide product variety in the market place, on the other hand they can be confronted with frustrating experience in selecting the right product that would exactly fit their expectations among a multitude of alternatives available. Similarly, manufacturers face the trade off between attracting more customers by providing them with large product variety and the need to manage this variety in design and fulfillment in such a way that operational aims like low cost, short lead times and high quality are met. The product design process (setting the solution space) plays a major role when planning and implementing mass customization. Providing value for customers by highly differentiated products without increasing the prices beyond customers' affordability is influenced heavily at the design level.

Thus, Part III of this book addresses the design issues of being customer centric. Managing the variety in the design domain is a challenging problem for manufacturers. The use of product families and modularization techniques are important means of dealing with this variety issue. Designing a family of products using a common platform approach instead of designing single products has gained momentum in various industries. Product families and common product platforms should help mass customizing companies to ensure economies of scale (on the level of modular components and platforms) while serving all customers differently (on the product level). In Chapter 8 *Du, Jiao and Tseng* present how an Architecture of Product Families (APF) contributes to generating families of products efficiently. APF is a logical organization of the product family covering the whole value chain from both a sales and an engineering perspective. Customer requirements in the functional domain are mapped with the variety parameters of a generic data structure for such a product family. Instantiation of the generic data structure determines the product structure and bills of materials, specific to customer specification.

Siddique and Rosen extend this discussion in Chapter 9 and present an approach to identify common platform architectures for a set of existing similar products. This is a major challenge faced by companies becoming more customer centric as it requires the development of product and process models and tools to facilitate configuration reasoning. The authors present an approach called 'Common Platform Identification (CPI)' focusing on the configuration aspect. Given different platforms for similar products, CPI first identifies the common modules. These modules are then re-modularized to enhance commonality further by breaking the modules that are not fitting to a common platform.

The topic of product design for modularity is also addressed in Chapter 10. *Cox, Roach and Teare* discuss how to increase productivity in the product development process by using reconfigurable models and product templates. In the last three decades, significant investments have been made in process technologies to increase productivity and efficiency in product development. But often the return on investment in these technologies has not yielded the gains in productivity that were expected, as new process tools were integrated into old product development processes. The authors argue that investments in new customer centric manufacturing tools and technology will be fruitful only when the product development process is made reconfigurable correspondingly. The authors show that the keys to increasing productivity are reconfigurable artifacts and product templates. By doing so, they provide important input to set up product development processes in mass customization systems which are characterized by the need for fast and efficient new product development processes.

Case-Based Reasoning and TRIZ are significant methodologies that, although not originally developed in a mass customization setting, can improve the design and set- up of customization systems. Estimating the cost of customization precisely, without exactly knowing all the manufacturing parameters and conditions, is an important means of providing the right quotations to customers during the order process. The ability to generate a quick and accurate quotation brings a significant advantage to mass-customized production companies. Unlike existing parametric cost estimation techniques that compute an estimate based on a mathematical relationship between product specification and cost, the approach presented by *Wongvasu, Kamarthi and Zeid* in Chapter 11 uses case-based reasoning to model the relationship between product configuration, resources requirement, and costs. Their approach can be further extended to estimate cycle times. The TRIZ methodology has been proposed for solving the contradiction or trade-offs in different areas. TRIZ is a tactic for inventive problem solving. Its basic philosophy is to challenge the contradictions accepted as fundamentals. As the previous chapters of this book have shown, mass customization is character- ized by trade-offs and the need to counterbalance these contradictions. *Mann and Domb* discuss in Chapter 12 the application of TRIZ in the area of mass customization based on four paradigm shifts. The authors analyze how systematic innovation methods are beginning to be used to successfully overcome rather than accept the trade-offs and compromises often held to be inherent. Understanding the trade-offs and contradictions of mass customization provides an important contribution to designing an appropriate "solution space" for a customer centric enterprise.

8 Product Families for Mass Customization

Understanding the architecture

Xuehong Du[1], Mitchell M. Tseng[2] and Jianxin Jiao[3]
[1] Artesyn Technologies Asia-Pacific Ltd, Hong Kong
[2] Department of Industrial Engineering & Engineering Management,
The Hong Kong University of Science & Technology; Hong Kong
[3] School of Mechanical and Production Engineering, Nanyang
Technological University, Singapore

The rationale of developing product families with respect to satisfying diverse customer needs with reasonable costs, *i.e.*, mass customization, has been well recognized in both industry and academia. Earlier research often highlights isolated and successful empirical studies with limited attempt to explore the theoretical foundations surrounding this economically important class of engineering design problem. In this chapter we investigate the fundamental issues underlying product family development. The concept of Architecture of Product Family (APF) is introduced as a conceptual structure and overall logical organization of generating a family of products. APF constructs – including common bases, differentiation enablers, and configuration mechanisms – are discussed from both a sales and an engineering perspective. Further, variety generation methods are evaluated in regard to producing custom products based on the modular product architecture and configure-to-order product development. To support APF-based product family design, a Generic Product Structure (GPS) is proposed as the platform for tailoring products to individual customer needs and generating product variants. At the end of the chapter, we present a case study of an industrial example to illustrate the feasibility and potential of our proposed framework.

8.1 Introduction

Mass customization has recently received much attention and popularity from both industry and academia alike [1, 2, 3]. It has been considered as a new battlefield for manufacturing enterprises [4]. Mass customization aims at delivering an increasing product variety to satisfy diverse customer needs while maintaining near mass production efficiency [5]. Essentially, it is an oxymoron of variety for catering to customization and low costs of variety fulfillment.

Developing product families has been recognized as effective means of achieving economies of scale in order to accommodate an increasing product variety across diverse market niches. In addition to leveraging the costs of delivering variety, product family design can reduce development risks by reusing proven elements in a firm's activities and offerings [6, 7].

As the backdrop of product families, a well-planned architecture – the conceptual structure and overall logical organization of generating a family of products – will provide a generic umbrella to capture and utilize commonality, within which each new product instantiated and extends so as to anchor future designs to a common product line structure. The rationale of such a product family architecture lies in not only unburdening the knowledge base from keeping variant forms of the same solution, but also in modeling the design process of a class of products that can widely variegate designs based on individual customization requirements within a coherent framework.

With this view, this chapter discusses the Architecture of Product Family (APF) for the effective implementation of mass customization. An extensive survey of related work and the background leading to this research are given in section 2, followed by an exploration of fundamental issues underlying product family in section 3. In connection to the challenges of product family design, a framework of APF is presented in section 4. Technical details of APF-based product family design are discussed in section 5. In section 6, an application of APF to an industrial example is presented to illustrate the feasibility and potential of the proposed framework.

8.2   Related work

Over the past five years, it has been witnessed an exponentially increased number of literature reporting mass customization, which involves many, if not most, aspects and issues of manufacturing enterprises and tackles mass customization from various perspectives. In the field of product development and engineering design, approaches to and strategies for variety design and product families for mass customizing products are prevalent in the literature, which can be classified by three themes, namely product architecture, product platform, and product family modeling. An overview of related work falls into five sub-groups, as depicted below.

(1) Product Platform Development and Representation. A number of perspectives on product platform exist in literature. A review suggests that product platform has been defined diversely, ranging from being general and abstract [8, 9, 10] to being industry and product specific [11, 12, 13]. In addition, the meaning of platform differs in the scope. Some definitions and descriptions focus mainly on the product/artifact itself [14, 15], whereas others try to explore the platform concept in terms of a firm's value chain [7].

There are two streams of research prevailing in the field of developing product platforms to support product family development. One perspective refers to a platform as a physical one, namely a collection of elements shared by several related products. Accordingly, the major concern is how to identify common denominators for a range of products [16, 11, 12, 17, 18, 19]. The effort is geared towards the extraction of those common product elements, features, and/or subsystems that are stable and well understood, so as to provide a basis for introducing value-added differentiating features [8, 20].

Meyer and Lehnerd's work is the representative of another dominating perspective to product platform [21]. They defined a product platform as "a set of subsystems and interfaces developed to form a common structure from which a stream of derivative products can be efficiently developed and produced" [21, p. 39]. The major issue is to exploit the shared logic and cohesive architecture underlying a product platform. One endeavor towards product platform development is to design product families in the way of "stretching" and/or "scaling" [22]. The development of Boeing 7XX aircrafts epitomizes such a design practice [23]. The robust design of product families is discussed in [24, 25, 26]. Simpson *et al.* proposed a product platform concept exploration method to facilitate the synthesis and exploration of certain common product platform concepts that can be scaled to generate an appropriate family of products [27]. Siddique *et al.* employed graph grammars to identify the common platform for a set of similar products and to specify possible product portfolio supported by the platform [28]. To facilitate platform-based product family development, interface management is reported as a distinct process of defining the physical interfaces between subsystems [29]. Set-based model is an attempt to the formal representation of product platform design and manufacturing processes [30].

(2) Product Architecture. Product architecture can be defined as the way in which the functional elements of a product are arranged into physical units and the way in which these units interact [31]. One approach directly relevant to product family design is the development of modular product architecture. According to Ulrich and Tung, the modular product architecture involves one-to-one mappings from functional elements in the function structure to the physical components of a product, where decoupled interfaces between components can be specified [32]. Ulrich pointed out that the modular product architecture allows each functional element of the product to be changed independently by changing only the corresponding component, thus advantageous to produce custom-built product from standard models [6]. It also makes standardization possible, which is essential to achieve the economy of scale. Therefore, using the modular product architecture, variety can be created by combinations of component building blocks. Pahl and Beitz also discussed the advantages and limitations of modular products [33]. Newcomb et al. and Rosen investigated methods to reason about sets of product architecture and to perform configuration design of modular products [34, 35].

While Ulrich and Tung defined five categories of modularity (i.e., component swapping, component sharing, fabricate-to-fit, bus, and sectional modularity) [32] and most extensions built their efforts upon these modularity types [36], current practice mostly refers to the product architecture as physical structures in terms of physical parts or components [37]. Little effort has been devoted to the implications of architecture with respect to functional features and design parameters [38], especially in terms of systematic planning of modular architecture starting from the early conceptual design stage. In addition, current research mostly focuses on product architectures and modular product design in the context of a single product. Since manufacturing companies increasingly develop product families to offer a large variety of products with low development and manufacturing costs, the architecture for product families becomes a major concern [39]. So far, only limited amount of literature has been devoted to addressing issues regarding the product architecture in the context of product families.

(3) Product Family Architecture. Fujita and Ishii pointed out one important characteristic to discern the architecture of a family of products from that of a single product, *i.e.*, the simultaneous handling of multiple products [40]. Erens and Verhulst adopted various product models to describe the architecture of product families [41]. The functional and physical architectures for product families are outlined. Essentially, they modeled the architecture of product families as a packaging of single product models, instead of a unified product family model. Yu *et al.* and Zamirowski and Otto approached product architectures from a functional perspective, that is, defining the architecture of product portfolio based on customer demands [42, 43].

Tseng and Jiao recognized the rationale of a product family architecture (PFA) with respect to design for mass customization [5]. They pointed out the development of PFA involves systematic planning of modularity and commonality in terms of building blocks and their configuration structures across three consecutive domains, namely the functional view (as seen from customer, sales and marketing viewpoints), the behavioral view (as seen from the product technology or design engineer perspective), and the structural view (as seen from the fulfillment or manufacturing and logistic perspectives) [44]. Zamirowski and Otto also perceived the necessity to develop the product architecture and platform with the synchronization of multiple views such as those from customer needs, function structures and physical architectures [43]. The leveraging of modularity and commonality in PFA development is also observed by Siddique *et al.* and Siddique and Rosen [28, 45]. While PFA approaches product families from both design content and design process perspectives, some basic issues such as variety fulfillment, product customization, and variant derivation have not been tackled in detail.

(4) Variety Design and Variety Fulfillment. Ishii and his group developed metrics for evaluating the importance and costs of product variety [46, 47, 48]. Their work is largely based on one-to-one correspondences between functionality and components and assumes product variety results from the combination of components, thus suitable only for dealing with simple products where functional differentiation can be directly embodied by specific components, *i.e.*, "seeable" variety. Fujita and Ishii discussed design for variety in terms of structuring essential tasks and issues associated with variety design [40]. Fujita *et al.* (1998; 1999) tried to optimize the system structure and the configuration of product families simultaneously [49, 50]. Their focus is on computational approaches to

the variety design process, such as design synthesis, optimization, representation, system constraints, and so forth. As pointed out by Fujita and Ishii and Ishii *et al.* product variety optimization is extremely difficult and far from successful practical applications [40, 47]. Simpson *et al.* applied goal programming and statistical analysis techniques to design optimization of product families [25, 26]. Similarly, their work focuses on computational support for the design process and excels in parametric design instead of product architecture planning, which is related to the conceptual design stage.

Jiao and Tseng observed the difference between customer-perceived variety in terms of functionality and technical variety in terms of fulfillment, which results in different variety design themes [44]. As an effort towards customer-perceived variety, the work in product line selection focuses on product functional attributes/features with only implicit engineering concerns [51]. Conjoint analysis is widely used in exploring customer preferences and shedding light on product line rationalization [52].

While the current practice of optimal variety design mostly emphasizes those physical components that can be shared by a series of predefined products, there is a paucity of published work and a lack of good, well-documented and well-structured case studies in connection to the construction of flexible product architectures for customized product development, in terms of the process of variety propagation across different domains, as well as mechanisms of variety generation [40].

As for variety fulfillment, a few methods have been reported, including (a) collaborative, adaptive, cosmetics and transparent customizations (called four faces of mass customization) [53]; (b) component selection and attached features [40]; (c) scaling from a common product platform [22, 27]; and (d) combination of components [6, 32]. While four faces of mass customization suggest strategic guidelines for variety design, component selection and attached features reveal two important aspects of variety fulfillment. One is the form of variety – attached features, and the other is the generation method of variety – component selection. Component combination based on modular product design suggests itself as a practical approach to variety generation.

(5) Product Family Modeling. As pointed out by Fujita and Ishii, one of the challenges of product family design is the representation of families of products other than single products [40]. The most significant effort towards representing products with options probably is the generic bill-of-materials (GBOM) product model [55]. Built upon the bill-of-material (BOM) structure, which is commonly used in industry, GBOM defines a generic product as a set of variants that can be identified through specifying alternative values for a set of parameters. Many researchers tackle product family representation based on the GBOM model. Erens and his colleagues developed a GBOM-based product modeling language for product variety [41, 56, 57]. Wortmann and Erens proposed generic product modeling (GPM) to represent a product family from both commercial and assembly views [57]. Mckay *et al.* intended to merge the description of detailed

product data with product variety [58]. Van Veen investigated GBOM-based product family modeling from the production perspective [59].

Jiao and Tseng observed different data relationships underlying product family modeling, including product-to-product, product-to-family and family-to-family relationships that involve commonality, instantiation and evolution issues, respectively [60]. They also identified two types of constraints to deal with variety mapping from sales to engineering domains and variety configuration within each domain. A generic variety structure (GVS) consisting of the common product structure, variety parameters, and configuration constraints is also proposed to characterize variety and its derivation [61].

As an attempt to address the complexity of design data management, Baldwin and Chung developed a discipline-independent data model to provide constructs for modeling products with optional contents [62]. In their framework, meta-data is introduced along with some dimensions for organizing design data such as the component hierarchy, the classification hierarchy, configuration management, and so on. With rigorous formulations of optional features, restrictions, include conditions, and constraints, the message-handling mechanism is developed to allow users to extract design information from a representation.

8.3 Fundamental issues underlying product family

8.3.1 Concept implication

In an increasingly diversified market, product design needs to be tailored to various customer needs. The customer requirements of a product are characterized by a set of (functional) features along with the associated feature values. Designing families of products has proved to be an effective means of maintaining economies of scale while satisfying the scope of customers' needs [13, 18, 31, 63, 64, 65, 66].

A *product family* refers to a set of similar products that are derived from a common platform and yet possess specific features/functionality to meet particular customer requirements. Each individual product within a product family, *i.e.*, a family member, is called a *product variant*. While a product family targets a certain market segment, each product variant is developed to address a specific set of customer needs of the market segment. All product variants share some common structures and/or common product technologies, which form the platform of the product family.

The interpretation of product families depends on different perspectives. From the marketing/sales perspective, the functional structure of product families exhibits a firm's product line or product portfolio and thus is characterized by various sets of functional features for different customer groups. The engineering view of product families embodies different product technologies and associated manufacturability and thereby is characterized by various design parameters, components, and assembly structures.

8.3.2 Modularity and commonality

There are two basic issues associated with product families, namely modularity and commonality. Table 1 highlights different implications of modularity and commonality, as well as the relationship between them.

Table 1: A comparison of modularity and commonality.

	Modularity	*Commonality*
Focused Objects	Type (Class)	Instances (Members)
Characteristic of Measure	Interaction	Similarity
Analysis Method	Decomposition	Clustering
Product Differentiation	Product Structure	Product Variants
Integration/Relation	Class-Member Relationship	

The concepts of modules and modularity are central in constructing product architectures [6]. While a module refers to a physical or conceptual grouping of components that share some characteristics, modularity tries to separate a system into independent parts or modules that can be treated as logical units [34]. Therefore, *decomposition* is a major concern in modularity analysis. In addition, to capture and represent product structures across the entire product development process, modularity is achieved from multiple viewpoints, including functionality, solution technologies, and physical structures. Correspondingly, there are three types of modularity involved in product realization, *i.e.*, functional modularity, technical modularity, and physical modularity.

What is important in characterizing modularity is the *interaction* between modules. Modules are identified in such a way that between-module (inter-module) interactions are minimized whereas within-module (intra-module) interactions may be high. Therefore, three types of modularity are characterized by specific measures of interaction in particular views. As for functional modularity, the interaction is exhibited by the relevance of functional features (FFs) across different customer groups. Each customer group is characterized by a particular set of FFs. Customer grouping lies only in the functional view and is independent of the engineering (including design and process) views, that is, to be solution-neutral. In the design view, modularity is determined according to the technological feasibility of design solutions. The interaction is thus judged by the coupling of design parameters (DPs) to satisfy given FFs regardless of their physical realization in manufacturing. In the process view, physical interactions between components and assemblies (CAs) are derived from manufacturability, *e.g.*, physical routings of CAs on a PCB, and thus become the major concern of physical modularity related to physical product structures.

It is commonality that reveals the difference of the architecture of product families from that of a single product. While modularity resembles decomposition of product structures and is applicable to describing module/product types, commonality characterizes the grouping of similar module/product variants under specific module/product types characterized by modularity. Corresponding to the three types of modularity, there are three types of commonality in accordance with functional, design and process views. Functional commonality manifests itself through functional classification, that is, to group similar customer requirements into one class, where similarity is measured according to the Euclidean distance among FF instances. In the design view, each technical module, characterized by a set of DPs corresponding to a set of FFs, exhibits commonality through clustering similar DP instances by chunks [31]. Instead of measuring similarity among CA instances, physical instances (instances of CAs for a physical module type) are grouped mostly according to appropriate categorization of engineering costs derived from assessing existing capabilities and estimated volumes, *i.e.*, economic evaluation.

The correlation of modularity and commonality is embodied in the class-member relationship. A product structure is defined in terms of its modularity where module types are specified. Product variants derived from this product structure share the same module types and take on different instances of every module type. In other words, a class of products, *i.e.*, a product family, is described by modularity and product variants differentiate according to the commonality among module instances.

8.3.3 Product variety

Product *variety* is defined as the diversity of products that a production system provides to the marketplace [6]. Two types of variety can be observed, namely functional variety and technical variety. *Functional variety* is used broadly to mean any differentiation in the attributes related to a product's functionality, from which the customer could derive certain benefits. On the other hand, *technical variety* refers to diverse technologies, design methods, manufacturing processes, components and/or assemblies, *etc.*, which are necessary to achieve specific functionality of a product required by the customer. In other words, technical variety, though may be invisible to customers, is required for engineering in order to accommodate certain customer-perceived functional variety. Technical variety can be further categorized into product variety and process variety. The technical variety of products is embodied in different components/modules/parameters, variations of structural relationships, and alternative configuration mechanisms, whilst process variety involves those changes related to process planning and production scheduling, such as various routings, fixtures/setups, and workstations. While functional variety is mostly related to the customer satisfaction from the marketing/sales perspective, technical variety usually involves manufacturability and costs from the engineering perspective.

Even though these two types of variety have some correlation in product development, they result in two different variety design strategies. Since functional variety directly affects customer satisfaction, this type of variety should be encouraged in product development. Such a *design for "functional" variety* strategy aims at *increasing functional variety* and manifests itself through vast research in the business community, such as product line structuring [13,51], equilibrium pricing [67], and product positioning [68]. On the contrary, *design for "technical" variety* tries to *reduce technical variety* so as to gain cost advantages. Under this category, research includes variety reduction program [69], design for variety [46, 48], design for postponement [70], design for technology life cycles [47], function sharing [17], design for modularity [71], *etc.*

The above implication of variety has practical impact on the variety fulfillment. While exploring functional variety in the functional view through customer needs analysis, product family development should try to reduce technical variety in the design and process views by systematic planning of modularity and commonality so as to facilitate plugging-in modules that deliver specific functionality, reusing proven designs, and reducing design changes and process variations. Such a setup, therefore, presents manufacturers with a challenge of ensuring "dynamic stability", which means that a firm can serve the widest range of customers and changing product demands while building upon existing process capabilities, experience, and knowledge [72].

8.4 Architecture of Product Family (APF)

With the understanding of above-mentioned challenges, it has been realized the importance of exploring the architecture of product families in order to build up the theoretical foundation of product family development [5, 7, 39, 40, 41]. As to this architectural aspect of product families, this research differentiates the architecture of product family from that of product families, because the family-to-family relationship is involved in the architecture of product families and excluded in that of product family. As an incipient endeavor, this research focuses only on the architecture of product family, in which the part-to-part, part-to-product, product-to-product and product-to-family relationships are included.

8.4.1 Principles

The implication of the architectural perspective originates from the meaning of *architecture* in computing – *"The conceptual structure and overall logical organization of a computer or computer-based system from the point view of its use or design, also referring to a particular realization of this"* [73]. Two points are conveyed by this explanation – the "logic organization" and "from the viewpoint of use". Accordingly, the Architecture of Product Family (APF) can be defined as the logical organization of a product family from both sales and engineering viewpoints. In the context of mass customization, there must be a balance between commonality and distinctiveness within the APF [8]. In other words, the APF defines both what are common and what are different among family members, as well as mechanisms with which some product variants can be derived. Therefore, there are two correlated aspects associated with the APF, namely the compositional and the generative aspects.

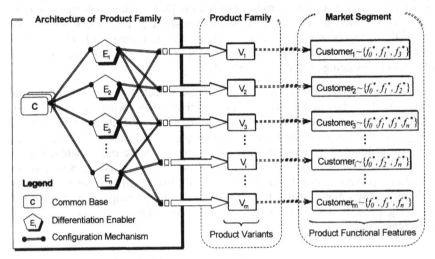

Figure 1: An illustration of APF

Figure 1 illustrates the principle of APF with respect to product family development for mass customization. From the sales point of view, customers are characterized by combinations of functional features, $\{f\}$, and associated feature values, $\{f^*\}$. A product family, $\{V_1, V_2, V_3, ..., V_i, ..., V_m\}$, is designed to address the requirements of a group of customers in a market segment, $\{Customer_1, Customer_2, Customer_3, ..., Customer_i, ..., Customer_m\}$, in which customers share certain common requirements, f_0^*, along with some similar and/or distinct requirements, $\{f_1^*, f_2^*, f_3^*, ..., f_n^*\}$. From the engineering perspective, the product variants of the product family, $\{V_i\}$, are derived from configuring common bases, $\{C_i\}$, and differentiation enablers, $\{E_i\}$, that are predefined for the product family. It was configuration mechanisms that determine the generative aspect of APF.

While the product-to-family relationship is embodied in common bases, product-to-product and part-to-product relationships manifest themselves through differentiation enablers and configuration mechanisms, respectively. In the following section, the compositional aspect of APF is introduced, the granularity and multiple-view issues of APF are highlighted, and then the mechanism of variety generation is elaborated.

8.4.2 Composition

The APF consists of three elements, namely the common base, the differentiation enabler, and the configuration mechanism, as depicted below.

(1) Common Base. Common bases (CBs) refer to certain shared elements within a product family. These shared elements may be in the form of either common (functional) features from the sales or customers' perspective, or common product structures and common components from the engineering perspective. Common features indicate the similarity of customer requirements related to the market segment. Common product structures and components are determined by product technologies, manufacturing capabilities and the economy of scale. In practice, a collection of CBs is usually called the base product of the family, which comprises the basis for product family design.

(2) Differentiation Enabler. Differentiation enablers (DEs) are basic elements making products different from one another. They are the source of variety for a product family. From customers' perspective, DEs may be in the form of optional features, accessories, or selectable feature values. Taking computers as an example, while a CD drive is specified as an optional feature (yes/no), the specification of RAM must be one instance of a set of selectable feature values, such as 64K, 128K, and 256K bytes. In the engineering view, DEs may be embodied in distinctive structural relationship (structural DEs) and/or various modules with different performance (constituent DEs). Each engineering DE usually has more than one alternative applicable to product variant derivation for a specific application.

(3) Configuration Mechanism. Configuration mechanisms (CMs) define the rules for and means of product variant derivation. Three types of configuration mechanism can be identified for the APF, namely selection constraints, include conditions and variety generation.

Selection constraints specify restrictions on optional features because certain combinations of options (*i.e.*, alternative values of optional features) are not allowed/feasible, or, on the contrary, are mandatory. An example of the selection constraint on a car might be: "If cylinder (feature) is 1.3 liter (option) and fuel (feature) is diesel (option), and a 5-speed (option) gear-box (feature) is manda-tory". Selection constraints eliminate those technically-infeasible or market-unwanted products from all possible combinations of the offered options [62, 74]. The theoretical number of combination is the Cartesian product of all possible feature values (options).

Include conditions are concerned with the determination of alternative variants for each differentiation enabler. The include condition of a variant defines the condition under which the variant should be used (or not) in order to achieve the required product characteristics. It may be in the form of a logic function with parameter values of the differentiation enabler or with its parent constituent as independent variables. For example, an office chair (a parent) consists of one supporting module (a child), which performs as a differentiation enabler.

Supposed there are two variants for this supporting module – using wheels or pads. The include condition of "using wheels" is "the office chair is drivable" while the include condition of "using pads" is "the office chair is not drivable". This include condition is defined in the form of a logic function of the parent's (office chair) variable (drivable or not). Essentially, include conditions are related to the engineering definition stage of product development.

Variety generation refers to the way in which the distinctiveness of product embodiment can be created. Its focus is on product structures. Such variety fulfillment is related to each differentiation enabler. In light of the rationale of modular product architecture [6, 32], this research identifies three basic methods of variety generation, namely attaching/removing, swapping and scaling. More complex variety generation methods can be composed by employing these basic methods recursively, called nesting, with reference to the hierarchical decomposition of product structures, called the Generic Product Structure (GPS). This issue will be discussed in detail following the next section.

8.4.3   Granularity

While sharing common bases across products allows for an economy of scale, differentiation enablers facilitate the delivery of variety. Such a trade-off of economies of scale and scope depends on appropriate identification of common bases and differentiation enablers. These APF constructs are usually selected based on (a) current and future customer needs; (b) commonality in design and fulfillment; (c) ease of configuration, and (d) appropriate level of aggregation. If the construct is at too low a level of aggregation, such as at the nuts and bolts level, then the number of constructs may be too many and the configuration becomes difficult. On the other hand, if the aggregation is at a very high level, such as complete modules or products, then the commonality may not be sufficient.

Figure 2 exhibits such a granularity issue of APF constructs. Product family, P, has a number of product variants. Each of them consists of three modules, M_1, M_2, and M_3. At the first decomposition level, M_1 is shared, whereas M_2 and M_3 have more than one variant (*i.e.*, alternative). There is no part of M_3 can be shared among its variants, thus no further decomposition. However, certain parts of M_2 are common among its variants. To take advantage of commonality, M_2 is further decomposed into M_{21}, denoting the common parts, and M_{22}, which characterizes the origin of variety for M_2. In such a way, M_1 and M_{21}, along with the associated structural relationships (*i.e.*, goes-into relationships), comprise the common bases of family, P, and M_3 and M_{22} become the differentiation enablers for all variants within P. The right part of Figure 2 illustrates APF constructs may be in the form of either constituent modules or structural relationships. It is obvious that different DEs imply different granularity (*e.g.*, M_3 and M_{22}), which is also true for CBs (*e.g.*, M_1 and M_{21}). Considering granularity for CBs and DEs benefits not only achieving economy of scale but also reducing the complexity in product and production management through identifying the source of variety. The later advantage is discussed by Wortmann, Erens and Verhulst [57, 41]. We will also demonstrate this in latter sections.

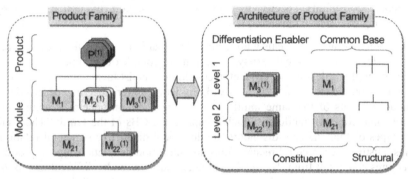

Figure 2: Granularity of APF constructs

8.4.4 Multiple views

Increasing variety has far-reaching influence on many organizational functions, such as sales and marketing, product engineering, and manufacturing. It is a common practice that different functional departments in a company interpret product families from different perspectives and for different purposes. For example, for sales and marketing, it must be able to represent product specifications in terms of functional parameters and constraints, and meanwhile be able to describe end-products, subassemblies, and components, along with constituent and structural information for engineering purpose. Therefore, CBs, DEs, and CMs imply different entities for sales and engineering, as summarized in Table 2.

Table 2: Multiple views of APF constructs

View APF Construct	Sales	Engineering
Common Base	Common features	Product technology Product structure Common modules
Differentiation Enabler	Optional features Selectable feature values	Distinctive modules Distinctive structural relationships Scaleable design parameters
Configuration Mechanism	Selection constraints	Include condition Variety generation

What customers order and perceive are in fact "product (functional) features" [57]. Therefore, the CBs of product family in the sales view should be those common features required by the whole customer group. Within the customer group, it was optional features and selectable feature values that make products different for individual customers. Therefore, from the sales or customers' perspective, optional features and selectable feature values constitute the DEs. Selection constraints specify the feasible combination of options, thereby comprising the CMs in the sales view.

The engineering view deals with technical realization of those features required in the sales view. Technical variety is much more complex than functional variety defined in the sales view. Product technologies, constituent components, and structural relationships are all involved. In terms of engineering descriptions, product variants of the same family exhibit one common general structure and share some common modules, which comprise the CBs of the family. There are two types of DEs, namely distinctive modules and distinctive structural relationships. Distinctive structural relationships are inter-module relationships in product structures and result from the fact that product variants can be different due to the occurrence of certain modules in product structures, such as different goes-into relationships in a BOM structure. A distinctive structural relationship in conjunction with the associated modules produces variety in terms of structural variation. Distinctive modules are employed to realize product differentiation within the same product structure, such as different performances in terms of parametric variation. Distinctive modules may be completely different designs or be variants of the same design [62]. Include conditions and variety generation methods become the major concerns of CMs for deriving engineering descriptions of product variants.

8.4.5   Variety generation

Ulrich and Tung summarized five types of modularity based on prevalent industrial practice towards component standardization and product variety [32]. Component swapping modularity and cut-to-fit modularity characterize the variation of component types used in one identical product architecture. Scaling can be regarded as a special case of cut-to-fit modularity when the module becomes the whole product and a closed-form system model can be formulated. Bus modularity and sectional modularity allow for variations of the number and/or locations of components, *i.e.* the variation of product structures. Nevertheless, all current practice only emphasizes physical modularity - referring modules to physical parts/components along with their structural relationships, without considering functional and technical modularity. In addition, how to draw clear boundaries between different modularity types and how they can shed light on modular design are still not addressed, especially with respect to producing custom-built product in certain customization circumstances.

In this research, aiming at producing custom-built products, some methods of variety generation widely used in practice are identified. The assumption is that the modular product architecture has positive impact on product family design [6].

Modules in this context can be a set of physical parts or a group of logic units that provides certain functions. Product variety can thus be achieved through combinations of modules. It is also assumed that any change of modules in order to produce a custom product will not induce a high cost or substantial changes of the other parts of the product

. Based on the above assumptions, three basic variety generation mechanisms can be distinguished, including attaching/removing, swapping, and scaling, as shown Figure 3.

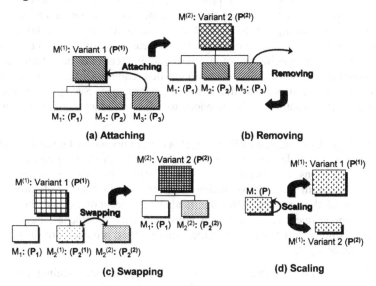

Figure 3: Basic methods of variety generation

(1) Attaching/Removing. In Figure 3(a), a particular module (M_3) carrying out certain additional functions (P_3) can be attached to a base product ($M^{(1)}$) to create a new product variant ($M^{(2)}$). These attachable modules should possess appropriate interfaces to the base product. A variety of end-products can thus be generated through attaching different attachable modules. Attaching appears in many situations, such as electronic devices with optional accessories and computer systems with peripherals, where optional accessories and peripherals exemplify attachable modules. Therefore, attaching is a popular method of product customization. In the practice of attaching, a postponement strategy is very useful, that is, to postpone product differentiation late to manufacturing stages or even to the point where products are delivered [70]. Opposite to attaching, *removing* can also be defined for variety generation. It means variant products can be derived through removal of certain modules. In most cases, a module carrying out certain functions that are redundant for a particular customer's application can be removed to have a simplified model with cost savings.

(2) Swapping. Swapping is mostly related to variety due to different perform-ance requirements for the same function. For example, the output voltage for a power supply may be either ±12V or ±15V. In Figure 3(b), both module $M_2^{(2)}$ and module $M_2^{(1)}$ carry out the same function (P_2) and each of them produces different performance levels of P_2 $(P_2^{(2)}$ and $P_2^{(1)})$. If we use $M_2^{(2)}$ to substitute its counterpart $(M_2^{(1)})$ in the base product $(M^{(1)})$, a new variant $(M^{(2)})$ can be obtained with a different performance $(P_2^{(2)})$ from that $(P_2^{(1)})$ of the original base product $(M^{(1)})$. Usually, substitutable modules $(M_2^{(1)}$ and $M_2^{(2)})$ should possess common interfaces to the other parts of the product. In most cases they are the variants of the same module. In fact, swapping can be regarded as a combination of removing and attaching. Swapping can be applied in situations where custom products are derived by modifying standard products. In this case, a standard product that most closely satisfies the specific set of customer requirements can be selected as the base product and those modules related to unfilled requirements are substituted by the modules that can achieve the required performance. Swapping is also an approach to give customers the freedom to adjust product performance while using [53].

(3) Scaling. As illustrated in Figure 3(c), a certain operational parameter (P) of the product or a module can be changed from $P^{(1)}$ to $P^{(2)}$, following certain system models or functions. For example, the size of office chair can be changed, and thus is scaleable. In this context, scaling is related to either the whole product or only a part of the product. Also, the fundamental technology and the system structure remain unchanged. Scaling is consistent with approaches suggested by Rothwell/Gardiner and Simpson/Mistree, where whole products are regarded as scaleable objects [22, 27].

(4) Variety Nesting. Attaching/Removing, swapping and scaling are basic mechanisms of variety generation. More complicated mechanisms can be composed by employing these basic methods recursively. This is referred to as variety nesting. Several representative scenarios of variety nesting are listed in Table 3. As illustrated in the table, attaching/removing, swapping, and scaling are only meaningful with respect to a particular module at a certain decomposition level. For instance, for variety nesting, SWAP(ATTA), at the lower level, variety of module M_2 is produced by attaching M_{23} to $M_2^{(1)}$, whereas at the higher level, variety of M is achieved by swapping between $M_2^{(1)}$ and $M_2^{(2)}$.

In summary, there are three aspects associated with the APF. The composition aspect deals with necessary elements of product family design. Multiple views of APF are concerned with the synchronization of different perspectives to variety and product families among different organizational departments. The granularity APF constructs emphasizes trade-offs of commonality and distinctiveness in product family design. A holistic understanding of these aspects reflects both the compositional and generative characteristics of the architectural perspective. With this architectural perspective, product variants can be derived through combina-tions of CBs and DEs in accordance with CMs.

Table 3: Variety generation through nesting

Variety Nesting / Notation	Product Variants	Building Blocks
Swapping-Scaling SWAP(SCAL)		$M_1: (P_1)$ $M_2: (P_2)$
Swapping-Swapping SWAP(SWAP)		$M_1: (P_1)$ $M_{21}: (P_{21})$ $M_{22}^{(1)}: (P_{22}^{(1)})$ $M_{22}^{(2)}: (P_{22}^{(2)})$
Swapping-Attaching SWAP(ATTA)		$M_1: (P_1)$ $M_{21}: (P_{21})$ $M_{22}: (P_{22})$ $M_{23}: (P_{23})$
Attaching-Swapping ATTA(SWAP)		$M_1: (P_1)$ $M_{21}: (P_{21})$ $M_{22}^{(1)}: (P_{22}^{(1)})$ $M_{22}^{(2)}: (P_{22}^{(2)})$

8.5 Product family design

Under the umbrella of APF, product family design manifests itself through the derivation processes of product variants based on APF constructs. Figure 4 illustrates the principle of APF-based product family design. Customers make their selections among sets of options defined for certain distinctive functional features. These distinctive features are the differentiation enablers of APF in the sales view. Selection constraints are defined for presenting customers with only feasible options, that is, both technically-feasible and market-wanted combinations. A set of selected distinctive features together with those common features comprise the customer requirements of a customized product design. As shown in Figure 4, in the sales view customized products are defined in the form of functional features and their values (options), whereas in the engineering view product family design starts with product specifications in the form of variety parameters. Within the APF, variety parameters correspond to distinctive functional features, and the values of each variety parameter correspond to the options of each functional feature.

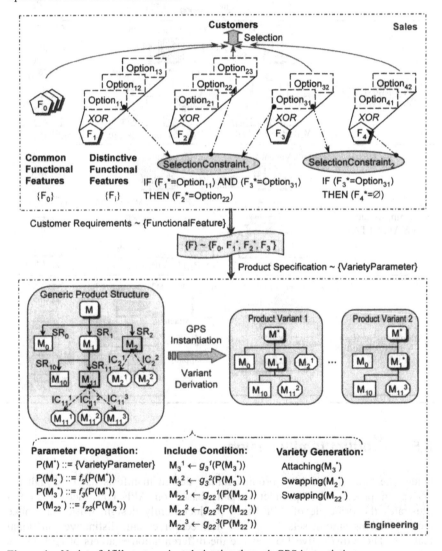

Figure 4: Variety fulfillment: variant derivation through GPS instantiation

To support APF-based product family design, a Generic Product Structure (GPS) is introduced as a generic data structure of a product family (Figure 5). In the engineering view, the derivation of product variants embodies the instantiation of GPS. While a GPS characterizes a product family, each instance of GPS corresponds to a product variant of the family. Each item in the GPS (either a module or a structural relationship) is instantiated according to certain include conditions that are predefined in terms of variety parameters. Variety parameters originate from functional features specified by the customers and propagate along the hierarchy of GPS. Variety generation methods, such as attaching/removing, swapping, and scaling, are implemented through different instantiations of GPS items. The GPS provides a platform (common base) for product family design, where distinctive items of GPS, such as distinctive modules and distinctive structural relationships, perform as the differentiation enablers of the family. Distinctive items are embodied in different variants (instances) and are identified by associated include conditions. Therefore, include conditions and variety generation methods embody the configuration mechanisms of APF in the engineering view.

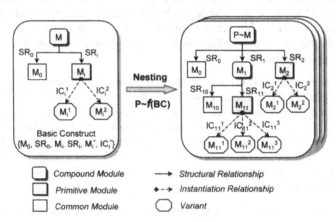

Figure 5: Generic product structure (GPS): basic construct and nesting

8.5.1 Generic product structure

To characterize the source of variety and its impact on product differentiation, this research introduces a concept of Generic Product Structure (GPS). As schematically illustrated in Figure 5, the GPS is proposed based on the hierarchical decomposition of product structures. Within a product family, all variants share a common structure. GPS performs as a generic data structure for such a product family. Figure 5 exhibits the basic construct of GPS, and that any complex GPS can be composed by employing basic constructs recursively (called nesting). There are three aspects associated with the basic construct, as depicted below.

(1) Constituent Item. Essentially, the basic construct is a hierarchy consisting of constituent items at different levels of abstraction, where items can be either abstract or physical entities. Two types of items are involved, namely modules, $\{M_i\}$, and structural relationships, $\{SR_i\}$, where i is the index of a module. A module is a set of physical or logical components grouped together to carry out certain functions. It may be at any level of abstraction - either an end-product or a sub-assembly or a component.

GPS distinguishes two types of modules, namely primitive modules and compound modules. A primitive module is one that can not be further decomposed, thus becoming leaf nodes of the decomposition structure, and possesses several variants (*i.e.*, instances of the same module type). A compound module is composed of primitive modules and/or other compound modules. The nesting of basic constructs is achieved by introducing compound module(s) as the component(s) of another compound module. In this sense, a nested GPS can be regarded as a multi-level decomposition structure of compound modules. In GPS, all variants of modules are controlled at leaf nodes. That is, the variants, $\{M_i^*\}$, of a module, M_i, are defined for and related to primitive modules only. This is because the variety of a compound module can be achieved through its primitive modules. Therefore, the relationship between variants and the corresponding module can be observed as instantiation of the module according to certain include conditions, $\{IC_i^*\}$ (to be explained below).

A structural relationship, SR_i, refers to the parent-child relationship between a parent module, M, and a child module, M_i. With respect to product structures, it is equivalent to the goes-into relationship defined for BOM structures (van Veen, 1992). There are only two variants associated with a structural relationship, *i.e.*, if exists ($SR_i = 1$) or not ($SR_i = 0$). The existence of SR_i means that the child module, M_i, is included as the component of the parent module, M. Otherwise ($SR_i = 0$), M_i is excluded. Different variety generation methods can be implemented through defining such SR variants.

A common module, M_0, means that there exists a fixed structural relationship ($SR_0 \equiv 1$) linked to a parent module and that all of its variants are same as itself ($M_i^* \equiv M_0$) and free from any include conditions. All other modules ($M_i \neq M_0$) and those variable structural relationships ($SR_i = 1$ or 0) comprise two sources of variety: variation resulting from different variants of modules and that due to possible existence of certain structural relationships.

(2) Variety Parameter. Usually, there is a set of attributes, A_i, associated with each module M_i. Among them, some variables are relevant to variety, and thus are defined as variety parameters, $\{P_{i,j}\} \subset A_i$, where j is the index of a variety parameter related to module, M_i. Like attribute variables, parameters can be inherited by child/primitive modules from a parent/compound module. Different instances of a particular $P_{i,j}$, *i.e.*, parameter values, $\{P_{i,j}^*\}$, embody the diversity resembled by, and perceived from, product variants, that is, $\{M_i^* | P_{i,j} = P_{i,j}^*\}$,

where M_i^* denotes a variant of M_i. In this way, all variants of primitive modules are in fact organized by different instantiations of variety parameters. However, common modules are supposed to have no variety parameter.

The instantiation of GPS is accompanied by the propagation of variety parameters from higher-level to lower-level modules. At the top level, all variety parameters are transformed from functional features and options (so called functional variety) and conveyed by the end-product. The allocation of these variety parameters to lower-level modules involves the relationships of variety parameters and differentiation enablers, and thus requires domain knowledge about mappings from functional requirements to design parameters [38]. Some research has been conducted towards such variety fulfillment issues. Functional variety (commercial view characteristics) is linked to component families according to direct correspondence [57]. Each child node inherits parameters from its immediate parent node [61]. In both work, inheritance is assumed to be direct correspondence. However, more complicated scenarios exist. Functional variety may be in a different form from technical variety defined in terms of engineering parameters. For example, "Office chairs should be durable" is functional variety, whereas the corresponding engineering parameters may be a set of parameters related to materials and/or design concepts. In addition, certain parameters of a lower-level module may be derivatives of its parent's parameters. For instance, "Feedback depth" is a parameter of the feedback module in power supply design. It must be introduced in response to different "Transient times" (original variety parameter) of its parent – the power supply. Furthermore, parameter values of a child may be a function of the same parameter of its parent. Malmqvist studied the latter two scenarios in the context of parametric design [75]. In a framework of managing product data with multiple customer-selected options, Baldwin and Chung highlighted that "the option combinations of each component must be specified in binding as a function of option combination of the aggregate design" [62]. Therefore, it can be concluded that parameters propagate along the GPS hierarchy and obtain their values at each node in light of variety parameters of its immediate parent nodes.

Parameter propagation in GPS can be explained using a simple example depicted in Figure 6. A compound module, M_1, is composed of several lower-level modules, one of which is M_2. M_2 is also a compound module and can be further decomposed into primitive modules, M_{21} and M_{22}. In the figure, the following facts can be observed:

(a) Every module has its own parameter set, for example, $\mathbf{P}_{M_1} \sim \{P_{1,1}, P_{1,2}, P_{1,3}, P_{1,4}, P_{1,5}\}$, $\mathbf{P}_{M_2} \sim \{P_{2,1}, P_{2,2}\}$, $\mathbf{P}_{M_{21}} \sim \{P_{21,1}\}$, and $\mathbf{P}_{M_{22}} \sim \{P_{22,1}, P_{22,2}\}$.

(b) A few parameters of the parent node propagate to its child nodes. For example, $P_{1,4}$ and $P_{1,5}$ propagate from M_1 to M_2, and then $P_{2,1}$ propagates from M_2 to M_{21}, and both $P_{2,1}$ and $P_{2,2}$ of M_2 propagate to M_{22} simultaneously. Any change of parameters in a parent module must result in changes of relevant parameters in the influenced child modules.

(c) A child module may have parameters that are same as those of its parent module and thus all parameters take the same values. Such parameter propagation is referred to as inheritance (direct correspondence), which reflects Wortmann and Erens' observation [57]. For example, relationships between $P_{1,4}$ and $P_{2,1}$, $P_{2,1}$ and $P_{21,1}$, and $P_{2,2}$ and $P_{22,2}$ belong to this category.

(d) Certain parameters of a child module may not be the same as those of its parent module. Instead, its values result from functions of its parent's parameters. This constitutes parameter derivation. For example, $P_{2,2}$ is derived from $P_{1,5}$, and $P_{22,1}$ is derived from $P_{2,1}$ and $P_{2,2}$.

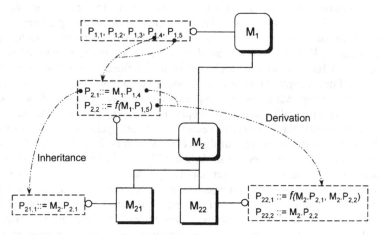

Figure 6:   Parameter propagation in GPS

(3) Include Condition. Include conditions are logical expressions defined for the instantiation of constituent items of GPS. The general form of a include condition is as following:

IF (antecedent) THEN (consequent)

where logical operations, such as AND, OR, and XOR, can be applied to both antecedent and consequent. The include condition of a structural relationship, IC_{SR_i}, is a binary variable with a 1/0 or Yes/No value, which indicates the existence of SR_i, meaning if or not the child module, M_i, is included in the parent module, M. On the other hand, the include condition of a module, IC_i, denotes a selective type instantiation, that is, a module, M_i, possesses several (≥ 1) variants, $\{M_i^*\}$, and the relationship of variants is exclusive all (XOR), meaning that only one out of the many variants can be included. Usually, IC_i is defined in terms of different combinations of variety parameter values of related modules, for example, IF $(P_{i,1} = P_{i,1}^1)$ AND $(P_{j,2} = P_{j,2}^1)$ THEN $(M_i^* = M_i^1)$ AND $(M_j^* = M_j^1)$. These include conditions describing the compatibility of various module variants implying the technical feasibility of module combinations.

8.5.2 Variant derivation

The generation of product variants has been dealt with in some product family models, including such useful concepts as parameter inheritance [57, 61], configuration constraints [57, 60, 62] and selection conditions [57, 62]. Nonetheless, the generative capability of these models is limited due to unclear handling of the underpinning structures of constraints across different views.

In this research, four elements are proposed for variant derivation, namely selection constraints, parameter propagation, include conditions, and variety generation. The considerations are: (a) Selection constraints are used to specify the applicable option combinations of functional features; (b) Both correspondence and derivation types of parameter propagation are modeled; (c) Include conditions are bounded to variants of primitive modules and to variable structural relationships as well; and (d) Variety generation defines the engineering realization of product variety. The last three are closely related to the GPS. The variant derivation process is depicted in Figure 7, where four stages are involved.

(1) At the stage of product specification, customers select available options of functional features subject to some selection constraints. The result is a set of customer-selected options, which reflects customers' requirements.

(2) At the second stage, selected functional features are transformed to the variety parameters of the end-product and then propagated down the hierarchy of the GPS. Through this parameter propagation, all parameters of modules in the GPS obtain specific values.

(3) The third stage, variant instantiation, is closely related to the second stage (parameter evaluation). With GPS, in fact, these two processes can be executed automatically [76]. The instantiation process for deriving the variant of a compound module can be summarized as the following:

Step 1. Obtain a value for each parameter from its parent according to its own propagation function;

Step 2. Check whether there is any variable structural relationship at the first level of GPS. If no variable structural relationship exists, proceed to step 4;

Step 3. If there is a variable structural relationship, check the include conditions of the associated structural relationship one by one. Delete those structural relationships along with the related modules, if their include conditions do not hold true; and

Step 4. In case a child module is a primitive module, select (specify) a variant from its alternatives, if its include condition holds true;

Step 4'. In case a child module is a compound module, proceed to Step 1.

By the end of this stage, we can obtain a product structure, *i.e.*, a BOM structure, specific to the product specification for the customer. The primitive variants suitable for implementing the required variety are then determined.

(4) The fourth stage is to generate the product design of the instantiated variant, where variety generation is involved. Variety generation methods are defined for each compound module and can be modeled in the GPS as the following:

- Attaching/Removing can be represented by a variable structural relationship that is of binary-type;

- swapping of primitive modules can be embodied in the include conditions of their alternatives. Swapping of compound modules resembles variant instantiation; and

- scaling is modeled by propagation functions of the scaled modules.

According to the product structure instantiated from the GPS, the custom product can be designed from a base product. Components of the base design are compared with the components of the same type in the instantiated product structure. If a component in the base product is not the one specified in the instantiated product structure, this component will be modified according to predefined variety generation methods. The manipulation of modules can be elaborated using graph grammar formalisms [77].

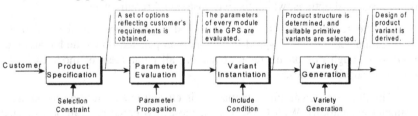

Figure 7: An illustration of the variant derivation process

8.6   A case study from the power supply industry

Power supplies are widely used in any product running on electrical energy. Few subsystems are as unappreciated as power supplies while a system is in the process of design. Usually, specification or design of a power supply is not considered seriously until the design of its host system reaches substantial maturity. Consequently, a custom power supply is always required to fit into the host system. With more than 1,200 product offerings, an electronics company under our investigation has been constantly challenged to accommodate an increasing variety of custom power supplies while leveraging costs. There is a significant amount of engineering expenses for meeting diverse customer applications. This case study is summarized from the practice of the proposed APF-based product family design approach in the company to deal with the challenge of increasing variety. A switching-mode power supply family, called FB65-T, is used here.

8.6.1   FB65-T product family

Power supply product families are usually identified according to the power conversion topology and the power rate with respect to diverse customer requirements. The characteristics of the FB65-T family include universal input, triple output, 65W normal output power, and flyback conversion topology. Based on market research, three categories of most required customization are identified as follows.

(1) Value of output voltage. The variation range is $0 \sim +15\%$ normal output voltage because customers may need higher output voltage to compensate the voltage drop in wires. The degree of the required voltage is dependent on the electrical distance between the power supply and the load. To change the value of an output voltage, the number of turns of the secondary coil of the transformer has to be recalculated based on the following equation [78]:

$$N_{sec} = \frac{N_{pri}(V_{out} + V_{fwd})(1 - \tau_{(max)})}{V_{in(min)}\tau_{(max)}},$$

where, V_{fwd} denotes forward voltage drop of the anticipated output rectifier, $\tau_{(max)}$ means maximum "on" time duty cycle (usually 50%), and $V_{in(min)}$ stands for the minimum expected input voltage. This equation can be solved for the required number of secondary turns at the lowest anticipated input voltage. At any input voltage lower than that, the regulator will fall out of regulation.

(2) The accuracy of output voltage. The accuracy of output voltage is deter-mined by the feedback circuit design. Cross-regulation is a term associated with the regulation of more than one output in multiple-output power supplies. It is the amount by which the output changes in voltage when one or more of the outputs experience a change in the load. In real-world switching power supplies, it is uneconomical to sense the voltages of all outputs. Some outputs thus keep unsensed. As a result, when a sensed output experiences an increased load, the unsensed output rises in voltage. When an unsensed output is loaded, it decreases and the sensed output remains at its rated voltage. In a typical power supply with +5V and ±12V outputs, when +5V output goes from half-rated load to full load, the +12V one would typically go to +13V and the −12V one would go to −14.5V. The technique of multiple output sensing can reduce such voltage variation among all outputs and thus should be used whenever it is possible. In this regard,

customers have to make tradeoffs between costs and performance when specifying the accuracy of output voltage.

(3) Output protection scheme. In the design of a switching power supply, it is important to consider the protection of the load and the protection of the power supply from failures in the load. To determine suitable protection methods, both the end-user of the product/system and its functionality should be considered. There are two prevalent protection methods, i.e., the short-circuit (SC) and the over-voltage (OV) protections. Customers can select either one or two of them in accordance with their protection philosophies. In addition, an AuR feature, meaning auto-recovery after the fault is removed, can be selected pending upon the selection of SC protection.

To meet the above demands for variety, the FB65-T family is designed as the following:

(a) Base product (specification): +5V, ±12V, and normal output accuracy;

(b) A variety of output voltages can be achieved using the scaling method, in which secondary turns are calculated according to Equation (2);

(c) A variety of output accuracy specifications can be fulfilled using the swapping method. Either multiple-output sensing (MOS) or single-output sensing (SOS) can be selected as the feedback circuit in accordance with the accuracy level specified by customers; and

Table 4:   Customization and variety generation methods for the FB65-T family

Customization		*Module (variety generation)*			
		Transformer	*Feedback Circuit*	*Protection Circuit*	*Start-up Circuit*
Output Voltage		Scaling (N_{sec})	Nil	Nil	Nil
Output Accuracy	Normal	Nil	Swapping (MOS/SOS)	Nil	Nil
	High				
Protection Scheme	SC	Nil	Nil	Attaching (SC, OV, SC+OV)	Nil
	OV				
	SC&OV				
	AuR	Nil	Nil	Nil	Swapping (BS/SDOO)

(d) Two concerns are associated with the protection scheme. At a higher level of decomposition, the protection circuit is a module that is relatively independent of other parts of the system. However, the module itself can be further decomposed into the SC protection, the OV protection and the compensation sub-modules. In power supply realization, a protection module may consist of the compensation sub-module and any one or two of the protection sub-modules. If the AuR feature is specified, a shut-down-on-overcurrent (SDOO) start-up circuit is required to replace the boot-strap (BS) circuit. Table 4 summarizes the customization and variety generation methods used in the FB65-T family.

8.6.2 Multiple views of APF

The APF of the FB65-T family can be described from both the sales and the engineering perspectives. Table 5 gives the sales view of APF constructs, including functional features, and options. There are seven features, among which the power rate, voltage of primary output, and RFI are common features. Sets of options are available for such features as the voltage of secondary output, output accuracy, the protection scheme, and the function of auto recovery. These features along with their options become the differentiation enablers of the family from customers' viewpoint. A selection constraint is specified for the selection of the protection scheme and the AuR function because the selection of AuR depends on the selection of the SC protection. This selection constraint exhibits the configuration mechanism in the sales view.

Table 5: The sales view of APF for the FB65-T family

APF Construct	*Feature*	*Option (Feature Value)*
Common base, common features	F_1 : Power rate	$Option_1 = 65W$
	F_2 : Primary output	$Option_2 = 5V$
	F_6 : RFI	$Option_6 = VDE$
Differentiation enabler, optional features	F_3 : Secondary output	$Option_3 \in [\pm 12*(1+15\%)]$
	F_4 : Output accuracy	$Option_4 \in \{Normal, High\}$
	F_5 : Protection scheme	$Option_5 \subseteq \{SC, OV, SC\&OV\}$
	F_7 : Auto recovery	$Option_7 \in \{Yes, No\}$
Configuration mechanism, selection constraint	If ProtectionScheme $\notin \{SC, SC\&OV\}$, then AuR = No.	

The engineering view is described in Tables 3 and 4. Table 6 lists the common bases from the engineering perspective. The generic product structure and the flyback topology are those technologies determining the ultimate functionality and capability of the family, such as the maximum output rate, the regulation level, switching frequency, *etc.* Accordingly, the input rectifier, power switch, output rectifier, controller, RFI, and the compensation module in a protection circuit become common modules to be used by every family member. It is obvious that the common bases of APF in the engineering view are determined by product technologies and their embodiments are those common modules of the family.

Table 6: The engineering view of APF for the FB65-T family: common bases

Technology	Common Module
Generic product structure	Input rectifier
Flyback topology	Power switch
-	Output rectifier
-	Controller
-	RFI
-	Compensation

As shown in Table 7, the differentiation enablers in the engineering view are categorized into distinctive modules and distinctive structural relationships. They are described in terms of variety parameters, variants, include conditions, and associated variety generation methods. There are three distinctive modules, namely the transformer, the feedback module, and the startup module.

Table 7: The engineering view of APF for the FB65-T family: differentiation enablers

Differentiation Enabler		Variety Parameter	Variant	Include Condition	Variety Generation
Distinctive modules	transformer	N_{sec}	StandardT	$N_{sec} = 40$	scaling
			CustomT	$N_{sec} \neq 40$	
	feedback	sensing	SOS	sensing=single	swapping
			MOS	sensing=multiple	
	startup	CircuitType	BS	CircuitType=BS	swapping
			SDOO	CircuitType=SDOO	
Distinctive structural relationship	SR_{SC}	Nil	SR_{SC}^{1}	WithSC = No	attaching
			SR_{SC}^{2}	WithSC = Yes	
	SR_{OV}	Nil	SR_{OV}^{1}	WithOV = No	attaching
			SR_{OV}^{2}	WithOV = Yes	

Each of the three modules has more than one variant that can be identified by certain include conditions described in terms of appropriate variety parameters. For instance, two variants of the feedback module, SOS and MOS, can be identified by the sensing method. Thus Sensing is defined as a variety parameter. When this parameter takes a value of Single, SOS should be included in the designed power supply. Otherwise, when Sensing takes a value of Multiple, MOS should be included. In addition, two distinctive structural relationships are introduced to differentiate a variety of protection schemes. They are identified using associated include conditions defined in terms of variety parameters of the protection circuit. Finally, different product variants are derived according to appropriate variety generation methods.

8.6.3 Generic product structure

The Generic Product Structure (GPS) of the FB65-T family is given in Figure 8. All modules and structural relationships that may occur are shown in the structure, including distinctive structural relationships and distinctive modules, as well as fixed structural relationships, common modules, compound modules, and primitive variants. In this GPS, the end-product itself is a compound module, which consists nine child modules. The protection circuit is also a compound module, which is decomposed into three child modules: the compensation, the SC protection and the OV protection circuits. Such a decomposition results in a variety of protection schemes. Two distinctive structural relationships, SR_{OV} and SR_{SC}, are introduced to characterize this kind of differentiation.

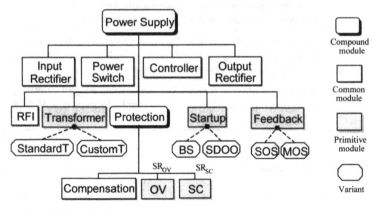

Figure 8: The GPS for the FB65-T family

8.6.4 Parameter propagation

Parameter propagation reveals the relationship between parameters of parent modules and those of child modules, thus characterizing the origin of variety and the fulfillment of functional variety. Variety parameters of a child module may be the equivalent of those of its parent module, or be defined as functions of the parameters of its parent module. First, functional features described in the sales view are transformed to the variety parameters of the end product. Then, the variety parameters of each module in the GPS are evaluated according to the parameters of its immediate parent. Table 8 illustrates parameter propagation in the case of FB65-T family. Four variety parameters are used to differentiate variants of the end-product. Each of them is bound to a functional feature and obtains its value according to customer-selected options. Such parameters as N_{sec} for the transformer, Sensing for the feedback module, CircuitType for the startup module, and WithSC and WithOV for the protection module are all defined as functions of the parameters from their parent modules.

Table 8:   Parameter propagation in the GPS for FB65-T family

Item	Parameter Propagation
End-Product	Eng.PowerSupply.SecOutput ::= Sales.SecondaryOutput Eng.PowerSupply.OutputAc ::= Sales.OutputAccuracy Eng.PowerSupply.ProtSch ::= Sales.ProtectionScheme Eng.PowerSupply.AuR ::=Sales.AutoRecovery
Module	Transformer.N_{sec} = f(Parent.SecOutput) = f(PowerSupply.SecOutput)
	Feedback.Sensing = f(Parent.OutputAc) = f(PowerSupply.OutputAc) Feedback.Sensing = Single, if PowerSupply.OutputAc = Normal Feedback.Sensing = Multiple, if PowerSupply.OutputAc = High
	Startup.CircuitType = f(Parent.AuR) = f(PowerSupply.AuR) Startup.CircuitType = BS, if PowerSupply.AuR = No Startup.CircuitType = SDOO, if PowerSupply.AuR = Yes
	Protection.WithSC = f(Parent.ProtSch) = f(PowerSupply.ProtSch) Protection.WithSC = With, if PowerSupply. ProtSch ∈ {SC, SC&OV} Otherwise Protection.WithSC = Without
	Protection.WithOV = f(Parent.ProtSch) = f(PowerSupply.ProtSch) Protection.WithOV = With, if PowerSupply. ProtSch ∈ {OV, SC&OV} Otherwise Protection.WithOV = Without

8.6.5   Variety generation

The FB65-T family serves a market where custom products are frequently required to meet customers' special needs. For this product family, custom designs start from a base product, which consists of all common bases of the GPS, as shown in Figure 9. A variety of end-products are generated by changing the number of turns of the secondary coil of transformer, replacing different variants of the startup, the feedback and the protection modules. The variants of protection scheme are created by attaching different circuits of SC protection and/or circuits of OV protection to the compensation module, which performs as the base product of protection module. In other words, the protection module itself is a family and possesses its own APF. Figure 10 illustrates variety generation for the protection circuit family.

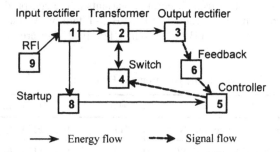

Figure 9: Base product for the FB65-T family

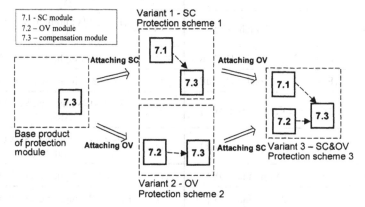

Figure 10: Variety generation for protection circuits

8.6.6 Variant derivation

Upon receiving a set of selected options from a customer, all variety parameters are evaluated according to appropriate parameter propagation functions. Based on the variety parameter(s) of each module, a suitable variant for the module can be instantiated according to associated include conditions. Then these instantiated variants of the module are manipulated to modify the base product using predefined variety generation methods. After all applicable modifications are performed, a particular product variant that meets the customer's requirements is derived. For the FB65-T family, because the protection module itself is a compound module, its variant should be derived first.

Afterwards, this protection variant is used in the derivation of the variant of the end-product, in which the protection module becomes one of the constituent modules of the end-product (compound module). This scenario illustrates node nesting in GPS instantiation. Node nesting is particularly useful for complex products, whose GPS is characterized by multiple layers and more than one compound module. Table 9 lists all possible variants at the end-product level that can be derived from the base product of the FB65-T family. All data entities and relationships involved in the FB65-T product family design are summarized in Figure 11.

Table 9:   List of product variants derived from the FB65-T family

Variant	Description*	Variant	Description*
#1	{$F_3 = \pm 12V$; normal; SC}	#11	{$F_3 = \pm 12V$; high; OV}
#2	{$F_3 \neq \pm 12V$; normal; SC}	#12	{$F_3 \neq \pm 12V$; high; OV}
#3	{$F_3 = \pm 12V$; high; SC}	#13	{$F_3 = \pm 12V$; normal; SC&OV; AuR}
#4	{$F_3 \neq \pm 12V$; high; SC}	#14	{$F_3 \neq \pm 12V$; normal; SC&OV; AuR}
#5	{$F_3 = \pm 12V$; normal; SC; AuR}	#15	{$F_3 = \pm 12V$; high; SC&OV; AuR}
#6	{$F_3 \neq \pm 12V$; normal; SC; AuR}	#16	{$F_3 \neq \pm 12V$; high; SC&OV; AuR}
#7	{$F_3 = \pm 12V$; high; SC; AuR}	#17	{$F_3 = \pm 12V$; normal; SC&OV; AuR}
#8	{$F_3 \neq \pm 12V$; high; SC; AuR}	#18	{$F_3 \neq \pm 12V$; normal; SC&OV; AuR}
#9	{$F_3 = \pm 12V$; normal; OV}	#19	{$F_3 = \pm 12V$; high; SC&OV; AuR}
#10	{$F_3 \neq \pm 12V$; normal; OV}	#20	{$F_3 \neq \pm 12V$; high; SC&OV; AuR}

*   For illustrative simplicity, those features common to every variant are not listed, including 65W, universal input, triple output, +5V output, RFI, and so on.

8.7   Summary

Motivated by the rationale of designing families of products to satisfy diverse customer requirements with reasonable costs, *i.e.*, to realize mass customization, we have discussed in this chapter the necessity to explore the fundamental issues of product family development. The concept of Architecture of Product Family (APF) is proposed to capture the underlying logic of a product family and thus to support product variant derivation. A product family refers to a set of similar products that are derived from a common platform and yet possess specific features/functionality to meet particular customer requirements.

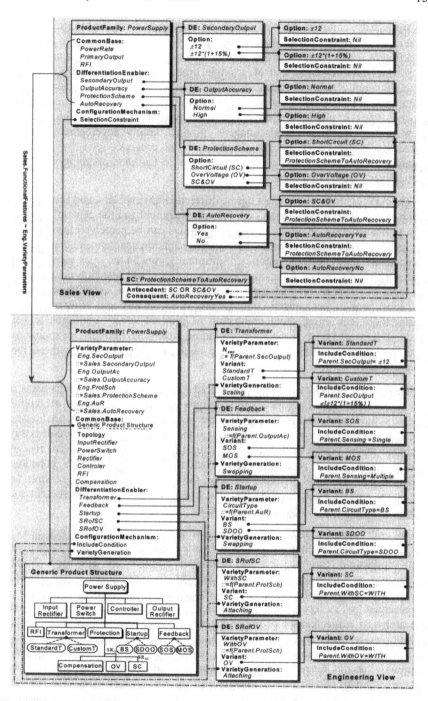

Figure 11: Data modeling for power supply product family design based on APF

Modularity, commonality and variety are important concerns in product family development. Modularity provides the flexibility for a (base) product to be tailored to different requirements. Commonality reveals the difference of the architecture of product families from that of a single product. While modularity resembles decomposition of product structures and is applicable to describing module types, commonality characterizes the grouping of similar module variants under specific module types. In addition, while encouraging functional variety in the functional view through customer needs analysis, product family development needs to reduce technical variety in the design and the process views by systematic planning of modularity and commonality so as to facilitate plugging-in modules that deliver specific functions.

The APF consists of common bases, differentiation enablers, and configuration mechanisms. They characterize both the compositional and the generative aspects of a product family. In addition, product families can be perceived from different perspectives. Therefore, common bases, differentiation enablers and configuration mechanisms imply different entities in different views. In the sales view, the APF is described in terms of functional features, options, and selection constraints on the options. Product technologies, structural relationships, constituent modules, and configuration mechanisms comprise the APF from an engineering perspective. Furthermore, the engineering realization of product variety conforms to some basic patterns of variety generation, including attaching/removing, swapping, and scaling. Variety fulfillment can also be achieved through nesting these basic mechanisms.

The Generic Product Structure (GPS) and variant derivation mechanisms are the core of APF-based product family design. The GPS encompasses all modules and structural relationships that may occur in the family. Structural relationships are identified as one type of constituent item in the GPS. Common modules can be included in the product using fixed structural relationships. Distinctive modules can be manipulated using variable structural relationships. With constituent modules and structural relationships, the GPS is capable of representing both structural and parametric variations of a product family, which represent most practice of customized product development.

All APF constructs perform as the configuration mechanisms necessary to derive engineering definitions of product variants. Selection constraints guarantee only applicable combinations of functional features are specified. Variety parameters are introduced to synchronize variety propagation across different views and along different levels of abstraction. Functional variety can be represented as variety parameters, which in turn propagate along the GPS hierarchy by a mechanism of either inheritance or derivation. Defining include conditions for distinctive structural relationships can capture the variation caused by the existence or non-existence of a module. Relating include conditions to primitive variants reveals the variation caused by different variants of the same type of modules. Basic variety generation methods, such as attaching/removing, swapping and scaling, can be modeled by using variable structural relationships and defining appropriate include conditions of modules. In this sense, the GPS provides a platform for APF-based product family design. Therefore, product variant derivation actually becomes a GPS instantiation and variety generation process, which can be implemented automatically provided that the module manipulation capability is fully supported.

References

[1]   Pine II, B.J.: Mass Customization: The New Frontier in Business Competition, Boston 1993.

[2]   IEE Colloquium on Manufacturing 'Mass Customization', London, 1996.

[3]   Proceedings of the Think Custom Conference, New York 2000, on www.think-custom.org/presentation.htm.

[4]   Wortmann, H.C.; Muntslag, D.R.; Timmermans, P.J.M.: Customer driven Manufacturing, London 1997.

[5]   Tseng, M. M.; Jiao, J.: Design for mass customization, in: Annals of the CIRP, 45 (1996), pp. 153-156.

[6]   Ulrich, K.: The Role of Product Architecture in The Manufacturing Firm, in: Research Policy, 24 (1995), pp. 419-440.

[7]   Sawhney, M. S.: Leveraged High-Variety Strategies: From Portfolio Thinking to Platform Thinking, in: Journal of the Academy of Marketing Science, 26 (1998) 1, pp. 54-61.

[8]   Robertson, D.; Ulrich, K.: Planning for Product Platforms, in: Sloan Management Review, Summer 1998, pp. 19-31.

[9]   Corso, M.; Muffatto, M.; Verganti, R.: Multi-Product Innovation: Emerging Policies in Automotive, Motorcycle and Earth-moving Machinery Industries, in: 3rd International Product Development Management Conference on New Approaches to Development and Engineering, Fontainebleau 1996.

[10]   Wheelwright, S. C.; Sasser, W. E. J.: The New Product Development Map, in: Harvard Business Review, 1989.

[11]   Ericsson, J.; Karlsson, P. O.; Mercer, G.; Robertson, D.: Sharing Parts across Car Models: Lessons from The Manufacturers, in: Europe's Automotive Components Business, 1 (1996), pp. 150-171.

[12]   Jandourek, E.: A Model for Platform Development, in: Hewlett-Packard Journal, 1996.

[13]   Sanderson, S.; Uzumeri, M.: Managing Product Families: The Case of the Sony Walkman, in: Research Policy, 24 (1995), pp. 761-782.

[14]   Meyer, M.; Utterback, J.: The Product Family and the Dynamics of Core Capability, in: Sloan Management Review, Spring 1993, pp. 29-47.

[15]   McGrath, M.: Product Strategy for High-Technology Companies, New York 1995.

[16]   Wilhelm, B.: Platform and Modular Concepts at Volkswagen – Their Effects on the Assembly Process, in: Shimokawa, K.; Jürgens, U.; Fujimoto, T. (Eds.): Transforming Automobile Assembly, Berlin and Heidelberg 1997.

[17]   Ulrich, K. T.; W. P. Seering: Function sharing in mechanical design, in: Design Studies, 11 (1990), pp. 223-234.

[18]   Kota, S.; Sethuraman, K.: Managing Variety in Product Families through Design for Commonality, in: 1998 ASME Design Engineering Technical Conferences, Atlanta 1998.

[19]   McAdams, D. A.; Stone, R. B.; Wood, K. L.: Understanding Product Similarity using Customer Needs, in: 1999 ASME Design Engineering Technical Conferences, Las Vegas 1999.

[20]   Moore, W. L.; Louviere, J. J.; Verma, R.: Using Conjoint Analysis to Help Design Product Platforms, in: Journal of Product Innovation Management, 16 (1999), pp. 27-39.

[21]   Meyer, M.; Lehnerd, A. P.: The Power of Product Platform – Building Value and Cost Leadship, New York 1997.

[22]   Rothwell, R.; Gardiner, P.: Robustness and Product Design Families, in: Oakley, M. (Ed.): Design Management: A Handbook of Issues and Methods, Cambridge 1990, pp. 279-292.

[23]   Sabbagh, K.: Twenty-First Century Jet: The Making and Marketing of Boeing 777, New York 1996.

[24]   Chen, W.; Simpson, T. W.; Allen, J. K.; Mistree, F.: Use of Design Capability Indices to Satisfy a Ranged Set of Design Requirements, in: Dutta, D. (Ed.): Advances in Design Automation, Irvine 1996.

[25]   Simpson, T. W.; Chen, W.; Allen, J. K.; Mistree, F.: Conceptual Design of a Family of Products Through the Use of the Robust Concept Exploration Method, in: 6th AIAA/USAF/NASA/ISSMO Symposium on Multidisciplinary Analysis and Optimization, 2 (1996), Bellevue, pp. 1535-1545.

[26]   Simpson, T. W.; Rosen, D.; Allen, J. K.; Mistree, F.: Metrics for Assessing Design Freedom and Information Certainty in the early Stages of design, in: Proceedings of 1996 ASME Design Engineering Technical Conferences, Irvine 1996.

[27]   Simpson, T. W.; Maier, J. R. A., Mistree, F.: A Product Platform Concept Exploration Method for Product Family Design, in: Proceedings of 1999 ASME Design Engineering Technical Conferences, Las Vegas 1999.

[28]   Siddique, Z.; Rosen, D. W.; Wang, N.: On the Applicability of Product Variety Design Concepts to Automotive Platform Commonality, in: Proceedings of 1998 ASME Design Engineering Technical Conferences, Atlanta 1998.

[28]   Sundgren, N.: Introducing Interface Management in Product Family Development, in: Journal of Production Innovation Management, 16 (1999), pp. 40-51.

[30]   Finch, W.W.: Set-based Models of Product Platform Design and Manufacturing Processes, in: Proceedings of 1999 ASME Design Engineering Technical Conferences, Las Vegas 1999.

[31]   Ulrich, K.; Eppinger, S. D.: Product Design and Development, New York 1995.

[32]   Ulrich, K.; Tung, K.: Fundamentals of Product Variety, in: Issues in Design Manufacturing/Integration, 39 (1991), pp. 73-77.

[33]   Pahl, G.; Beitz, W.: Engineering Design: A Systematic Approach, 2th Ed., New York 1996.

[34]   Newcomb, P. J.; Bras, B.; Rosen, D. W.: Implications of Modularity on Product Design for the Life Cycle – Design Theory and Methodology, in: Wood, K. (Ed.): DTM'96, Irvine 1996.

[35]   Rosen, D. W.: Design of Modular Product Architecture in Discrete Design Spaces Subject to Life Cycle Issues, in: Proceedings of ASME 1996 Design Engineering Technical Conferences, Irvine 1996.

[36]   Kusiak, A.; Huang C. C.: Development of Modular Products, in: IEEE Transactions on Components, Packaging, and Manufacturing Technology, 19 (1996) 4, pp. 523-581.

[37] Henderson, R. M.; Clark, K. B.: Architectural Innovation: The Reconfiguration of Existing Product Technologies and the Failure of Established Firms, in: Administrative Science Quarterly, 35 (1990), pp. 9-30.

[38] Suh, N. P.: Axiomatic Design – Advances and Applications, 2000.

[39] Meyer, M. H.: Revitalize Your Product Lines through Continuous Platform Renewal, in: Research-Technology Management, 40 (1997), pp. 17-28.

[40] Fujita, K.; Ishii, K.: Task Structuring toward Computational Approaches to Product Variety Design, in: Proceedings of 1997 ASME Design Engineering Technical Conferences, Sacramento 1997.

[41] Erens, F.; Verhulst, K.: Architectures for Product Families, in: Computers in Industry, 33 (1997), pp.165-178.

[42] Yu, J. S.; Gonzalez-Zugasti, J. P.; Otto, K. N.: Product Architecture Definition Based Upon Customer Demands, in: ASME Journal of Mechanical Design, 121 (1999), pp. 329-335.

[43] Zamirowski, E. J.; Otto, K.N.: Identifying Product Portfolio Architecture Modularity Using Function and Variety Heuristics, in: Proceedings of 1999 ASME Design Engineering Technical Conferences, Las Vegas 1999.

[44] Jiao, J.; Tseng, M. M.: Fundamentals of Product Family Architecture, in: Integrated Manufacturing Systems, 11 (2000) 7.

[45] Siddique, Z.; Rosen, D. W.: Product Platform Design: A Graph Grammar Approach, in: Proceedings of 1999 ASME Design Engineering Technical Conferences, Las Vegas 1999.

[46] Ishii, K.; Juengel, C.; Eubanks, C. F.: 1995a, Design for Product Variety: Key to Product Line Structuring, in: Proceedings of the 1995 DETC, 83 (1995) 2, pp. 499-506.

[47] Ishii, K.; Lee, B. H.; Eubanks, C. F.: Design for product retirement and modularity based on technology life-cycle, in: Manufacturing Science and Engineering, 2-2 (1995), pp. 921-933.

[48] Martin, M.V.; Ishii, K.: Design for Variety: Development of Complexity Indices and Design Charts, Proceedings of the 1996 ASME Design Engineering Technical Conferences, Sacramento 1997.

[49] Fujita, K.; Akagi, S.; Yoneda, T.; Ishikawa, M.: Simultaneous Optimization of Product Family Sharing System Structure and Configuration, Design for Manufacturing –DFM'98, Atlanta 1998.

[50] Fujita, K.; Sakaguchi, H.; S. Akagi: Product Variety Deployment and Its Optimization Under Modular Architecture & Module Commonalization, Processing of DETC'99, Las Vegas 1999.

[51] Page, A. L.; Rosenbaum, H. F.: Redesigning Product Lines with Conjoint Analysis: How Sunbeam Does It, in: Journal of Product Innovation Management, 4 (1987), pp.120-137.

[52] Green, P. E.; Srinivasan, V.: Conjoint analysis in marketing: new developments with implications for research and practice, in: Journal of Marketing, 54 (1990) 4, pp. 3-19.

[53] Gilmon, J. H.; Pine II, B. J.: The Four Faces of Mass Customization, in: Harvard Business Review, 75 (1997) 1, pp.91-101.

[54] Hegge, H. M. H.; Wortmann, J. C.: Generic Bill-of-Material: A New Product Model, in: International Journal of Production Economics, 23 (1991), pp. 117-128.

[55] Erens, F.; McKay, A.; Bloor, S.: Product modeling using multiple levels of abstraction instances as types, in: Computers in Industry, 24 (1994) 1, pp. 17-28.

[56]   Erens, F. J.; Hegge, H. M. H.: Manufacturing and Sales Co-ordination for Product Variety, in: International Journal of Production Economics, 37 (1994) 1, pp. 83-99.

[57]   Wortmann, H. C.; Erens, F. J.: Control of Variety by Generic Product Modeling, in: Processing of The First World Congress on Intelligent Manufacturing Processes and Systems, 2 (1995), University Puerto Rico, San Juan, pp. 1327-1342.

[58]   Mckay, A., Erens, F.; Bloor, M.S.: Relating Product Definition and Product Variety, in: Research in Engineering Design, 2 (1996), pp. 63-80.

[59]   van Veen, E.A.: Modeling Product Structures by Generic Bill of Materials, Amsterdam 1992.

[60]   Jiao, J.; Tseng, M. M.: An Information Modeling Framework for Product Families to Support Mass Customization Production, in: Annals of the CIRP, 48 (1999) 1, pp. 93-98.

[61]   Jiao, J.; Tseng, M; Ma, Q.; Zou, Y.: Generic Bill of Materials and Operations for High-Variety Production Management, in: Concurrent Engineering: Research and Application, 9 (2000) 4.

[62]   Baldwin, R. A.; Chung, M. J.: Managing Engineering Data for Complex Products, in: Research in Engineering Design, 7 (1995), pp. 215-31.

[63]   Paula, G.: Reinventing a Core Product Line, in: Mechanical Engineering, 119 (1997) 10, pp. 102-103.

[64]   Yamanouchi, T.: Breakthrough: The Development of the Canon Personal Copier, in: Long Range Planning, 22 (1989) 5, pp. 11-21.

[65]   Schonfeld, E.: The Customized, Digitized, Have-It-Your Way Economy, Fortune, 1998, pp. 115-124.

[66]   Wheelwright, S. C.; Clark, K. B.: Leading Product Development: the senior manager's guide to creating and shaping the enterprise, New York 1995.

[67]   Choi, S. C.; Desarbo, W.S.: A Conjoint-based Product Designing Procedure Incorporating Price Competition, in: Journal of Product Innovation Management, 11 (1994), pp. 451-459.

[68]   Choi, S. C.; Desarbo, W.S.; Harker, P. T.: Product Positioning under Price Competition, in: Management Science, 36 (1990), pp. 175-199.

[69]   Suzue, T.; Kohdate, A.: Variety Reduction Program: A Production Strategy for Product Diversification, Cambridge 1990.

[70]   Feitzinger, E.; Lee, H. L.: Mass customization at Hewlett-Packard: the Power of Postponement, in: Harvard Business Review, 75 (1997), pp. 116-121.

[71]   Erixon, G.: Design for Modularity, in: Huang, G. Q. (Ed.): Design for X - Concurrent Engineering Imperatives, New York 1996, pp. 356-379.

[72]   Boynton, A. C.; Victor, B.: Beyond flexibility: building and managing the dynamically stable organization, in: California Management Review, 34 (1991) 1, pp. 53-66.

[73]   Oxford English dictionary, New York 1989.

[74]   Giesberts, P. M. J.; Tang, L. V. D.: Dynamics of the Customer Order Decoupling Point: Impact on Information Systems for Production Control, in: Production Planning & Control, 3 (1992) 3, pp. 300-313

[75]   Malmqvist, J.: Towards Computational Design Methods For Conceptual And Parametric Design, Ph.D. Thesis, Chalmers University of Technology, Goteborg 1993.

[76]   Du, X.; Tseng, M. M.; Jiao, J.: Graph Grammar Based Product Variety Modeling, in: Proceedings of 2000 ASME Design Engineering Technical Conferences, Baltimore 2000.

[77]   Du, X.; Jiao, J.; Tseng, M. M.: Modeling Product Family: A Graph Grammar Approach, Working Paper, 2000.
[78]   Brown, M.: Power Supply Cookbook, Boston 1994.

Acknowledgments: This research is partially supported by Computer Products Asia-Pacific Ltd. (Artesyn Technologies) under grant CPI 95/96.EG01, the HKUST Research Infrastructure Grant (RI 93/94 EG08), and the Hong Kong Government Research Grant Council (HKUST 797/96E and HKUST 6220/99E). The authors would like to express their sincere thanks to Dr. M. Eugene Merchant, Professors Stephen C-Y Lu, Num P. Suh, Gunnar Sohlenius and Martin Helander for their valuable advice.

Contact:

Prof. Mitchell M. Tseng, Ph.D.
Department of Industrial Engineering & Engineering Management,
Hong Kong University of Science & Technology, Kowloon, Hong Kong
E-mail: tseng@ust.hk

Dr. Roger Jianxin Jiao
School of Mechanical and Production Engineering,
Nanyang Technological University, Singapore
E-mail: mjiao@ntu.edu.sg

9   Common Platform Architecture

Identification for a set of similar products

Zahed Siddique[1] and David W. Rosen[2]
[1] School of Aerospace and Mechanical Engineering, University of Oklahoma, Norman, Oklahoma, USA
[2] School of Mechanical Engineering, Georgia Institute of Technology, Atlanta, USA

Designing a family of products using a common platform approach instead of designing single products has gained momentum in various industries. One of the challenges faced by companies include identifying common modular platform configuration for a set of existing similar products, which requires development of product and process models and tools to facilitate configuration reasoning. The purpose of this chapter is to present a configuration reasoning framework that can be applied to identify common platform architectures. To accomplish this objective, first representations are developed for product architecture, then mathematical tools are developed to identify common platform from these representations along with a re-modularization scheme to investigate alternative architectures. Product architectures are discrete, as such graph and set representations are used to model product architectures. Discrete mathematical tools such as isomorphism are then adopted and applied to identify common platform architecture. The application of this configuration reasoning framework is illustrated using identifying common platform architecture for a set of automotive underbody front structures.

9.1   Introduction

Companies are being faced with the challenge of providing as much variety as possible for the market with as little variety as possible between products, which has introduced the concept of developing a product family utilizing a platform approach. "A product platform is a collection of the common elements, especially the underlying core technology, implemented across a range of products" [1]. Focusing product strategy at the platform level simplifies the product development process because there are fewer platforms than products and major platform decisions are only made every few years. Product family development problems can be divided into two broad categories: (1) Aggregation implies determining the common platform for a set of similar existing product. (2) Differentiation implies increasing product portfolio using the existing platform. The aggregation approach is *a posteriori* approach characterized by redesign and/or consolidation of existing product groups to create more competitive product families. The reasoning behind the aggregation approach is to (1) simplify the product offering and reduce part variety by (2) standardizing components so as to (3) reduce manufacturing and inventory costs and (4) reduce manufacturing variability (i.e., the variety of parts that are produced in a given manufacturing facility) and thereby (5) improve quality and customer satisfaction [2]. The diversification approach is *a priori* approach to product platform design characterized by a conscious up-front decision to develop and manage a product family based on a common core or platform. In this chapter the focus is on aggregation approach to simplify the product offerings using configuration design, primarily modular design. In configuration design, we are interested in identifying the components in an artifact and the relationships among the components [3]. To address these questions, it is necessary to identify common platform configurations, for a set of products, considering structure and assembly process information.

Activities involved in the aggregation approach are to identify the common platform configuration for a set of similar products and to establish a common assembly process. Both platform structure and assembly process reasoning have similar steps associated with them. Conceptually identifying common platform involves:

- Representation of product architectures and process information to facilitate product platform design.
- Identification of the common platform elements and assembly process.

We will present a Common Platform Identification (CPI) method to identify common platform structure and process. The method starts with a set of existing similar products to identify a common platform and assembly process. The CPI method only focuses on configuration; scaling and dimensional synthesis are not performed. The CPI method consists of a set of mathematical tools that can be applied towards common platform reasoning. Product architectures are composed of discrete components with relationships among them. As such, graph and other discrete representations are used to represent the product configuration and assembly process information. Mathematical tools are then applied on these representations to identify common platform structure and assembly process. The product architectures are then reconfigured to consider alternative configurations to improve platform and common assembly process.

The remaining of this chapter is organized as follows: We start with a brief literature review of product representation schemes and common platform identification. In the following section, the CPI method is introduced, followed by a presentation of graph representation and steps for common platform identification for product architecture and assembly. This method is then applied to common front structure identification for a set of automotive front structures. We will end with some concluding remarks.

9.2 Literature review

9.2.1 Research in product family design

Research done in the areas of mass customization, design for product variety, designing family of products, etc. has stressed the need for developing a common platform. Martin and Ishii identified commonality, modularity and standardization as the core characteristics of product families [4]. Chen *et al.* suggests designing flexible product architectures to enable small product changes to increase product variety [5]. Pine suggests modularity to achieve maximum individual customization while minimizing cost. Stone *et al.* proposed a method to identify modules from a functional description of the product [6]. Kota *et al.* suggested standardization of all components that does not need to be varied to satisfy the varying function requirements of the product family [7]. Simpson *et al.* presented a Product Platform Concept Exploration Method (PPCEM) to facilitate the design of a common scalable product platform for a product family [8]. Gonzalez-Zugasti *et al.* focus on modular product family development based on customer demands [9, 10]. Newcomb et al., use Steward's Design Structure Matrix to define modular product architecture(s) from different life cycle viewpoints [11, 12].

This research in product family design and modular architecture development does not systematically investigate all feasible platform and product family member architectures. The first step towards achieving this goal is to generate all feasible platform and product family configurations given a set of product and process constraints.

9.2.2 Product architecture representation

Identification of common platform for a set of similar products requires representation of product configurations. The representation is required to capture and represent platform and product family constraints. In order to model product architecture, graphical approaches are useful due to their inherent representative power. The use of graphs provides a somewhat intuitive and highly visual context for such representation. In the graph representations the nodes denote the individual components and the arcs denote dependencies between the components. Different types of graphs are available for varying representational requirements of the node interactions [13, 14]. An added benefit of using graphical representations to model product structure is that a large amount of associated research exists in the area of graph theory [12, 15]. In this chapter product architecture will be represented using graphs and other discrete models similar to that of. Modeling product architectures using graphs allow concepts from graph theory to be applied for platform reasoning to identify common platform from the representations for a set of similar products. The graphical representation will be modified and extended to represent product architectures.

9.2.3   Identification of common platform

Although many researchers have identified the need for development of common platform for set of similar products, in most cases their efforts do not specify approaches that can be used for common platform identification and development. Simpson et al. use the Product Platform Concept Exploration Method (PPCEM) for designing scalable common product platform for a family [8]. Pederson has developed a framework for more *effective* realization of large and complex made-to-order systems, coined the Hierarchical Product Platform Realization Method (HPPRM) [16]. The HPPRM is founded on *Decision-Based Design* acknowledging that humans are at the center of decision making, and an *Evolutionary Approach* acknowledging the importance of historicity (heredity), variation (population thinking), and probabilistic decisions (natural selection) in product development.

9.3   Common platform identification method

Our interest is to identify common platform configurations for a set of similar products, which involves determining the common elements and relationships for a set of similar products. To perform aggregation of similar set of products, tools and models need to be developed and identified. Activities required to identify common platform structure and assembly process are shown in Figure 1.

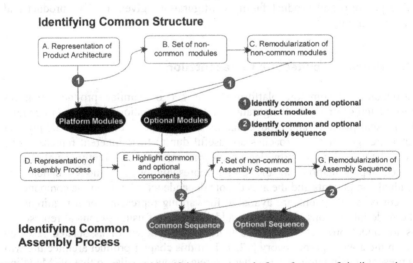

Figure 1:   Modeling related to identifying common platform for a set of similar products

Identifying Common Platform Structure involves representing product structure information, which includes initial modularization and connection information (A). Applying algorithms and methods on these representations to identify common and optional product modules (B). Remodularizing the non-common modules and applying algorithms and methods to identify alternative configurations of the product, that will increase commonality (C). Identifying common assembly process representing assembly process information (D) and highlighting common and optional components (E) identified as a result of Step C. Applying algorithms and methods on the representations to identify common and optional sequences (F). Remodularizing the non-common assembly sequence and applying algorithms and methods to identify alternative assembly sequences, that will increase commonality (G). Mathematical concepts used to model the product and platform architecture and assembly processes along with tools for common platform identification will be presented in the next two sections.

9.4 Identifying common platform structure

9.4.1 Representation of product structure

Configuration design involves determining components that are in the product, relations among these components, and module information, hence this information needs to be modeled. To identify the common platform for a product family, product structure information need to be represented and the representation scheme should also facilitate the identification of viewpoint specific common platform. Product architectures are discrete structures; as such, discrete representations are well suited for platform representation. The representation scheme used to model product information is based on graphs and subgraphs. Several preliminary definitions are presented first.

Definition 1: A *graph* is a pair G= (V,E)

Where the elements of V are the vertices (or nodes, or points) of the graph G and E is a set of pairs of distinct vertices called edges. The vertex set of graph G is referred as V(G), its edge set as E(G).

Vertices u and v are said to be *adjacent* if there is an edge {u,v}. The usual way to picture a graph is by drawing a dot or a circle for each vertex and joining of these dots (circles) by a line if the corresponding two vertices form an edge. A graph G is called a *labeled graph* if its vertices are assigned labels from the vertex label alphabet, ΣV. In this research the vertices are assigned "names", which indicates the component/element of the product the vertex is representing. So, ΣV are the names of components/elements of the product. In the following we use the term graph to indicate labeled graphs with labeled vertices.

Definition 2: Let G=(V,E) and G'=(V',E') be two graphs. If V'⊆V and E'⊆E then G' is a *subgraph* of G and is denoted as G'⊆G and is said G *contains* G'.

To facilitate the representation of product module information a subgraph G' is considered to have the name G', which from an engineering perspective indicates the intent (function) of the module. In the pictorial presentation, the boundaries of the subgraphs are explicitly drawn and named. The subgraph representation provides an easy way to maintain correspondence between modules and components associated with them. It also provides means of performing reasoning at both module and component level.

An example representation of segments of an automotive front structure is shown in Figure 2. The vertices in the graph represent components of the front structure and the edges between the vertices represent existing connections among the components. The subgraphs (with their explicit boundary and name) in Figure 2 represent modules of the front structure: the subgraph named "Front Cross Member Assembly" specifies the 10 components and connections among these components that constitute the module.

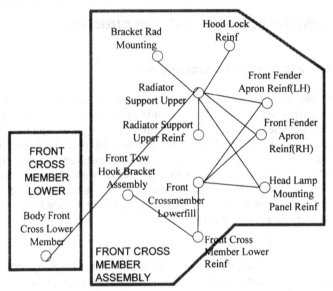

Figure 2: Labeled Subgraph Representation of a partial automotive front structure

9.4.2   Determining common structure platform from graph representation

Development of a modular common product platform to support family of products requires separating structures of the products into common and optional modules. This separation requires identification of common modules in all the product architectures - which is equivalent to identifying labeled subgraphs that are common in all representations, which is called identifying *isomorphic labeled subgraphs*.

Definition 3: Two labeled subgraphs are *isomorphic* if

1. the names of the subgraphs are equivalent and
2. the labeled subgraphs are isomorphic.

From an engineering perspective the first condition checks if the intent of the modules are equivalent. The second condition determines if the structure of the modules are same. Given two or more graphs with labeled subgraphs, identification of isomorphic labeled subgraphs is performed using the procedure described below. The procedure identifies the common labeled subgraphs and as a byproduct identifies vertices and edges that is not common. The procedure is as follows:

Labeled subgraph isomorphism

Input: Two or more labeled graphs with subgraphs representing modules.

G1,G2,....,Gn are the input labeled graphs, representing n similar products. gi,1,gi,2,....,gi,qi⊂Gi where i=1,2,...,n and qi is the number of subgraphs in the ith graph. Also L, is the label operator, L(gi,j) represents the label of gi,j labeled subgraph.

(1) Search and arrange subgraph labels that are equivalent for all graphs under consideration. This step corresponds to determining if the modules have the same intent.
If Q={$Q_1,Q_2,...,Q_p$} represent set of labeled subgraph, with Q_k representing set of labeled subgraphs with equivalent labels then:

Q_k={g_{ij}|L(g_{ij})≡L(Q_k) where L(Q_k) is the equivalent label for the current set Q_k, $g_{ij}∈G_i$ i=1,2,...,n and $g_{ij}∉Q_m$, m=1,2,..k-1} where k is the number of labeled subgraph sets with equivalent labels.

(2) For each set of labeled subgraphs with equivalent label determine if the vertices contained in the subgraphs have equivalent labels. This step corresponds to determining if the components in the modules are same.
For each set of subgraph in Q with equivalent label, determine if there exists a labeled vertex isomorphism for the subgraphs in the set, Q_k. The vertex isomorphism is determined by coloring the vertices with common labels for all the subgraphs in Q_k. If all the vertices are colored common for all subgraphs in Q_k, then there exists vertex isomorphism for all $g_{ij}∈ Q_k$

(3) For all subgraphs that have same set of vertices determine if an edge bijection exists. This step corresponds to determining if the relations among the components are same. If the edge bijection exists then the labeled subgraphs in the set Qk are said to be isomorphic.

Labeled Subgraph Isomorphism can be applied to labeled graphs to identify modules that are common in all the members of the product family. The intent of this chapter is to determine common platform for a set of similar products, keeping a module perspective. To accomplish this objective, in many cases, remodularization of the products is required to increase commonality among products, which will lead to a larger common product platform. The product can be remodularized based on intuition and experience. A more structured way is to first determine possible modularizations and then identify the potential modularization for increase in commonality. Possible modularization for a set of components with a set of constraints, using configuration design spaces, will not be presented here. For information on these configuration design spaces, readers are referred to [17].

Once the remodularization of the set of products has been determined, Labeled Subgraph Isomorphism can be reapplied to the modified remaining product architectures to determine common modules. The identified common modules make up the common platform, whereas the uncommon modules will be used to provide options for a product family.

9.5　Identifying common assembly process

9.5.1　Assembly modeling using family of vertex subsets

The primary assembly process information that is of interest here is the assembly sequence. In the assembly viewpoint, modules are developed by assembling components and then the modules are assembled with other modules/components to develop larger modules, until the final product is assembled. As products are composed of discrete components which are assembled in a sequence, discrete representation based on sets is used to represent the assembly sequence. A set is a collection of objects called the elements of the set. A set A is said to be a subset of a set B if every element of A is also an element of B. In this case, it is said that A is *contained* in B, and written as $A \subseteq B$. If $A \subseteq B$, it is also said that B is a *superset* of A. To represent an assembly sequence, the elements of the set are labeled vertices representing components of a product. Subsets of this vertex set represent assembly modules. Assembly sequences can be represented as a *family of vertex subsets*. In this representation the subsets are given a label, which represents the assembly workstation name. The boundaries and the label (name) of the subsets are explicitly drawn along with the label of the vertices. Assembly sequences are either parallel or sequential, which implies that (in this assembly representation) if B and C are two subsets of a set A then:

$$B \cap C = B \text{ or } C \text{ or } \varnothing \text{ ($\varnothing$ represents the empty set)}$$

This property of the subgraph representation makes the *family of vertex subsets* a partially ordered set under the $\subseteq$ relation. Representation of assembly sequence using *family of vertex subsets* is straightforward. In this representation subsets contain components and other subsets that are assembled at that particular assembly step. An example provides Figure 3: (b) shows the *family of vertex subsets* representation for the coffeemaker's assembly sequence (a). Focusing on the start of the assembly to illustrate the representation scheme: The labels of the subsets represent assembly workstations, as such the components "Heater 1" and "Base Plate" are assembled together in CM-1 workstation. In the next step two "tubes" are assembled with the existing assembly module in workstation CM-2. This continues until the complete coffeemaker has been assembled.

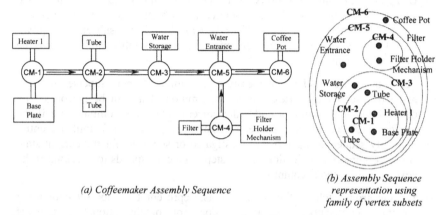

(a) *Coffeemaker Assembly Sequence*

(b) *Assembly Sequence representation using family of vertex subsets*

Figure 3:   Assembly sequence representations of a coffee maker

9.5.2   Identification of common assembly process

The common assembly sequence for a set of products can be determined from the *families of vertex subsets* for the set of products.

Definition 4: Let F1 and F2 be two *families of vertex subsets* and AF1 and AF2 be two subsets of F1 and F2 respectively. AF1 and AF2 are considered common if:

1.   The names of A_{F1} and A_{F2} are same.
2.   The sets of vertices for A_{F1} and A_{F2} are equivalent.
3.   All subsets of A_{F1} and A_{F2} satisfy the two conditions above.

From an engineering perspective the first condition checks if the workstations are equivalent. The second condition determines if the components (total) of the current assembly sequence are same. The third condition determines if the assembly sequences before the current assembly state are also equivalent. The common assembly sequence for a set of products is determined using the following procedure:

Identify common vertex subsets

Input: Two or more families of vertex subsets representing assembly sequences for two or more similar products.

(1) Determine the minimal subsets for each of the *families of vertex subsets* under considerations from their partial order. (A subset *b*, in a *family of vertex subsets*, is said to be minimal if no subset precedes *b*) This corresponds to determining the start of assembly sequence branches for the products.

(2) Using only the minimal subsets determine the set of common minimal labeled subsets. The minimal subsets do not have any subsets under them, as such this can be determined by checking if vertices included in the subsets are equivalent. This step corresponds to determining if the assembly modules have the same components.

(3) For each subset under consideration determine if their superset in their respective family of vertex subsets is common. This can be performed by determining if the subsets under the current subset are common and if the vertices included in each subset are same. Perform step 3 until the entire assembly sequences have been investigated or supersets for the current state are not common. Application of this step once corresponds to checking if the next assembly step is common.

(4) Starting from each of the final subsets, represent the superset containing optional components (if the superset does not contain optional components then represent the current subset) with *representative vertex* and apply step 3 to determine if the superset is common (considering the representative vertex as common). Repeat Step 4 until the maximal subset of the family of vertices has been reached. Application of this step corresponds to ignoring sequences that are not common and determining if the rest of the sequence is common.

Procedure Identify Common Vertex Subsets can be used to identify a common assembly process for two or more similar products. In the next section representations and procedures that have been presented will be applied towards identifying common platform and assembly process for automotive front structures.

9.6   Automotive front structure common platform

Balancing the need to customize products for target markets while enabling the economies of scale of a "world car" is a challenge faced by every automotive manufacturer. A proliferation of options and model derivatives leads to increased tooling cost and production line complexity. Careful commonization of automotive platforms can be used to increase product variety while reducing the number of components and the product line complexity. Automobile manufacturers are changing how they design, engineer and manufacture their products. Most, if not all, of the major players in the automotive industry are moving towards developing common platforms.

Imagine that an auto manufacturer has two platforms for small size cars and wants to develop a common platform. The common platform is required to support the configuration of both the platforms, while using maximum number of common components/modules and assembly process. The designer is focusing on the front structure of the platform and wants to identify the common platform configuration. The focus is on modularity to develop a common front structure that can support requirements for both automobiles. Automotive underbody along with the main modules of the front structure is shown in Figure 4. A common front structure is required to support the configuration of front structures for both the platforms, while using maximum number of common components/modules. The two front structures were designed for the same class cars with similar requirements, as such similarities exist between them. Even with these similarities, the two platforms are manufactured and assembled in different platform plants. Recently the automotive manufacturer has decided to develop a common platform for the same class automobiles to take advantage of common platform thinking.

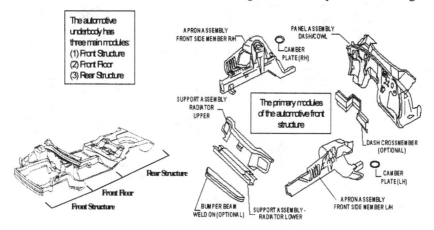

Figure 4: Automotive underbody front structure and its main modules

The designer has information about which components are in both the platforms, the current modular structure of the platform architectures (which components belong to which modules), and the connections among the different components of the platform. The objective is to: *Develop a common front structure from the existing two front structures, by reasoning at the configuration level, that can support both the front structures.*

The front structures are named Front Structure #1 and Front Structure#2. The two front structures under consideration are from the same class cars. Application of Platform Commonization steps on these two front structures is presented in the rest of the section, starting with the representation of the structure.

9.6.1 Identification of common modules for structure viewpoint

From the information provided for the two platforms, the structure of the automotive front structures is modeled using graphs and sets. Module and connection information is captured using the representations to facilitate the development of the common front structure. Components of the front structures are represented as vertices with connections among the components represented using edges. Subgraphs are used to identify components that are in a module. The labels of the subgraphs indicate intent of the modules. The initial modularization (components, modularization, and connections) of the two front structures were also provided. The initial modularization of Front Structure#1 and Front Structure #2 are shown in Figure 5.

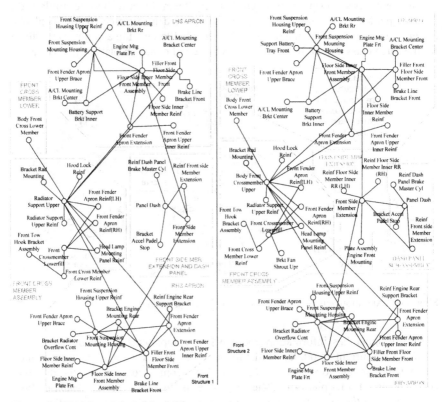

Figure 5: Initial modularization of front structures

As can be seen from Figure 5, Front Structure #1 has LHS-Apron, RHS-Apron, Front Cross Member Assembly, Front Cross Member Lower, and Front Side Member Extension and Dash Panel as modules. In the case of Front Structure #2 the Front Side Member Extension and Dash Panel are separate modules. Other differences exist in the two front structures due to components that are not common. In the next three sections the common structure platform for the two front structures will be identified from the labeled subgraph representation, keeping a module perspective. The result of applying labeled subgraph isomorphism on the two automotive front structure representations is shown in Figure 6.

In the figure subgraphs that have common label and components across the two front structures have been colored common (Shaded). The two modules that have been colored common (red) are: RHS Apron and Front Cross Member Lower modules. Front structure components that are not common are also colored black in Figure 6.

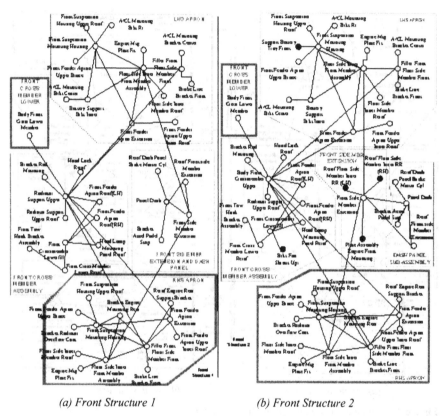

(a) *Front Structure 1*   (b) *Front Structure 2*

Figure 6: Identifying common and optional modules/components between front structures

The product modules that have the potential to be common have not been shaded. In these modules some of the components are common while others are not. Remodularization of each of these modules is performed to determine if any alternatives exist that might increase commonality between the two front structures. As an example consider the LHS Apron module for the two front structures – except for the Support Battery Tray Front component in Front Structure#2 all other components are same. From the set of possible remodularizations for these 15 components, based on experience and intuition, the designer decided that the modularization with the Support Battery Tray Front as a new module, Front Battery Support, and the rest of the components as a module will be chosen. This option is feasible because the Support Battery Tray Front component can be separated from the LHS Apron. Remodularization of Front Cross Member Assembly, and Dash Panel was also performed in a similar manner. Remodularization of the other non-common module was not feasible, as such it was not modified. The new modularization of the two front structures is shown in Figure 7.

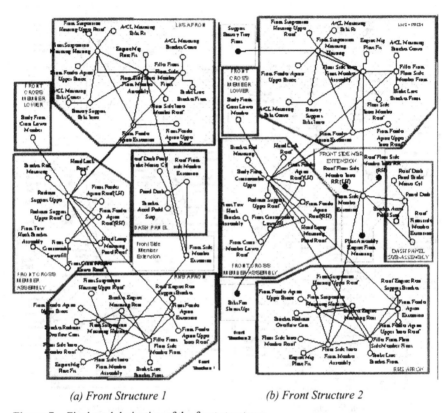

  (a) Front Structure 1   (b) Front Structure 2

Figure 7: Final modularization of the front structures

With the remodularization performed on the front structures the next step is to identify the common elements and options for these new modules and to specify the available options. Applying labeled subgraph isomorphism on the modules that have not been shaded yet, it is determined that the LHS Apron, Front Cross Member Assembly and Dash Panel Assembly are common. The Front Battery Support, Front Side Member Extension and Fan Bracket modules are optional. The final coloring of the two front structures is shown in Figure 7. Identification of the common platform for the two front structures is straightforward from the colored graphs. Modules that have been colored common are part of the common front structure. The common front structure platform (Figure 8) includes the following modules: RHS Apron, LHS Apron, Dash Panel, Front Cross Member

Assembly, and Front Cross Member. The options for the two front structures include two types of Front Side Member Extension, Front Battery Support and Fan Bracket. The identified common platform for the structure viewpoint and the labeled graphs will be used as input to identify common assembly process from the family of vertex subsets representation for the two front structures.

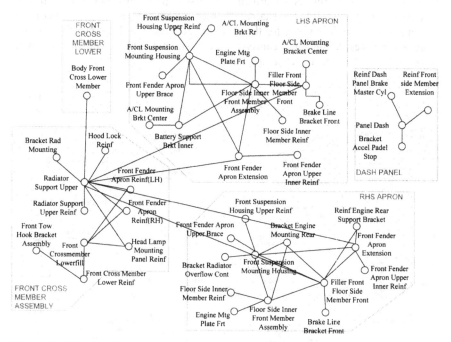

Figure 8: Common front structure for structure viewpoint

9.6.2 Identification of common assembly process

In this section the common assembly process will be identified from the *family of vertex subsets* representation of the two front structures. The approach for identifying common assembly process was described in the previous section. Identification of the common assembly process is performed separately from the structure viewpoint as indicated in Figure 1. The activities involved in this step will be performed and results for each will be presented. The initial assembly process representations, using family of vertex subsets representation, for the two front structures are shown in Figure 9.

Applying the Identify Common Vertex Subsets procedure on the two front structure representations, the common assembly process can be identified. The shaded subsets represent assembly sequences that are common for both front structures. The shaded nodes in Figure 9 represent components that are not common. As can be seen in Figure 9, although many components are common among the two front structures, the assembly processes are very different. To increase the commonality of the assembly process, assembly sequences that have common components but do not have common sequences will be reorganized. As an example, remodularization of the component for the subset (RA-2) containing Rein Engine Rear Support Bracket, Filler Front Floor Side Member Front, and Brake Line Bracket Front is performed. The first step towards remodularization is to determine the possible alternatives, which is determined by constructing a combinatorial assembly space for the elements contained in the subset [17]. From these alternatives the designer can choose a feasible assembly sequence that will be common for both of the front structures.

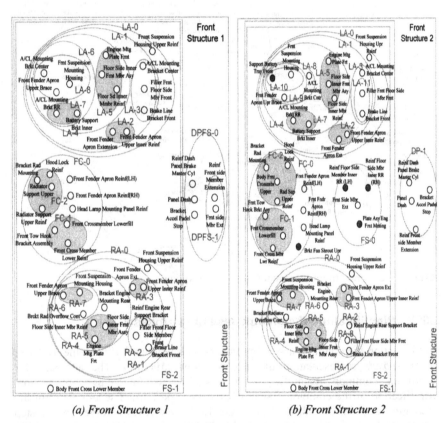

(a) Front Structure 1   (b) Front Structure 2

Figure 9: Identifying common assembly process for two front structures

Remodularization was also performed for FS-0, DPFS-0, LA-8, RA-2, FC-0. After remodularization of all the assembly modules the resulting assembly processes for both front structures are shown in Figure 10. As before the shaded areas in the graphs represent the common assembly sequence.

Applying the Identify Common Vertex Subsets procedure on the final remodularized families of vertex subsets, the common and optional assembly sequences can be colored, which can be used to identify the common assembly process for the two front structures (Figure 11).

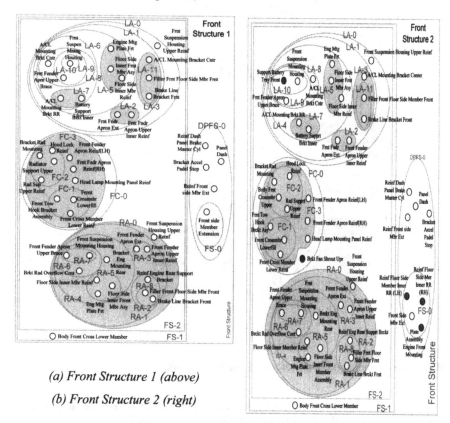

(a) Front Structure 1 (above)

(b) Front Structure 2 (right)

Figure 10: Final remodularization of assembly process for the two front structures

9.7 Conclusions

To facilitate the identification and development of common platform configurations and assembly process for a set of existing products (1) product structures were modeled using labeled subgraphs and labeled subgraph isomorphism was applied on these graphs to identify the common platform (2) families of vertex subsets were used to model assembly sequences for products and common vertex subsets were identified, which was the common assembly process. The significance of this work is in systematically identifying common platforms and processes for a set of similar product architectures keeping a modular perspective based on graph theory.

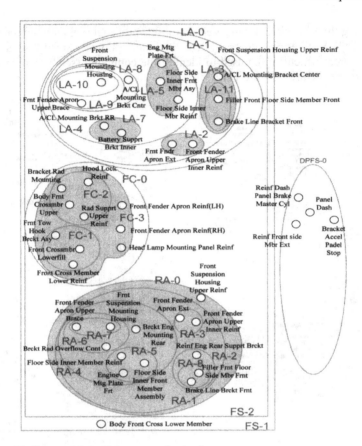

Figure 11: Common Assembly process for the front structures

Based on the process of identifying common platforms and the automotive front structure example, the following statements can be supported:

- Graphs can be used to model product and platform architectures, with vertices representing components and edges between the nodes representing relationship between the components.

- Labeled subgraphs have the capability to explicitly represent module information for products.

- Labeled subgraph isomorphism can be used to identify common modules from a set of labeled graphs representing product structures.

- Family of vertex subsets can be utilized to model assembly process information. Algorithms and procedures can be applied on these representations to identify the common assembly process.

In this chapter we have identified the common product architecture and assembly process separately. To identify an overall common platform both component and assembly process viewpoints need to be combined in future work.

References

[1]   McGrath, M. E.: Product Strategy for High-Technology Companies, New York 1995.

[2]   Simpson, T. W.: A Concept Exploration Method for Product Family Design, Atlanta 1998.

[3]   Dixon, J. R. et al.: A Proposed Taxonomy of Mechanical Design Problems, in: Computers in Engineering Conference, San Francisco 1988, pp. 41-46.

[4]   Martin, M. V.; Ishii, K.: Design for Variety: Development of Complexity Indices and Design Charts, in: 1997 ASME Design for Manufacturing Conference, Sacramento 1997.

[5]   Chen, W.; Rosen, D.; Allen, J.; Mistree, F.: Modularity and the Independence of Functional Requirements in Designing Complex Systems, in: Concurrent Product Design, 74 (1994), pp. 31-38.

[6]   Stone, R. B.; Wood, K. L.; Crawford, R. H.: A Heuristic Method to Identify Modules from a Functional Description of a Product, in: 1998 ASME Design Engineering Technical Conference, Atlanta 1998.

[7]   Kota, S.; Sethuraman, K.; Miller, R.: A Metric for Evaluating Design Commonality in Product Families, in: ASME Journal of Mechanical Design, 122 (2000) 4, pp. 403-410.

[8]   Simpson, T.W.; Maier, J.R.A.; Mistree, F.: A Product Platform Concept Exploration Method for Product Family Design, in: 1999 ASME Design Enginnering Technical Conferences, Las Vegas 1999.

[9]   Gonzalez-Zugasti, J. P.; Otto, K.N. Otto; Baker, J.D.: A Method for Architecting Product Platforms with an Application to Interplanetary Mission Design, in: Advances in Design Automation, Atlanta 1998,

[10]   Gonzalez-Zugasti, J. P.; Otto, K. N. Otto; Baker, J. D.: Assessing Value for Product Family and Selection, in: Advances in Design Automation, Las Vegas 1999.

[11]   Newcomb, P.J.; Bras, B.; Rosen, D.W.: Implications of Modularity on Product Design for the Life Cycle, in: ASME Journal of Mechanical Design, 120 (1998) 3, pp. 483-490.

[12]   Steward, D.V.: Systems Analysis and Management: Structure, Strategy, and Design, New York 1981.

[13]   Rosen, D. W.: Design of Modular Product Architectures in Discrete Design Spaces Subject to Life Cycle Issues, in: 1996 ASME Design Automation Conference, Irvine 1996.

[14]   Feng, C. X.; Huang, C. C.; Kusiak, A.; Li, P. G.: Representation of functions and features in detail design, in: Computer Aided Design, 28 (1996) 12, pp. 961-971.

[15]   Roth, J.; Hashimshony, R.: Algorithms in Graph Theory and Their Use For Solving Problems in Architectural Design, in: Computer Aided Design, 20 (1988) 7, pp. 373-381.

[16]   Pedersen, K.: Designing Platform Families An Evolutionary Approach to Developing Engineering Systems, Georgia Institute of Technology, Atlanta 1999.

[17]   Siddique, Z.; Rosen, D. W.: On Discrete Design Spaces for the Configuration
        Design of Product Families, in: Artificial Intelligence in Engineering, Design,
        Automation, and Manufacturing, 15 (2001), pp. 1-18.

Contacts:

Dr. Zahed Siddique
School of Aerospace and Mechanical Engineering,
University of Oklahoma, Norman, Oklahoma, USA
E-mail: zsiddique@ou.edu

Prof. David W. Rosen
The George W. Woodruff School of Mechanical Engineering,
Georgia Institute of Technology, Atlanta, Georgia, USA
E-mail: david.rosen@me.gatech.edu

10   Reconfigurable Models and Product Templates

Means of increasing productivity in the product development process

Jordan J. Cox[1], Gregory M. Roach[1] and Shawn S. Teare[2]
[1] College of Engineering & Technology, Brigham Young University, Provo, USA
[2] Lockheed Martin, USA

Over the past 25 years significant investments have been made in process technologies in an effort to increase productivity and efficiency in product development. However, the return on investment in these process technologies has not yielded the gains in productivity that would be expected. This chapter asserts that this is due to the injection of new process technologies and tools into old product development processes. Globalization and mass customization are also demanding higher levels of productivity. The solution to achieving effective returns in this area of product development lies in designing and implementing strategies that integrate and optimize the new process tools and technologies. Two primary product development strategies are emerging that integrate and optimize the new process tools; first, reconfiguration of product knowledge, artifacts and data, and second, the product continuum. The gathered data supports two conclusions. One is that product design processes that take advantage of reconfigurable models enjoy significant savings in time and cost in the model creation segment of the design process. The instantiation time for the reconfigurable models averages 3% of the time to create conventional models. As the size and/or complexity of the product increases, the potential benefit of reconfigurable models also increases. Another observation is that the reconfigurable template is an effective tool for organizing and administering design information in the process. The extra time spent setting up the reconfigurable models in the first design cycle is easily repaid through faster future design cycles, easier design information maintenance, and effective reconfiguration of previous work. Productivity as measured by reduced cycle times and increased throughput must be designed and built into the process. Our experience is showing that the keys are reconfigurable artifacts and product templates.

10.1   Introduction

Investments in process technologies during the past two decades to increase productivity and efficiency in product development have not yielded the gains in productivity that would be expected. This lack of return on investment is due to injecting new process technologies and tools into old product development processes. Globalization and mass customization, termed 'the new business imperative', are also demanding higher levels of productivity [1]. The solution to achieving effective returns on investments in this area of product development lies in designing and implementing product development process (PDP) strategies that integrate and optimize the new process tools and technologies.

In this chapter, we present new strategies for product development processes that increase productivity and facilitate the implementation of mass customization. These strategies are organized around two concepts: reconfigurable models, defined as parametric scalability of product knowledge and product continua, defined as functional forms of the PDP. Using these concepts we propose constructing new processes as functional forms of reconfigurable product artifacts which can generate continua of product configurations within a specified customization envelope. This approach leverages the new process technologies with a new PDP designed from the outset for productivity and mass customization.

10.2 Background

Companies are investing large amounts of money to upgrade process technologies such as CAD/CAE/CAM expecting that the return on investment will be large. Considering the fact that most companies have invested in process technology for the past 25 years and have upgraded, on average, three times during this period, this represents a significant investment. According to G. Moore, "the billions of dollars invested in office automation have not improved the productivity of the office place one iota "[2].

Many companies have not taken full advantage of the process technology investments they have made. They have not used new technology to the fullest extent and miss out on many of the gains in efficiency that such technology can provide. For example, in the early 1990's CAD companies introduced parametric solid modeling. This provided new capabilities to develop updatable solid models. However, most engineering companies continue to use these new CAD systems in the traditional way to create solid models for creating drawings. In one example the authors reviewed, a scroll for a turbo-charger was being created the traditional way by creating cross-sections at defined intervals along the scroll and then "skinning" these cross-sections. Many problems were encountered including oscillations in the surfaces and inaccuracies in matching the profiles. This however was the process for defining scrolls used in manual drafting. Once it was discovered that all the design engineers wanted was a linear blend from the inlet to the outlet, the new CAD system was able to accomplish this mathematically in a single step. However, unless every task involving CAD is reviewed for possible improvements or mismatches, the new CAD software often goes underutilized. Also, slight changes in a product design can initiate total redesign of process plans or manufacturing artifacts because the ambiguities in the PDP require that new artifacts be made. Inefficiencies abound in our current processes due to the structure of the process and mismatches between the process and new process tools and technologies.

The structure of the PDP has a tremendous impact on the overall efficiency and productivity of the company. Companies are trying to insert new process technologies into old processes. The typical "stage gate" PDP was developed by NASA in the late 1960's and forms the basis for the processes in use today [3]. Companies are still working with the same basic philosophy for product development that they were using in the 1960's regardless of the fact that the technologies and tools are so dramatically different [4].

There are other forces that are driving the need for change in product development processes. Globalization and its drive towards mass customization have as large of a potential impact as these productivity issues. As consumers become more educated about market options the demand for custom products increases. This increase is forcing market shifts towards mass customization [5]. Mass customization demands mass production efficiencies and costs for producing custom products. Though not fully achievable, it is possible to dramatically increase the customizable aspects of a product and do it in a manner that retains efficiency in design and production. Product development processes however, must be much more carefully planned and customized to meet these requirements. Using one standard PDP for all types of products is not sufficient. Careful mapping of the product offering space is required to design for an envelope of solutions as opposed to single product solutions.

For years, little research has been done in developing new product development processes that take advantage of advancements in technology. M.E. McGrath reported that in meetings with corporate executives when the question 'How much do you invest improving the product development process?' was posed, they frequently replied, 'little or none' [6]. We postulate that new processes need to be developed that take advantage of new technology and tools to fully leverage the productivity gain throughout the entire PDP. We present two strategies for addressing these issues. The first strategy deals with using new technologies to structure product knowledge and artifacts for parametric reconfigurability. This strategy focuses on using parametric capabilities and process tools to format product artifacts so that they can be used again and again for generation of new product development projects. The second strategy focuses on restructuring the overall PDP as a product function capable of producing new forms of the product in a continuum. This involves orchestrating the process tools and technologies for optimal effectiveness. We have entitled this 'template strategies' [7, 8]. These strategies address the issues of parallelization, standardization and automation, and address the problems of globalization and mass customization.

10.3   Parametric reconfiguration of product knowledge and artifacts

A key concept in addressing productivity in product development is the concept of parametric reconfiguration. New tools provide the ability to create artifacts in a parametric and reconfigurable way. This requires formats for artifacts that allow for parametric updating so that the generation of artifacts becomes a simple instantiation process. Product development processes are collections of tasks, decisions, knowledge and models that are captured in artifacts such as solid model files, sets of drawings, reports, technical publications, hardware, assembly instructions, etc. The tasks within the PDP are usually structured to create the various artifacts. Decisions that are made during the PDP are typically focused on selecting the values of the defining design or manufacturing parameters. The timing of the decisions coupled with the known lead time requirements for artifact creation has dictated the structure of the PDP. This approach assumes that all the steps used to create the artifacts must be followed each time. This means that even though former versions of the artifact are available, they are used for reference only. Efforts to shorten cycle time focus on reducing the time required to perform the standard set of steps used to create the artifacts [9, 10, 11]. We maintain that this relies too much on the assumption that the process for creating the artifacts must remain the same, just done in less time. Orders of magnitude reductions in cycle time require completely new processes for creating artifacts and possibly the development of new forms of artifacts.

Artifacts usually have a consistent format. The decisions, knowledge and models associated with the artifact determine this format. A significant portion of the steps required to build artifacts are repeatable and require little or no new knowledge. Because of this, the artifacts can be created parametrically and the software and tools needed to construct them can be structured to allow derivative forms that all use the same base artifact. The implementation of parametric strategies and reconfigurability requires rethinking the steps in the creation of artifacts and how they are created (for further discussion on parametric strategies see [12]). If the artifacts are formatted for reconfiguration, the overall PDP can become reconfigurable as well. The tasks are structured to be parametric yet they remain tied to the reconfigurable artifacts.

The process of constructing artifacts for parametric reconfigurability is composed of five steps. Focus is placed on planning the envelope of instantiability to create derivative artifacts and the structure of the artifacts so that their creation is a simple instantiation. Consider a parametric solid model that can be instantiated to a multitude of different known configurations by simply varying the values of its input parameter list, or a technical report that can be updated by varying its defining parameters. A set of parameter values can be input to the reconfigurable model and it updates in a semi-automated fashion. This requires far more planning before creating the reconfigurable model to ensure that it will be robust. A discussion of the method for artifact reconfiguration and the implications of reusablility follows. The five steps for constructing reconfigurable artifacts and an example of their application are provided.

10.3.1 Steps for building parametrically reconfigurable product artifacts

Develop the instantiation envelope: This step requires the development of a conceptual design envelope to represent the variability of the artifact model. The envelope is a specification of key parameters and their limits. This includes items such as geometry, manufacturing capabilities and limitations, convergence criteria for analysis models, life cycle, etc. This requires planning and work with marketing, manufacturing and others to understand the product family offering and the company capabilities and limits.

Design: The artifact model is planned out in terms of its feature structure, parameter scheme, and the relations, i.e. mathematical equations that will be used to construct the model. The planning step is accomplished by sketching or detailing the features that will be used to construct the model and determining the ordering of these features. The parameter scheme of each of the features is then determined and finally the parametric relationships which determine how the model will update are developed.

Creation: The planned model is created using the appropriate tool (i.e. CAD/CAM/CAE, word processing software, etc.).

Debug: The model is debugged by "flexing" it to the limits of the envelope developed in step 1. Flexing of the model is accomplished by the instantiation of various sets of parameter values or designs that are encapsulated within the "design space" of the envelope. This may require adjustments to both the model and the envelope.

Production: The model may now be used as a reconfigurable model where product instantiation is accomplished by selecting appropriate parameters within the envelope governing the updating of the model.

10.3.2 Example development of a reconfigurable model

We will now examine the application of the five steps to create a reconfigurable model. CAD models are the artifacts within the PDP that are most readily formatted for reconfiguration. Consider the pump impeller represented in Figure 1. Reconfigurability in CAD models takes advantage of the history file capability coupled with the parametric capability of the specific tool. In order to truly make a CAD model reconfigurable it requires more than simply making the dimensions parametric. It requires careful model planning and coordination. The steps in the construction of the reconfigurable model differ significantly from traditional methods.

Figure 1: Example instantiations of a reconfigurable CAD model for an impeller

In traditional CAD model development the designer works with engineers to develop the preliminary geometry. Next, design iteration further refines the geometry. Finally, in the last stages of design, features such as fillets, chamfers, etc. are added for a complete geometric representation. In contrast, the reconfigurable CAD model is planned out in terms of its features, parameters and governing relations before the CAD system session is commenced. The five steps in the creation of a reconfigurable model are applied to the impeller CAD model as follows:

1. Developing the instantiation envelope. The envelope is a specification of key parameters and their limits. This step requires the development of the design envelope to represent the variability of the geometry of the part [13]. This requires planning and work with marketing, manufacturing and others to understand the product family offering and the manufacturing capabilities and limits. For example, the outer diameter of the impeller may vary between the limits of 3.0 and 20.0 cm.; the number of blades in the impeller may vary between 2 and 15; the height of the outer rim of the impeller may vary between 0.1 and 1.0 cm; and the height of the impeller may vary between 2.0 and 25 cm.

2. Design. The artifact model is planned out in terms of its feature structure, parameter scheme, and the relations that will be used to construct the model. Figure 2 shows the feature structure for an impeller.

Figure 2: The planning of the feature structure and sequence for the design of the reconfigurable CAD model

The planning phase is accomplished by designing the features that will be used to construct the model and determining the sequencing of these features. The first feature is a profile that is then revolved to create the solid. Feature two represents the xy spline of a single vane. In feature three, a cut is made to create the proper z profile of the vane. In feature four, rounds are added to the vane. Finally the proper number of vanes is patterned around the impeller hub. The parameter scheme of each of the features is then determined and finally the parametric relationships which determine how the model will update are developed. The parameters and relations are provided in Figure 3.

inner_radius	0.125
outer_radius	1.5
hub_height	2
wedge_angle_bottom	43.86197
wedge_angle_top	8.069017
vane instance angle	30
top_outer_rad	0.4
vane_thickness	0.1
datum_angle for spline in PROE	-30
vane_sp_point0x	0.125
vane_sp_point0y	0
vane_sp_point0z	2
vane_sp_point1x	0.532
vane_sp_point1y	-0.01849
vane_sp_point1z	2
vane_sp_point2x	0.893
vane_sp_point2y	-0.0536
vane_sp_point2z	1.56
vane_sp_point3x	1.181
vane_sp_point3y	-0.255
vane_sp_point3z	1.12
vane_sp_point4x	1.299038
vane_sp_point4y	-0.75
vane_sp_point4z	0.63
hub_sp_point1x	0.4
hub_sp_point1y	0
hub_sp_point1z	2
hub_sp_point2x	0.5
hub_sp_point2y	0
hub_sp_point2z	1.56
hub_sp_point3x	0.75
hub_sp_point3y	0
hub_sp_point3z	0.93
hub_sp_point4x	1.5
hub_sp_point4y	0
hub_sp_point4z	0.125

Relations

Instance_angle = 360 / num_vanes

hub_height > 0.75 * outer_radius

vane_sp_point2y > 0.1

hub_sp_point3x > top_outer_rad

The parameter scheme and governing relations for the impeller CAD model.

Figure 3: The parameter scheme and governing relations for the impeller CAD model

3. Creation phase. The model is now created using a CAD tool. Pro-Engineer was used to create the solid model of the impeller according to the design plan following the relations and design rules established in the previous steps.

4. Debug. The model is debugged by scaling it to the limits of the envelope developed in step 1, some instances are shown in Figure 1. Scaling of the model is accomplished by the instantiation of various sets of parameter values or designs that are encapsulated within the "design space" of the envelope. Note that for the impeller the debugging step resulted in a reduction of the limits on the xy profile of the vane. At the lower end of the limit for the coordinate of the hubspline the model would fail to update. This resulted in the addition of another relation to govern the model.

5. Production. The impeller model may now be used as a reconfigurable model where instantiation is accomplished by selecting appropriate values of the parameters governing updating of the model and importing them into the CAD package and refreshing the model. This model creation strategy can be used throughout the PDP for artifacts typically generated using CAD/CAE/CAM tools, but there are other artifacts: reports, technical publications, test documentation, etc. that also represent a significant portion of artifacts in the PDP, and they can also be made parametrically reconfigurable.

The strategy for reconfiguration can be applied to these artifacts as well. Consider a technical publication where sections in the document are the equivalent of features. Changeable numbers and text correspond to parameters. Parametric relationships are represented by links to text and picture files. In the impeller example, the five step method for creating reconfigurable models was used to create reconfigurable flow and stress analysis models, a reconfigurable manufacturing model, and a reconfigurable technical document for the impeller. Samples of the models are provided in Figure 4.

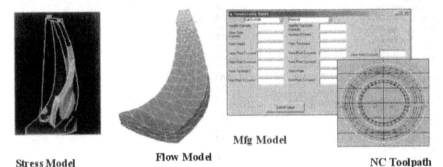

Stress Model   **Flow Model**   **Mfg Model**   **NC Toolpath**

Figure 4:   Examples of additional reconfigurable models for the impeller

In this manner, it is possible to construct all of the artifacts within the PDP as reconfigurable artifacts. This means that generation of a specific instance is a matter of instantiating the reconfigurable models which, with the values of the key parameters, can take minutes to achieve instead of months. Achieving this reconfigurability for just one given artifact can be useful. An example of this occurred when a company that developed sophisticated products for aerospace applications implemented a parametric CAD system into their existing PDP:

Following their documented process steps they implemented the parametric CAD system and used it to create designs of the product. The geometry in the CAD model would then be sent to the analysis group for radiation heat transfer analysis. The analysis group could then translate the geometry into their system and build an analysis model. Once the analysis was complete, the design was reviewed. Changes required that the process be repeated. One iteration in this process typically required several weeks to three months depending on the complexity of the geometry.

This process was modified for reconfigurability by first determining the envelope of variability within the product configuration and planning the CAD model and the analysis model. The two models were then coordinated in terms of feature structures and parameter schemes in the two separate systems. Once these two separate models, one in the CAD system and one in the analysis system were created, they could be instantiated simultaneously. Since they were based on the same parameter scheme and design envelope, there was no need to update the CAD model until the analysts were comfortable with the parameters that controlled the design of the part. The CAD model could then be refreshed and the appropriate drawings generated. Typical cycle times for one iteration were reduced to three minutes! Not only does this represent a significant reduction in cycle time, but it opens the way for other advanced process tools such as optimization. Since both groups had reconfigurable models and understood the geometry and governing parameters, the analysis group could optimize the values of the parameters without instantiating the design model. This small change in the process eliminated hours of tasks associated with geometry translation and model building. These kinds of changes must be orchestrated throughout the overall PDP.

10.4 Parametric reconfigurability comparison

In order to further demonstrate the effectiveness of reconfigurability, a number of reconfigurable models were created for different products. The eleven products used in this analysis are listed in Table 1. It is inherently difficult to quantify the effect of reconfigurability and measure the performance versus the conventional method. However, despite these limitations, these strategies were analyzed and compared against the traditional approach.

The numbers represent best estimates based on experience, actual tests, and common sense. It is recognized that this cannot represent empirical data but it is helpful in measuring the potential for impact of this method as well as measuring return on investment. The method of quantification is explained in the next section.

Table 1: Example products

Products for which reconfigurable models were created	
Radial impeller	Turbine wheel
Orthoplanar spring	Constant force contact
Gearbox	Fly reel
Water pump	Bicycle rear derailleur
Compliant switch	Air turbine starter
	Clutch

10.4.1 Quantification of reconfigurable model performance

Table 2 presents a summary of the reconfigurable model performance metrics used in the sample products. Row 1 of the table lists the names of the products that were used. Row 2 lists the number of parts within each product. Rows 3, 4, 5, and 6 list the number of reconfigurable models each product contains in four different groups. The first group is geometry models. Geometry models are mainly solid CAD models but can also include two dimensional drawings and other models that define or record geometry. The next group is analysis models. Analysis models include those created for FEA and CAE as well as simpler models such a spreadsheets that help determine the product definition. The next group consists of manufacturing models. These are the models that define and control the manufacturing processes required to produce the product. The final group consists of other models. These are all the models that do not fit in one of the three aforementioned categories such as technical publications, bill of materials, and so forth. Row 7 shows the total number of models created for each product. It should be noted that the number of models created for each product is not necessarily the required number for full product definition.

Table 2:   Summary of reconfigurable model performance metrics

1	impeller	spring	gearbox	pump	turbine wheel	switch	contact	flyreel	derailleur	ATS	clutch
2 number of parts	1	1	11	5	1	1	3	2	7	46	1
3 number of geometry models	0	1	11	5	2	1	3	2	4	46	1
4 number of analysis models	1	1	0	1	1	1	1	1	1	0	1
5 number of manufacturing models	1	1	11	5	1	1	3	2	4	0	1
6 number of other models	1	1	0	1	0	1	1	1	0	0	1
7 total number of models	3	4	22	12	4	4	8	6	9	46	4
8 complexity of geometry	0.0	1.0	2.0	1.3	3.0	2.0	1.3	1.0	1.0	2.9	0.8
9 complexity of analysis	4.0	2.0	0.0	2.0	3.0	3.0	2.0	0.5	1.0	0.0	2.0
10 complexity of manufacturing	5.0	1.0	2.0	1.4	4.0	1.0	1.5	2.0	1.0	0.0	1.4
11 complexity of other models	0.8	0.8	0.0	0.8	0.0	0.8	0.8	0.8	0.0	0.0	0.8
12 conventional model creation time (days)	10	5	44	16	13	7	11	7	9	132	5
13 number of engineers required	6	3	27	10	8	5	7	5	6	80	3
14 conventional model cost	$47,040	$11,520	$950,400	$130,400	$83,200	$27,200	$63,280	$29,200	$43,200	$8,448,000	$12,000
15 reusable model creation time (days)	12	6	55	20	16	9	14	9	11	165	6
16 number of engineers required	6	3	27	10	8	5	7	5	6	80	3
17 reusable model cost	$58,800	$14,400	$1,188,000	$163,000	$104,000	$34,000	$79,100	$36,500	$54,000	$10,560,000	$15,000
18 instantiation cycle time (days)	0.3	0.2	1.1	0.5	0.4	0.2	0.3	0.2	0.3	3.3	0.2
19 number of engineers required	6	3	27	10	8	5	7	5	6	80	3
20 instantiation cost	$1,440	$480	$23,760	$4,000	$2,560	$800	$1,680	$800	$1,440	$211,200	$480
21 reusable model creation & instantiation cost	$60,240	$14,880	$1,211,760	$167,000	$106,560	$34,800	$80,780	$37,300	$55,440	$10,771,200	$15,480
22 conventional model creation cost	$47,040	$11,520	$950,400	$130,400	$83,200	$27,200	$63,280	$29,200	$43,200	$8,448,000	$12,000
23 reusable cost/conventional cost	128%	129%	128%	128%	128%	128%	128%	128%	128%	128%	129%
24 instantiation cost/conventional cost	3.1%	4.2%	2.5%	3.1%	3.1%	2.9%	2.7%	2.7%	3.3%	2.5%	4.0%

template performance analysis results

To provide an indication of the complexity of each product and its associated reconfigurable models, a rating system was devised. Each model was rated for complexity on a scale from zero to five with five being the most complex. The average of all of the geometry model complexities within each product was calculated and that number provides a measure of the overall product geometry complexity. The geometry complexity averages are provided in row 8. The complexity ratings for the analysis, manufacturing, and other models were also averaged to provide a measure of complexity in each of these areas for the product. These averages are listed in rows 9, 10, and 11 respectively.

Part of the product design cycle is creating the necessary models to define the product. To provide an estimate for the resources required to create the models for each product used in this evaluation, the complexity rating is assumed to equal the number of days required to create a conventional, non-reconfigurable model. Therefore a simple model (a rating of less than or equal to one) is assumed to require one day to create and the most complex model (a rating close to and up to five) to require five days. This is, of course, a very conservative estimate since some complex models can take weeks to develop, but it serves for this evaluation. To provide an estimate of the total time required to create all of the models for the product, the number of each type of model is multiplied by that type's average complexity and the results are summed. This estimate represents the time required to develop the conventional models for the product and they are listed in row 12.

The number of engineers required to create the models is also a factor in resource consideration. Another assumption was made: a simple model (a complexity rating of less than or equal to one) was assumed to require one engineer and the most complex model (a rating close to five) to require three engineers. The total number of engineers required for the conventional model creation for each product according to this formula is given in row 13. Again this is a conservative estimate. Using the assumption that an engineer costs a company $200,000 per year in total costs and therefore $800 per day, a dollar figure can be placed on each product conventional model development cycle. This cost is listed in row 14.

The next set of numbers in Table 2 is the estimate of resources required to create reconfigurable models for each product. Based on experience planning and constructing these models, the average time penalty incurred in creating a reconfigurable model versus a conventional model is a factor of 1.25. The extra time is required because of the necessary planning phase (the definition of feature structure, etc.) and the parametric nature of the reconfigurable model. Once the planning has taken place, however, the actual model construction is usually quicker and easier because model design decisions are completed and not taking place during model construction, as is often the case in conventional model creation. The time required for creating reconfigurable models for each product is given in row 15 and is 125% of the time required for the conventional models (row 12). The same number of engineers is used (listed in row 16) and the total estimated cost for each product is provided in row 17.

Next, an estimate is made of the cost of instantiating the reconfigurable models. Based on experience with reconfigurable models, a complex model can be instantiated i.e. provided with parameter values, in less than one hour. Assuming 8 hours in a work day, a scaling factor of 1/40 was applied to the conventional cycle time resulting in models with a five rating being instantiated in one hour and so forth. Fractions of a day were rounded up to the nearest day. These times are listed in row 18. Again the same number of engineers was assumed for the instantiation process (row 19). The cost estimate of the instantiation process is listed in row 20.

Rows 21 through 24 in Table 2 contain some values useful for comparing the reconfigurable models and conventional models. Row 21 lists the cost of creating and instantiating the reconfigurable models i.e. the activities required in a single product design cycle with the template. Row 21 is the sum of the reconfigurable model cost (Row 17) and the instantiation cost (Row 20). Row 22 shows the cost of the creation of conventional models (same values as row 14). Row 23 provides a percentage representing the factor of the reconfigurable model cycle cost divided by the conventional model cycle cost. This number indicates the premium required during the first design cycle when using a reconfigurable template to create the reconfigurable models. The final row (24) contains the factor of the instantiation cost divided by the conventional model creation cost. This number represents the "savings" in the second and subsequent design cycles when the design models no longer need to be created but simply instantiated. As is seen in Table 2, instantiations of the reconfigurable models cost approximately 3% of the costs of one cycle of the regular PDP. Again it is important to point out that the quantification method employed here uses best estimates based on past experience. The conservative nature of the assumptions helps to provide meaningful results.

There are several factors that must be considered when examining these numbers. One is that the analysis only considers the model building phase of the product design process. There are other phases of the design process where using reconfigurability will have a different effect. The model creation phase is, however, a major part of the design process and is therefore a good indicator of the overall effect of cycle time reduction seen with the reconfigurable model method.

Another point to consider is because of the difficulty of the quantification, the assumptions made in the analysis were fairly subjective and for this reason very conservative. A complex model often takes weeks or even months whereas five days was the maximum assumed. A real life process should expect to see at least as great a benefit as described with more benefit certainly possible. The order of magnitude increase in productivity in the PDP can only be achieved when this type of reconfigurability is designed into the entire process and not just one aspect of it. This leads to the second key point, that of template strategies for the PDP.

10.5 Creating a new product development process

In the introduction we refer to one of the causes of the productivity problem as being the application of new productivity tools to old product development processes. The concept and application of reconfigurable models forms the backbone for the creation of a new process that is reconfigurable and maps a domain of customer specifications to a range of customized products called a product continuum. The rest of the chapter is devoted to the discussion of a new reconfigurable process. From the outset this new process is designed to be used like a mathematical function to map different customer specifications to new and unique products. The tasks within the PDP are organized into a reconfigurable template.

In conventional or single point design (the process of developing a functioning product for a given set of customer requirements), the process is defined as the product is developed. Tasks which map a specific customer request into a specific product are defined. In contrast, in a product continuum, the process is defined as a formal function where all elements of the domain of possible customer specifications are formally mapped to all possible elements of the range of functioning product. This requires a much more rigorous definition and construction of the PDP. The product function for a product continuum allows customer specifications to vary, within a predefined envelope, to produce customized product solutions. The process is designed from the outset to be parametric and repeatable. It converts the domain of customer specifications into a continuous function with a range of product solutions. The complex relationships associated with the process must be fully articulated for the process to be parametrically reconfigurable.

The greatest challenge in this approach is to construct a well-defined, parametric, repeatable process where all of the relationships are fully articulated. To do this, a method is presented for defining a fully-articulated product function for product continua using principles of abstract algebra. There are five steps involved in creating a reconfigurable PDP for supporting a product continuum. These steps are similar to the steps used to create reconfigurable models only now we are constructing a full process that can be instantiated. We will discuss the process abstractly, and then illustrate the application of the process through an example.

10.5.1 Steps in creating a product continuum

Step 1: Develop the customization envelope. This step requires the development of a conceptual design envelope to represent the variability of the product as opposed to a single product artifact. The envelope is a specification of key parameters and their limits. This includes items such as performance metrics, product costs, and quality metrics. It is the encapsulation of the design space within which mass customized product may be produced. This requires planning, work with marketing, manufacturing and others to understand the product family offering and the company capabilities and limits. For further discussion on the strategy for constructing the product offering domain space see [13].

Step 2: Specification of the domain and range sets. The development of a product continuum requires that the PDP be defined in the form of functions, sets and intermediate mappings or functions that make up the overall product function.

All of the inputs and outputs of the process must be accounted for. Using principles of abstract algebra to map the domain sets to the range sets, the PDP for a product function was developed. A schematic depiction of the complete process and a brief description of the sets are provided in Figure 5.

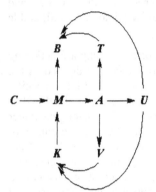

C – Customer specifications (product behavior metrics)

M – Master parameter list (parameters)

A – Product artifacts

U – Product in the field (product behavior metrics)

B – Product behavior (product behavior metrics)

T – Test and validation (product behavior metrics)

K – Company conventions (best practices)

V – company behavior (vaulted artifacts)

Figure 5:   Schematic depiction of the PDP for a product continuum and definition of the sets in the new PDP

Step 3: Enumeration of the sets. The next step is to enumerate the members of the sets within the product continuum. The definition of the membership of the sets C, U, and B is straightforward. Membership of the set C is provided by the customer as the product specifications. The set U is defined by the customer as the desired deliverables such as hardware, technical documentation, etc. The set B is composed of the members of the set C and the additional behavior metrics that may be required to determine the customer specifications. For example, suppose one of the customer specifications for a product is the number of loading cycles or the life of the product. The stress in the component is required to predict the life. Stress then becomes a behavior metric in the set B. The membership of the set A can be derived from the set U. A is defined by a process called backwards mapping. Backwards mapping identifies the map that maps A into F. By knowing the membership of the set F the membership of the set A can be inferred. The final deliverables in the set U are mapped backwards through all of the tasks required for their generation. This exercise identifies the product artifacts that may be converted into reconfigurable models. These artifacts comprise the set A.

The remaining sets to be defined are V, T, K, and M. The set V is defined by company convention. The company selects the artifacts that it chooses to archive such as drawing packages, analysis models, technical documents, etc. The members of the set T are generally defined by the need for validation and risk management. They comprise the results of tests that are performed on product artifacts chosen by the company. The set K is defined by company practices, rules, conventions, etc. that are used to provide elements of the set M not provided through the D mapping. For example, a company rule may specify that the thickness of a housing must equal 0.1 * length.

The set M is a special set because its members come from mapping from other sets within the PDP. The governing parameters from the reconfigurable artifact models define elements within the set M. The parameters required in the predictive models to produce the metrics in the set B are also members of the set M. The parameters that are related to the customer specifications are further included as elements of M. Finally the parameters generated from the company rules, and best practices set K complete the set M.

Step 4: Define the maps. The next step is to define the mappings between the sets that generate the outputs from the given inputs. The mappings are presented in the sequencing of the PDP for illustration. The process sequencing involves five phases: the predictive phase, the instantiation, validation, release, and support. These phases and the sequencing of the process are shown in Figure 6.

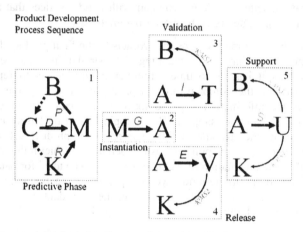

Figure 6: The sequence of the PDP for a product continuum

In the first phase, the predictive phase, three new maps need to be defined. The first map, P, maps the master parameter list to the set of behavior metrics though the predictive models. The second map is the map D. The D map takes the customer specifications and determines the appropriate values for parameters in the master parameter list. The third map, R, takes the company rules and best practices and generates the associated parameters in the master parameter list.

In the instantiation phase the master parameter list is mapped through the mapping G to generate the product artifacts. The mapping G is the aggregation of the parametric, reconfigurable artifact models. The validation phase involves the mapping I. The I map is the collection of the tests performed to validate the behavior metrics. The validation step also consists of a knowledge management strategy in the form of statistical models to validate and update the performance metrics and the predictive models.

In phase 4 the map E is the archiving and vaulting requirements for product definition. Most vaulting strategies include drawing packages, technical documentation, process sheets, the analyzes from predictive models and other artifacts vaulted for legal reasons. However, it is possible to only vault the master parameter list values for the particular instance. These parameters can then be recalled at a later date to reinstantiate the complete template and create all of the product artifacts as needed. Each instantiation of product artifacts represents a data point in what the product template has generated. This information is valuable in terms of modeling the actual customization envelope. As the product template is put into practice, the set V develops a characterization of the multi-dimensional design space of the customization envelope. V then becomes the set of product instantiations from various configurations of M. The release phase also consists of a knowledge management strategy in the form of statistical models that can be used to validate, update, and derive company rules and practices that are used to generate unknown parameters in the master parameter list.

In the final phase the map S maps the artifacts to the final product deliverables through support procedures. The mapping S is similar in some sense to the mapping I in that it has to document product failure. S has to represent the relationship between the company and the customer. The mapping deals with issues such as the information required for warranty, company procedures for product failure, and product support. This mapping is set up based on the set C because this is how the customer defines failure — the product not meeting the desired specifications. The mapping S has to document product failures. Thus S is a time dependent documentation of the members of C. The ideal for data gathering would be to have information on how well the product is meeting C in all aspects. Typically for reasons of cost and convenience, the only data that is gathered is data related to product failure.

Two knowledge management strategies are associated with this mapping. The first relates the field data to the product behavior metrics in B. Recall that the set C is a subset of B. Hence, the field-use data may be used to provide additional models of product performance. Once again this strategy takes the form of statistical models. The second knowledge management strategy relates the field data to K, the set of company rules and best practices. Data gathered from field performance is used to validate the company rules. It may also be used to derive new practices to eliminate the product failures.

Step 5: Storyboarding and creating a product template. With all of the mappings and sets captured in a reconfigurable format the final step is to define page by page the tasks, activities and sequencing of a web application. This includes setting up the links between the various sets and mappings and controls the passing of parameters and the instantiations of the reconfigurable models. This represents the manner in which the company or customer will navigate through the product continuum to design a custom product.

Web portal technology can be effectively used to automate the tracking of information, formats, locations of information, and the structure of processes. This leaves the tasks of reasoning, review and decision making to the users of the process, where these tasks should be. To do this, a web-based environment must be created that captures the structure of the product continuum. This requires the storyboarding of each of the screens in the web portal that will be used to walk the users through the PDP. Much like storyboarding a movie scene by scene, the process is storyboarded task by task with links to reconfigurable models, information databases and other necessary references designed into the screens. Reviews and decisions can be storyboarded into the process to ensure integrity. Figure 7 shows an example of storyboarding for a hard drive connector.

The result is an automated process template that walks the company employees through a consistent PDP that optimizes reconfigurability, keeps track of formats, manages information, etc. This "product template" is the electronic encapsulation of the product continuum capable of producing mass customized product solutions. Making all of the artifacts reusable impacts the PDP in two significant ways. The first is the realization that all the reconfigurable models must be coordinated and instantiate within a common parametric envelope. This is not a trivial task. Consider that in traditional processes the format and coordination of models is left as a task to be accomplished only if an engineering change request requires it. In the new approach all models are created for multiple instantiations and therefore must be coordinated within the same parametric envelope.

The second impact to the PDP is the need to orchestrate the instantiation of all the models to take advantage of inherent parallelism. In the example described above that reduced cycle time from several weeks to three minutes, this reduction was achieved because the models were formatted for reconfigurability and also because the process was restructured.

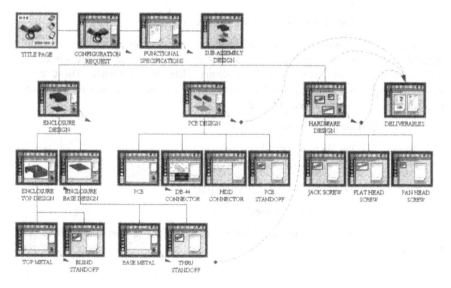

Figure 7: Storyboarding of the PDP for a hard drive connector

10.5.2 Example product template

As an example to aid in understanding the definition of the product continuum and the product template we will examine the development of a constant-force compression spring. A constant-force compression spring is a device that exerts a constant or near-constant force for a given displacement. These types of springs have application in robot end-effectors, electrical contacts, grinding operations, etc. For this product the configuration choice is to use a compliant constant-force mechanism. A compliant mechanism is a mechanism that gains some or all of its force and motion from the deflection of flexible segments. The constant-force mechanism configuration is a member of a class of such mechanisms developed by Howell et al. [14]. The chosen configuration has a compliant segment that is pinned on one end and fixed on the other. The flexure of the segment provides the near constant-force. The configuration is provided in Figure 8.

Figure 8:   System configuration for the constant-force compression spring (example)

The sets are represented as follows:

{Set Name:

Element member (min value, max value)}.

The project initiates with the customer request. In this case, the customer requests a constant-force compression spring and specifies the desired force, the desired displacement distance, and the desired life of the spring. These three customer requirements comprise the membership of the set C where the range specifications for the requirements are

{C (Customer Requirements):

member (min, max)

Force (0.5 N, 15.5 N)

Displacement (1.3 mm, 12.7 mm)

Life (1 cycle, infinite)}.

The ranges on the customer requirements make up the customization envelope for the spring.

The set U is composed of the final deliverables to the customer. In the case of the spring, the customer desires the hardware and the technical documentation. The set F is

> {U (Product in the Field):
>
>> member
>>
>> assembled spring
>>
>> technical documentation}.

The next set to be enumerated is B, the set of product behavior metrics. Recall that the set C is a subset of B. In order to predict the life of the spring the stress due to the displacement loading of the spring is required. Thus, the metric stress also becomes an element of the set B. The set B is represented as

> {B (Product Behavior):
>
>> member (min, max)
>>
>> Force (0.5 N, 15.5 N)
>>
>> Displacement (1.3 mm, 12.7 mm)
>>
>> Life (1 cycle, infinite)}
>>
>> Stress(0 MPa, 27.6 MPa)}.

The enumeration of the set U allows for the population of the set A through backwards mapping. The artifacts identified as members of the set A for the constant-force compression spring are as follows:

> {A (Product Artifacts):
>
>> member
>>
>> technical publication
>>
>> assembly specification sheet
>>
>> assembly drawing
>>
>> assembly solid model
>>
>> compression spring NC model
>>
>> compression spring solid model
>>
>> compression spring drawings
>>
>> compression spring specification sheet
>>
>> compression spring process plans
>>
>> compression spring fixturing
>>
>> compression spring tooling}.

It is important to remember that all of the listed artifacts can be constructed as parametric reconfigurable models. The members of the set V are the artifacts that the company chooses to archive and vault. In this case, the company maintains the set of drawing artifacts, the master parameter list, and the test results performed on the given configuration product request. The membership of the set V is provided below:

> {V (Company Behavior):
>
> > member
> >
> > Spring assembly drawing
> >
> > Ground link drawing
> >
> > Input link drawing
> >
> > Flexible link drawing
> >
> > M
> >
> > Force tested
> >
> > Cycles tested}.

For this product, the set T is comprised of two elements. The spring is tested to measure the force-deflection characteristics of the spring and the number of cycles to failure. The results are compared to the customer specifications and are correlated to the predictive models. The membership of the set T is

> {T (Test and Validation):
>
> > member
> >
> > Force-deflection plot
> >
> > Number of cycles}.

The set K for the spring is composed of company rules and best practices. For the constant-force spring there is only one element in the set K. This member of the set governs the out-of-plane thickness of the spring. The rule states that the design engineer is to specify the material thickness value:

> {K (Company Conventions):
>
> > member (min, max)
> >
> > thickness (1.6 mm, 19 mm)}.

The final set is the set of master parameters M. The set M is a very extensive set of elements. Only a few elements of this set are provided below. Other elements of M are further enumerated in the next step, the process of creating the mappings. The mappings are used to transform the domain set to the range set.

> {M (Master Parameter List):
>
> > member (min, max)
> >
> > link width (0.8 mm, 5.0 mm)
> >
> > link thickness (1.0 mm, 19.0 mm)

　　　　modulus of elasticity (1379 MPa)

　　　　nondimensionalized force term (0.4537)

　　　　nondimensionalized parameter R (0.8274)

　　　　characteristic radius factor (0.85)

　　　　ground link width(6.0 mm, 13.0 mm)

　　　　assembled spring graphic(spring.jpg)

　　　　ball endmill size (1.59 mm, 3.18 mm, 6.35 mm)

　　　　number of elements (5, 50)}.

The definition of the maps for the constant-force spring begins with the maps P, D, and R in the predictive phase. Recall that the map P maps elements of the master parameter list to the behavior metrics. The map P consists of three predictive models. The first model, the configuration of the constant-force spring, is modeled using pseudo-rigid-body theory for compliant mechanisms [15]. This model predicts the force-deflection characteristics of the spring. The second model is a parametric batch file for a commercial finite element modeling package. It calculates the stress in the compliant segment. The third model uses the stress prediction from the previous model to predict the life of the mechanism. The aggregation of the parameters associated with the three models forms the domain set for the P map, the subset mp of the master parameter list. The range set of the P map is all elements of the set B. The schematic of the P mapping is shown in Figure 9.

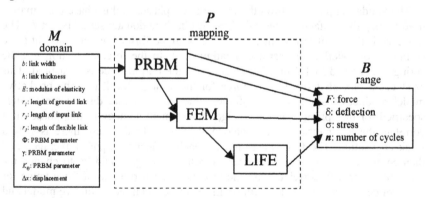

Figure 9:　The schematic of the P mapping for the constant-force spring

The mapping D maps the customer specifications to the master parameter list. The domain set for the D map is the set C, the customer specifications. The range set then becomes the subset of parameters within the set M. Notice that the D map takes a subset of the range set of the P map and maps those elements back into the domain set of the map P. Thus the D mapping is an inverse of the P mapping in some form. Because the P mapping involves non-invertible functions, in this case the finite element model, D is not constructible in a closed form. The mapping D is approximated by guessing the domain set in P and mapping it to B. The results are then compared to the customer specifications. This then becomes the iterative loop in design and continues until the proper set of parameters is chosen such that the elements in the set B satisfy the customer specifications.

The R mapping is used to translate the company rules and best practices into parameters in the set M. The R mapping is used to provide the parameters that are not provided for in the D mapping. The domain set is the set of company rules. The range set is the set of parameters in M generated by the mapping. For the spring, the out-of-plane thickness of the material is not provided by the D mapping. This parameter is provided by an engineer once the customer specifications for the product are received. This rule, that the design engineer is to provide the material thickness, is an element in the set K. This rule is mapped directly to the parameter thickness in the master parameter list. The mapping R may also take the form of relational equations between parameters.

In the instantiation phase the G mapping is constructed. Recall that the G mapping is the aggregation of the parametric reconfigurable models. The domain set for the G mapping is the set of all parameters required to instantiate the various artifacts. The range set is composed of all the parametric, reconfigurable artifacts required for product generation. Only one artifact model from the spring, the NC model, is presented. Similarly, the remaining artifacts each artifact have their own parametric model and list of parameters. The collection of parameters from all of the artifact models constitute mg, another subset of the master parameter list M.

The validation phase follows the instantiation phase. In this phase the I map is used to map the artifacts to the set of test results. The domain set is the set A. The range set is the set of test results, T. Two tests are performed on the constant force spring: a force-deflection test and a fatigue test. In the force-deflection test the spring is deflected in 1.0 mm increments and the output force is measured. This produces a plot of deflection vs. force for the spring. Recall that the predictive model produces a single constant-force value for the entire displacement range specified for the spring. The knowledge management strategy in this phase must correlate the test metrics to the behavioral metrics. In this case there is only a weak correlation. The force measured for each deflection input is averaged and then compared to the predicted force. A better strategy would be to change the predictive model to correlate with the test such that it provides a force value for each deflection increment. This allows the error between the predictive model and the actual measurements to be determined.

The second test is a fatigue test where the spring is repeatedly loaded to the full displacement provided in the customer requirements until the spring fails. This data can then be correlated directly to the prediction from the life model of the spring. As more data is gathered from further creation of product from the template, statistical models can be used to provide information to update the predictive models and in some cases the statistical models may be used to replace some or all of the predictive models.

The next phase is the release phase. The mapping associated with the release phase is the E map. The E mapping contains archiving and vaulting requirements for product definition. The domain set of the E mapping is the set A. The range set is a subset of A that is stored for company purposes. As the product function is used, the vaulting process begins to populate the set V with data points that are used in a statistical model for updating the set K.

For example, consider the scenario where after many design cycles for the spring are completed, the set V contains a large amount of useful data. Upon examination of the data it is noticed that the material thickness values that are most commonly specified are clustered around four thickness values. With this information, the rule in the set K where the material thickness is to be specified by an engineer is modified to allow the engineer to choose one of the four thicknesses. Standardizing on the thicknesses increases manufacturing performance and reduces material costs.

The final phase in the sequencing of the process is the support phase. Recall that this phase deals with the release of the product to the customer and the gathering of field performance data. The domain is the set A, and the range is the set of artifacts delivered to the customer, F. For the constant-force spring, the S map consists of two intermediate maps to address the two ways that the spring can fail. The first map addresses failures of the spring where a component of the spring breaks. For this case the mapping S is the warranty procedure. In the event the spring breaks, a new spring is sent out to the customer to replace the defective one. As part of the procedure, the customer is required to return the defective spring and fill out some additional warranty information. In the second map the failure of the spring to meet the force/deflection characteristics is treated. As in the previous case, a new spring is sent to the customer and the defective spring along with additional warranty information is required of the customer.

The knowledge management strategy relating the set F to the set B for the spring involves examination of the failed spring and using this information to correct the models such that they more accurately predict actual behavior. For example, it is determined that the acceptable stress in the spring is too high and is limiting the life. The predictive model on life is then corrected to change the acceptable stress value that is used in the life predictions.

For the second knowledge management strategy relating the set F to the set K, the field data is used to validate and correct the company rules and practices. Recall that in the example from the previous phase that the material thickness rule was changed to allow only four choices of thickness for the spring. The strategy is to use field performance data to validate how the product is performing when the spring thickness is restricted to the four choices. Field data may show that for many situations, the four thicknesses perform adequately. It may also identify some applications where the four thicknesses are all inadequate and a custom thickness is called for. This information may then be used to update the rule pertaining to material thickness.

The sequencing of the process is followed by the fifth step, the storyboarding and the creation of the product template. With all of the mappings and sets captured in a reconfigurable format, the tasks, activities and sequencing of the web application for the constant-force spring are defined. This includes setting up the links between the various sets and mappings and controls the passing of parameters and the instantiations of the reconfigurable models throughout the PDP. The process is depicted in Figure 10.

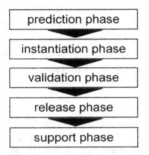

Figure 10: The sequence for storyboarding the product template

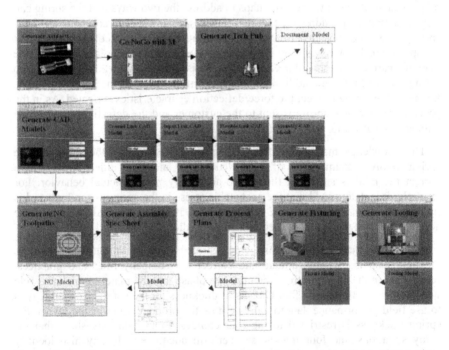

Figure 11: Storyboard of PDP for instantiation phase in the constant-force compression spring example (dashed arrows represent links to configurable models)

For each phase of the product template the proper pages are constructed for navigation through the tasks. The links to the proper reconfigurable artifacts, the master parameter list, the mappings, etc. are laid out and organized. The storyboard for one phase of the process, the instantiation phase, is provided in Figure 11. The instantiation phase begins with a decision to accept or reject the current master parameter list. If the list is accepted, the designer proceeds to instantiate the individual artifact models. The task pages and their associated models are shown in Figure 11. It is important to note that for clarity not all of the links are represented. For example, the technical publication has images that are linked to the assembly solid model. It also contains specifications that are linked to the predictive models and to the customer specifications. The storyboards for the rest of the phases are linked together to complete the product template.

10.6 Summary and conclusions

The construction of reconfigurable artifacts and storyboarding and construction of a product template for a product continuum can represent daunting tasks for a company. They require that the company employees coordinate and plan product offerings and processes to optimize throughput. Productivity as measured by reduced cycle times and increased throughput must be designed and built into the PDP. Early indicators are showing that the keys are reconfigurability and product continua.

The data provided with respect to the use of reconfigurable models supports the conclusion that product design processes that take advantage of reconfigurable models enjoy significant savings in time and cost in the model creation segment of the design process. There are many advantages to reducing design cycle time such as saving resources and enhancing the ability to be competitive in the product market. The instantiation time for the reconfigurable models averages 3% of the time to create conventional models. As the size and/or complexity of the product increases, the potential benefit of reconfigurable models also increases.

Another observation is that the reconfigurable template is an effective tool for organizing and administering design information in the PDP. With the functional forms of reconfigurable product artifacts it is possible to generate continua of product configurations within a specified customization envelope. The extra time spent setting up the reconfigurable models in the first design cycle is easily repaid through faster future design cycles, easier design information maintenance, and effective reconfiguration of previous work. The initial investment of time and resources to create the template can be returned after a single additional development cycle. This process would therefore be effective to use in any design that has potential for reconfigurability. This process may not be as helpful in the case where design information will not be reconfigured. However, even if a product design will not be reconfigured, the template is still helpful in clarifying and documenting the design and the design decisions.

References

[1]   Anderson, D.M.: Agile Product Development for Mass Customization, Chicago 1997.

[2]   Moore, G. A.: Crossing the Chasm, 1991.

[3]   Cooper, R.G.: Third-Generation New Product Processes, in: Journal of Product innovation Management, 11 (1994) 1, pp. 3-14.

[4]   Cha, J.; Gou, W.: The Methodology and Environment for Modeling and Implementation in Concurrent Engineering, in: Advances in Design Automation, 65 (1993) 1, pp. 51-56.

[5]   Pine II, J. B.: Mass Customization, Boston 1993.

[6]   McGrath, M. E.: Setting the PACE in Product Development, Boston 1996.

[7]   Cox, J.J.: Product Templates: A Parametric Approach to Mass Customization, CAD Tools and Algorithms for Product Design, in: Dagstuhl Seminar Report, 227 (1998).

[8]   Cox, J.J.; Zobott, K. D.: Product Templates – A Method of Implementing Mass-Customization, in: Proceedings from 32nd ISATA Automotive Mechatronics Design & Engineering, Vienna 1999.

[9]   Abernathy, W.J.: The Productivity Dilemma: Roadblock to Innovation in the Automobile Industry, Baltimore 1978.

[10]   Adler, P.S.; Mandelbaum, A.; Nguyen, V.; Schwerer, E.: Getting the Most out of Your Product Development Process, in: Harvard Business Review 75 (1996), pp. 134-152.

[11]   Brown, S.L.; Eisenhardt, K.M.: Product Development: Past Research, Present Findings, and Future Directions, in: Academy of Management Review, 20 (1995) 2, pp. 343-378.

[12]   Shah, J. J.; Somnath, S.; Sutanu, G.: An Intelligent CAD Environment for Routine Mechanical Design, in: Computer in Engineering, 1 (1991), pp. 111-117.

[13]   Hunsaker, D.: A Framework for Modeling Mass Customization Envelopes, Provo 1998.

[14]   Howell, L.L.; Midha, A.; Murphy, M. D.: Dimensional Synthesis of Compliant Constant-Force Slider Mechanisms, Machine Elements and Machine Dynamics, in: Proceedings of the 1994 ASME Mechanisms Conference, 71 (1994), pp. 509-515.

[15]   Howell, L.L.: Compliant Mechanisms, New York 2000.

Contacts:

Professor Jordan Cox
Mechanical Engineering, College of Engineering & Technology,
Brigham Young University, Provo, UT, USA
E-mail: cox@byu.edu

Gregory M. Roach, Ph.D.
Mechanical Engineering, College of Engineering & Technology,
Brigham Young University, Provo, UT, USA
E-mail: roachg@et.byu.edu

11   Case-Based Reasoning

Rapid cost estimation of mass-customized products

Naken Wongvasu, Sagar V. Kamarthi and Ibrahim Zeid
Department of Mechanical, Industrial, and Manufacturing Engineering,
Northeastern University, Boston, MA, USA

The ability to generate a quick and accurate quotation brings a significant advantage to mass-customized production companies. Cost estimation is essentially a process that attempts to predict the final cost of a product, even though not all of the manufacturing parameters and conditions are known when the cost estimation is prepared. This chapter presents a case-based reasoning methodology for rapid and accurate estimation of cost of mass-customized products. Unlike existing parametric cost estimation techniques that compute an estimate based on a mathematical relationship between product specification and cost, this approach uses case-based reasoning to model the relationship between product configuration, resources requirement, and costs. Empirical results show that this approach produces reasonably accurate quotes. Thus, we contend that case-based reasoning approach hold a good promise for providing rapid and effective response to customers' request for quotation.

11.1   Introduction

Any mass-customized production company could potentially use case-based reasoning to estimate the cost of its products. The data required for the cost estimation is often already stored in the company's database in the form of bill-of-material (BOM) and BOM-based detail cost report. If one looks at the BOM and the cost of a product, one can distinguishes two types of parts, viz., purchased parts and manufactured parts. For the cost of purchased parts, one can use standardized cost data of the most recent order. However, one cannot easily obtain the cost of manufactured parts (and therefore, the cost of the product) as they are usually not available until after a specified period of time, i.e., until after manufacturing planning stage or actual production/assembly stage. This prevents one from being able to quickly respond to a customer's request for quotation (RFQ). To be able to respond quickly and accurately to an RFQ, costs of manufactured parts have to be available when a customer requests for a quotation. Thus, the problem is how to obtain the cost of the manufactured parts and the cost of the product as specified by a customer in the absence of a process plan and other production data (which may not be needed if the order is not placed).

While several mass-customization companies rely on their engineers' knowledge and experience to manually estimate cost of a product, a better approach is to use *parametric cost estimation* technique [1, 2]. Parametric cost estimation computes an estimate based on the mathematical relationship (using either multiple regression model or neural networks) between the product specification and its final cost. Although both regression and neural network approaches in parametric cost estimation are widely used, as they are reliable and reasonably accurate, the approaches have shortcoming [3, 4, 5, 6, 7]. First, a regression model usually produces a low accuracy (70% accurate). Second, although neural networks are more accurate, there are no specific or clearly defined approaches for choosing their architecture or a learning algorithm for any given application. Third, it is usually difficult to select appropriate cost-driving parameters that truly represent the relationship between product characteristics and cost. Because parametric cost estimation is mathematical-based, both regression and neural network approaches cannot be used to estimate (1) the cost of a product that does not have numerical specifications (2) the cost of a product in which the majority of the cost drivers cannot be expressed numerically.

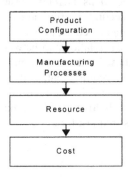

Figure 1:   Product configuration-based costing model

We have presented above a new approach based on case-based reasoning (CBR) to address the issues in the existing parametric cost estimation approaches. Unlike the existing parametric cost estimation approaches that compute an estimate based on the mathematical relationship between product specification and cost, this approach is based on an assumption that production cost is driven by production/assembly processes that, in turn, are driven by product configuration (see Figure 1); and compute an estimate based on the historical relationship between product configuration, resources requirement, and costs. This allows one to use not only numerical product specifications but also non-numerical product specifications, as cost drivers. In addition, it is much easier to explain cost estimation model using CBR. The next section describes the proposed case-based reasoning approach for estimating cost of mass-customization product. Section 3 presents a case study, an overview of the product configurator and cost estimator system, and analysis of the results. Section 4 concludes this chapter and suggests directions for future research.

11.2   Case-based reasoning for rapid cost estimation

11.2.1 Introduction into case-based reasoning

What is case-based reasoning (CBR)? CBR is a computerized method that attempts to solve the current problem by studying solutions that have been used in the past to resolve problems of a similar type [8]. The following are the general characteristics of CBR:

- CBR parallels actual human decision-making process. When a problem is presented, similar problems from the past are used as the basis for solving the new problem.

- CBR can incorporate new knowledge into the existing knowledge base without the need for a knowledge engineer to manually create and enter it as in the case of traditional knowledge-based systems.

- Creating a case-based system is usually more rapid than creating a traditional knowledge based system.

- In terms of speed, CBR takes less time to solve a problem as previously solved problems are either adopted or adapted for the new problem instead of starting from scratch.

- Because CBR uses previously solved problems to solve a new problem, less specific and less precise knowledge is required. This makes CBR an ideal candidate for approximating the cost of a product for RFQ response purposes.

- CBR provides better explanation and justification. CBR can easily justify a solution by pointing to similar solutions and describing the rules that are generated through the reasoning process.

- CBR does not require a large number of past data to be able to solve a new problem.

CBR brings several advantages to rapid cost estimation that differentiates this approach from the existing rapid cost estimation techniques. First, the cost estimation relationship between product specification/configuration, resources requirement, and cost are stored as cases instead of as rules or mathematical formulas. Second, decisions are made and justified based upon previous experience rather than pre-defined rules or mathematically derived formulas. Third, the most appropriate resources required to manufacture a given part are determined by means of similarity and analogical reasoning. Fourth, though not implemented in this work, the reasoning model can be seeded to compensate for an exceptionally high or low resources requirement. Fifth, CBR does not require an exact attribute match between the existing cases and the new part to be able to establish a cost estimation relationship. Consider part (a) and (b) in Figure 2 for example. With parametric estimation, the shape feature of these parts cannot be used as the cost driver as their feature descriptions are different. This is not the case with CBR that uses parts similarity instead of matching attributes to establish the cost estimation relationship. Last, CBR allows non-numerical product specifications, such as color, surface treatment, and product structure, to be used as cost drivers.

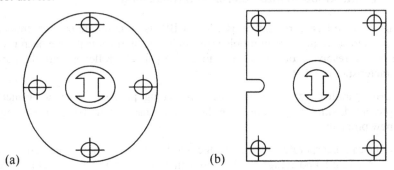

Figure 2: Example of different parts of the same type having different part description

Note that this approach needs to be tailored to specific characteristics of a mass-customized production company. For example, companies differ in terms of the type of product and services they offer to customers, e.g., product design, manufacturing, purchasing, assembly, and packaging. Thus, how CBR is implemented will depend on the nature of the product and service of the company. Hence, we will restrict our discussion to manufacturing cost estimation of make-to-order products. However, many of the ideas presented here are also applicable to rapidly estimating cost of other types of mass-customized products and services (such as assemble-to-order or engineer-to-order products). To ensure proper operation of CBR, the following guidelines are adopted:

- The estimation model is based on facts and real data, such as product configuration, corresponding resources and overhead data.

- Exceptional cases are not compensated for, unless it is necessary. If the compensation is necessary, it is implemented in such a manner that it does not affect the outcome of the normal cases.

- The estimation model uses only data that are available at the time of RFQ processing.

- The estimation model focuses on performance that is equivalent and consistent with product characteristics.

Figure 3 shows the process model of CBR for estimating cost of manufactured parts. In general, the process model of CBR for rapid cost estimation can be described in term of five tasks: matching, retrieval, adoption, repair, and integration.

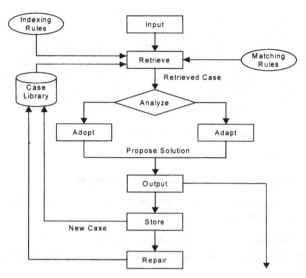

Figure 3: Process model of the CBR system for estimating the cost of manufactured parts

Matching: Retrieving the most appropriate case from a collection of past cases is a search-and-match problem in which configuration of a new product is used as a search criteria. The matching rule may be a straight match or it may be modified to improve the match (i.e., order of importance of different components may be included to influence the match result).

Retrieval: The retrieval task in CBR deals with searching past cases to find the best match between the new case and individual past cases using product configuration as matching criteria. For make-to-order products, retrieval is seldom based on a perfect match. Thus, a partially matched case that has highest match score is retrieved.

Adoption: This function is responsible for applying the information retrieved from the past cases to the new case. In most artificial intelligent (AI) problems, the information is usually modified to fit the new case's problem description. This modification requires additional "deep" knowledge about the application domain, and the modification process is referred to as case adaptation. However, in this work, previously used resources, overhead, and cost information are adopted for the new case.

Repair: Repairing a solved new case is a necessary step for preparing the new case to be stored for future reference. Before a new case can be used as a past case, one has to make sure that the new case mirrors the actual production information and that the product configuration is valid and final.

Integration: The integration task in CBR adds the new case to the collection of past cases. Additional knowledge that is relevant and useful for future reference may also be added.

In short, when the system receives input (i.e., specification of a product or part), it will retrieve the most appropriate past case from the case memory according to some pre-specified match rules. After an old case is retrieved for a given product or part, it is adopted and integrated into the case memory for future reference. During the cost estimation process or after actual production, if the proposed solution is found to be invalid, it can be manually repaired and re-integrated into the case memory.

In general, CBR system consists of three basic components:

- a collection or *library* of past cases,

- a means of *organizing* past cases in the case library and

- a means of using the key elements of new cases to *search and retrieve* the most similar case from the case library.

These are described in term of five components/tasks: (1) case library organization, (2) case indexing, (3) case initialization, (4) case search and retrieval, and (5) case adaptation.

11.2.2 Case library organization

The construction and organization of a library of past cases (case library) is a critical task. The organization of a case library affects CBR's performance and the ability to locate the most appropriate past case or part of a past case for solving a new problem. The effect of retrieving an improperly matched case is often more computationally expensive than selecting the wrong rule to execute in a rule-based system [9]. The case library may be organized as either a flat database or a hierarchically structured database, depending on the goal of the system. The representation of the cases in a case library is usually generalized for all cases in the case library, so that all cases are described by the same set of attributes or relationships between components of a case. This way, all cases in the case library are uniformly represented, thus, simplifying the search process, case retrieval, and integration process.

In earlier work the concept of generically representing BOMs of variants of a product family as a logical BOM of the product family is presented [10, 11]. The concept implies that there exist similarities among variants of a product family, thus, possessing similar cost-driving characteristics (i.e., production process and resource requirement). As a product is hierarchically built up from its immediate sub-assemblies and parts, one can classify these parts either as "purchased parts" or as "manufactured parts" (see Figure 4). In addition, specifications of both the purchased parts and the manufactured parts can further be classified either as being "modifiable" (by customer) or as "fixed" (not modifiable by customer). To produce a part or a sub-assembly, a set of manufacturing operations is required. Each operation's resources requirement (in term of materials, sub-components, and resource utilization) is governed by a set of product-related "drivers," which may be as simple as the part number or part type, or as complex as the part's feature description. Thus, for a given variant of a product family, one can use the relationship between the drivers, resources requirement, and that is captured in a case library for rapid cost estimation.

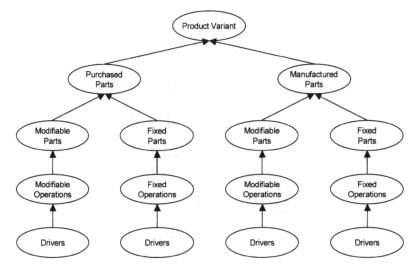

Figure 4: Representation of a make-to-order product

For example, consider the assembling of a personal computer. Let the operations be acquiring parts, assembling components, and testing. Further, let the driver for the operations be the desired computer configuration. Thus, whenever a personal computer is assembled, these operations need to be performed, and the resources required are determined by the value of the driver, in this case, computer configuration. Thus, the product configuration can be used to directly determine the resources requirement and cost for assembling a personal computer. As product configuration is presented by the BOM, one combines attributes from the logical BOM of a product family, the physical BOM of a product variant, corresponding resources requirement, and corresponding cost to represent the relationship between the driver (i.e., product configuration), resources requirement, and cost as a case in the case library. Thus, each case consists of two sets of interdependent objects:

- The *BOM of a product variant* and its attributes, such as logical part number, physical part number, part description, level of components in BOM, cost, and quantity of sub-components per parent component.

- *Resources* requirement and cost, such as resource number, resource description, usage rate, cost per unit, and overhead information, used for manufacturing parts and the product.

Figure 5 shows a schematic diagram of the organization of the case library and the content of each case. Table 1 shows the data representation scheme, e.g., case description, case property, attribute, base case, and operation of the case library.

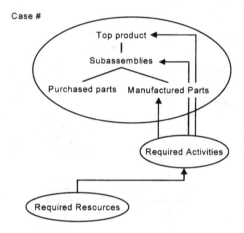

Figure 5: The representation of a case in the case library

Table 1: Representation of a case

Representation of a case	
Case description	Unique name of the case (e.g., product number)
Case property	A log file which log the reasoning process and indicate whether the case is a base case or a derived case.
Attributes	BOM of a product variant and case indices.
Base case	Derived case's parent as indicated in the log file. For the base cases, this attribute is the cases' product number and parts' number.
Operation	Corresponding resources requirement of the product variant and its manufactured parts.

11.2.3 Case indexing

The cases in the case library must be indexed to allow for efficient search. How the cases are indexed is a key issue in CBR. If one defines them too broadly, too many cases may be retrieved as similar cases. On the other hand, if they are defined too specifically, no similar cases may be found. In this work, each case in the case library represents product-configuration-based cost estimation relationships that model the cost of manufacturing a product as being driven by product configuration and the resources requirement.

Although it is a common practice to identify a product and its parts by part numbers, this alone will limit the ability to locate a past case. In previous papers we have presented a new product representation technique in which BOMs of variants of a make-to-order product family are generically represented by a logical BOM of the product family. By generalizing the BOMs, different top-product variants and its component-product variants are represented together as logical top-product and logical component-products, respectively. Implicitly, part numbers of both top-product variants and component-product variants are also generalized into logical top-product number and logical component-part numbers. Hence, one will use logical top-product number, logical part number, top-product number, and part number as case indices. This enables one to locate not only a case as a whole but also part of a case.

Specifically, the following attributes from logical BOM and physical BOM are used as case indices for each and every case in the case library (Figure 6 shows entity-relationship diagram of the representation of a case):

- *Product family number (PF\#):* Generic top-product number for distinguishing one product family from another.

- *Top-product number (P\#):* Specific top-product number for distinguishing variants of a product family.

- *Logical item number (LIN\#):* Generic component-part number for classifying component parts of different variants of a product family.

- *Part (Item) number (I\#):* Specific component-part number of each component-part in the BOM of the product variant.

- *Resource number (Res\#):* Specific resource number for each resource used to manufacture components.

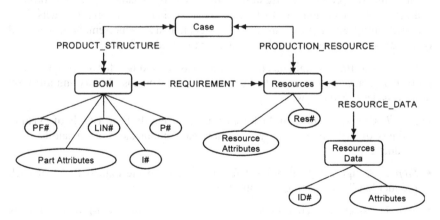

Figure 6: Entity-relationship diagram of the case representation

11.2.4 Case initialization

Initialization of a case library is needed to establish and represent the contents of each case. Before a CBR system can be used to solve a new problem (in this case, estimation of the cost of a product as specified by a customer), the case library must be initialized with past cases. According to Gupta and Veerakamolmal, the procedure to organize the structure of a case library consists of two principles, generalization and classification [12].

Generalization: The purpose of generalization is to list a general set of attributes that can precisely describe a product, resources requirement, and cost. The attributes are represented in the form of items from the logical BOM of the product family and resources requirement information from past orders. Since the process of using the relationship between past product specification, resources requirement, and cost to estimate the cost of new product CBR may require *decomposition* and *integration* of two or more product structures and the relationship between product specification, resources requirement, and cost of the products. In general, the process of breaking up the product into sub-assemblies and parts is called decomposition. An integration procedure is required to merge all the partial cost estimates for the sub-assemblies and parts to create final cost estimate.

Classification: For efficiency of the search and retrieval process, CBR needs to include knowledge for distinguishing the classes of parts. CBR makes extensive use of the classification scheme to initialize, append, and retrieve the resources requirement and cost data from the case library. To accommodate the representation for multiple product families, one employs the concept of logical BOM as the classification scheme. This classification is based on the fact that a family of make-to-order products consists of products, each of which is a modified version of a standard product configuration obtained by changing certain product characteristics. This implies that there exist similarities among products in a product family and that, by means of conceptual data representation, a logical (generic) product structure can be created for the product family.

CBR must first be initialized with past cases in order for it to solve new problems or accept new cases to the case library. The case library initialization process is described below:

- *Step 1:* Create a logical BOM for each product family from BOMs of existing variants of the product family and knowledge about the product family from product experts.

- *Step 2:* Assign a logical product family number to the logical BOM of each product family of existing variants.

- *Step 3:* Assign respective logical item numbers to component-product variants in BOMs.

- *Step 4:* List all resources (in the order of precedence) used in manufacturing each top-product variant and its component-product variants.

- *Step 5:* Create relationships between manufactured components and the resources data.

- *Step 6:* Store the BOM of a product variant, logical product family number, logical part numbers and the corresponding resources data in a case.

- *Step 7:* Repeat *Step 6* until all existing product variants are stored in the case library.

11.2.5 Case search and retrieval

Searching the case library is a part of retrieval process that directs the search to appropriate places in a case library, accessing potentially useful cases. In order to ensure that only relevant cases are retrieved, the extent to which two cases match each other requires one to consider (1) the selection of index, (2) how cases are matched, and (3) how matched cases are ranked.

The search and retrieval algorithms for ensuring that the most appropriate past resources requirement and cost are retrieved for a given product configuration is divided into two procedures according to the type of product in a BOM, viz., procedure for searching and retrieving top-product's resources requirement and procedure for searching and retrieving manufactured parts' resources requirement.

Procedure for searching and retrieving top-product's resources requirement

The search and retrieval algorithm for finding the most appropriate resources requirement and cost of manufacturing a new product variant uses product structure as the basis for determining similarity between cases. The search algorithm serially compares the BOM of the new product variant to each BOM of the past product variant in the case library. For each comparison, a match score is computed to rank the similarity and the one with the best score is chosen.

- *Step 1:* Let Λ be the representation of a new top product, Θ_i be the representation of a case in the case library, and σ_i be the match score, where $i = 1, 2, ..., n$ and n = number of cases in the case library.

- *Step 2:* Initially, let $i = 1$.

- *Step 3:* Compare Λ to Θ_I - for each similar component in Λ and Θ_i, increments σ_i by 1.

- *Step 4:* Repeat Step 3 for all i.

- *Step 5:* Retrieve resources requirement and cost from Θ_i in which σ_i is the highest. If there is more than one Θ_i in which σ_i is the highest, choose either the most recent Θ_i, arbitrary Θ_i, or a user chosen Θ_i from the set of Θ_i in which σ_i is the highest.

Procedure for searching and retrieving manufactured parts' resources requirement

The search and retrieval algorithm for finding the most appropriate resources requirement and cost of manufacturing a part also uses product structure as the basis for determining the similarity between cases. The search algorithm serially compares the BOM of the part to each BOM of a part in the case library. In addition to the product structure heuristic that is used as a basis to search and retrieve resources requirement and cost of a top-product, the search algorithm takes advantage of the logical BOM concept to ensure that the search only retrieves the most relevant resources requirement and cost for a given part. Specifically, the matching and ranking algorithm for component parts uses *domain expansion strategy* in which the algorithm first tries to match the part number. If no matching part number is found, the algorithm expands its search scope by attempting to find a matching logical part number. For each comparison, a match score is computed to rank the similarity, and the one with the best score is chosen.

- *Step 1:* Let λ_k represent a component part in Λ, θ_{ij} represent a component part in Θ_i in the case library, and σ_{ij} be the match score, where $i = 1, 2, ..., n; j = 1, 2,..., m; k = 1, 2, ..., p; n$ = number of case in the case library; m = number of component parts in Θ_i; and p = number of component parts in Λ.

- *Step 2:* Initially, $i = 1, j = 1$, and $k = 1$.

- *Step 3:* For each i and j, using part number as matching criteria, compare λ_k to θ_{ij}. If the part number matches, increment σ_{ij} by 1.

- *Step 4:* If there is no match, re-initialize i and j, and compare λ_k to θ_{ij} using logical part number and part structure as matching criteria - for each match between λ_k and θ_{ij}, increments σ_{ij} by 1.

- *Step 5:* Retrieve resources requirement and cost from θ_{ij} in which σ_{ij} is highest. If there is more than one θ_{ij} in which σ_{ij} is highest, choose either the most recent θ_{ij}, arbitrary chosen θ_{ij}, or a user chosen θ_{ij} from the set of θ_{ij} in which σ_{ij} is the highest.

- *Step 6:* Repeat *Step 3* through *Step 5* for all k.

11.2.6 Case adoption / adaptation

When the description (i.e., top-product part number, component-product part number) of a retrieved case is identical to that of a new product, the production and resource information from the retrieved case can be adopted for the new product. However, when the description of the retrieved case is not identical to that of the new product's configuration, the degree of difference between the predicted resources requirement and the actual resources requirement may vary from situation to situation.

Since this approach estimates cost of a product from the relationship between product configuration, resources requirement, and cost, the variation between the relationships stored in the case library directly affects the accuracy of the estimate. Without spending additional time and effort to plan the production process, several adaptation schemes can be used to modify certain information (e.g., resource usage, shrink factor, resource cost, overhead rate, overhead cost) of the retrieved resources requirement and cost data to a more suitable value. In this work, the authors use the following adaptation schemes:

- *Most frequently used:* The parameters of the retrieved set of resources requirement and cost are changed to the most frequently used values.

- *Most recently used:* The parameters of the retrieved set of resources requirement and cost are changed to the most recently used values.

- *Use maximum rate:* The parameters of the retrieved set of resources requirement and cost are changed to the highest values available.

- *Use statistical rate:* The parameters of the retrieved set of resources requirement and cost are changed to their statistically calculated values. Linear regression and neural network methods can be used for developing prediction models based on data from the past orders. This is useful when significant quantity of data are available, and more than two drivers can be described numerically.

- In addition, one may adapt the search and retrieval heuristics directly. Some of the possible matching schemes are as follows:

- *Most recently used:* Instead of retrieving resources requirement and cost from the most similar product, CBR retrieves the information from the most recently manufactured product.

- *Most frequently used:* Instead of retrieving resources requirement and cost from the most similar product, CBR retrieves the most frequently used set of resources requirement and cost from the case library.

- *Maximum number of resources required:* Instead of retrieving resources requirement and cost from the most similar product, CBR retrieves resource information from the past product that requires the highest number of resources.

- *Minimum number of resources required:* Instead of retrieving resources requirement and cost from the most similar product, CBR retrieves resource information from the past product that requires the least number of resources.

11.3 Case study

To validate the proposed CBR approach to approximate the cost of mass-customized products, a case study was conducted on a family of make-to-order products. The product under study was introduced by the company in 1996. At the time of this research, 28 varieties were produced by the company. Each product consists of approximately 40 to 48 parts, depending on the product configuration. Out of these parts, nine parts are manufactured and/or assembled in-house.

11.3.1 Product configurator and cost estimator system

As part of the experiment, an integrated product configurator and cost estimator system was developed to (1) assist engineers in configuring a variant of the product family and (2) estimate cost of the product variant based on the cost estimation model discussed in the previous section. It is designed to simplify the process of dealing with options and features on a particular variant of the product family. First, the system sets up a basic product structure based on the selected base-features of the product family. The system then lets the user select an option for each feature in the basic product structure. After each feature has been assigned an option, the system generates a "resulting" BOM of the product variant. The "resulting" BOM contains generic part numbers, part numbers, description, quantity, and other information about the product. It is used by the cost estimator as input for its case-based reasoner in estimating cost of the product variant.

Figure 7 shows the architecture of the system. The system is written in Visual Basic. It consists of seven modules, viz., user interface module, production configuration & manufacturability assessment module, cost estimation module, Logical BOM processor module, case-based reasoning (CBR) module, database module, and pricing module. The following subsections provide an overview of six of the modules to provide a better understanding of how a product variant is configured, how cost is estimated, and how the interaction between them is handled.

User interface module: The user interface module interacts with other modules as well as with the user to assist and simplify the process of dealing with options and features on a particular variant of the product family. Rather than having the user specify each component individually in steps, the user interface uses menus and events to drive the order of program execution during the product configuration process. This allows the user to interactively select options for the features associated with various combinations of the product's base-features. The user interface module is also responsible for displaying a variety of outputs passed by other modules, viz., indented BOM of a product variant, indented BOM/cost report, real-time CBR activity, and CBR activity log.

Note that, though the system is designed to be used by sales representatives and engineers, the user interface module can be easily extended to provide internet interface, thus allowing a customer to configure a product and receive an approximate price of the product in real-time.

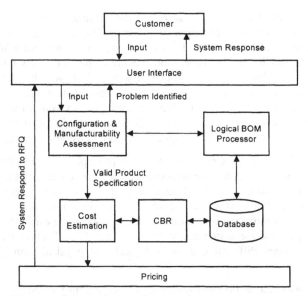

Figure 7: Architecture of the product configurator and cost estimation system

Product configuration & manufacturability assessment module: The function of the product configuration module is to evaluate the customer input against a set of rules that govern data validity, completeness, manufacturability, and the like, before it generates the BOM of the product and forwards it to the cost estimation and pricing modules for generation of the quotation. For example, if the module detects that the inputs are incomplete, it will prompt the user to make necessary additions or changes to the input. The module is also used to check whether the customer-specified product parameters and characteristics violate any manufacturability constraints of the product family. The module also assists the user by simplify the process of dealing with options and features on a particular variant of the caster. This is performed in two stages. The first stage is during user input where the module interacts with the user interface module and presents to the user only features with selectable options as well as pre-selects some options for the user. Which features to be presented or selected are constrained by the combinations of the product's based features. The second stage begins after options of the presented features are selected but before a BOM is generated. This is when the module begins to complete the product configuration by selecting the most appropriate parts for the product, based on the selected options.

Logical BOM processor module: Logical BOM processor module represents and processes product structure data for the product configuration & manufacturability assessment module. The module consists of logical BOM structure and representation of product configuration constraints.

Cost estimation module: Manufacturing cost estimation lies at the core of an RFQ response and work cooperatively with the case-based reasoning module. Indeed, the lack of a rapid RFQ processing system and the continued reliance on traditional, time-consuming and personal-judgment based RFQ processing procedures by a company unfortunately leads to a waste of valuable time and effort in RFQ processing. In PCCE, the cost estimation module employs the new CBR for rapid cost estimation methodology as described in section 2 above. The technique relies on (1) similarity between specification of the current product and specification of past products; and (2) the relationship between product configuration, resource requirement, and cost. With CBR, the module is capable of generating a reasonably accurate manufacturing cost estimate quickly, even for custom-made parts.

Case-based reasoning (CBR) module: The principle of CBR for rapid cost estimation is based on reusing previous cost calculations to estimate the cost of new products. For efficiency, the CBR module is designed with generalization and classification capability in order for the module to work in a multi-product environment. Generalization is a process that incorporates logical BOM structure of a product or part into the search and retrieval process. Classification allows CBR to distinguish between different classes of logical and physical items in the case library.

Case library organization and indexing: The case library consists of past cases; each represents cost estimation relationship in the form of relationship between past product's BOM, the product and its parts' respective resources requirement, overhead, and cost. Each case in the case library is indexed by five attributes:

1. Product family number.

2. Individual caster number.

3. Logical item number.

4. Part number.

5. Required resources number.

Case search and retrieval: Based on the five indices, after a product configuration is passed from the product configuration and manufacturability assessment module, CBR searches through the case library to retrieve (1) the most similar product's resources requirement and cost, and (2) the most similar part's resources requirement and cost for the manufactured part. Regarding cost of the purchased parts, they are retrieved in a similar manner during the product configuration process. The search and retrieval process can be summarized as follows:

1. Compare the BOM of a new product configuration to BOMs of past product variants in the case library, counting the number of similar components.

2. Retrieve resources requirement and cost data from the past product that has the highest number of similar component counts. If there is a tie, retrieve the information from the most recent product that has the highest number of similar component counts.

3. Compare the BOM of component-parts of the new product configuration to the BOMs of component-parts of the existing product configuration in the case library, counting the number of similar sub-components.

4. Retrieve resources requirement and cost data of component-parts from the past product that has the highest number of similar sub-component counts. If there is a tie, retrieve the information from the most recent product that has the highest number of similar sub-component counts.

Regarding the cost of the purchased parts, they are retrieved in a similar manner during the product configuration process. However, instead of comparing the BOM of component-parts, the search and retrieval process compares the individual part of a new product configuration to that of the past product configurations and retrieves the most recent cost of the purchased parts from the case library. The matching criteria for most similar product and part employ a domain expansion search as described in above. In short, the search first uses specific part and product numbers as matching criteria. If an exact match is not found, the search expands its scope by using logical part numbers as matching criteria and retrieves resources requirement and cost from the most recently manufactured part.

Case adoption: In this version, PCCE operates on the assumption that cost estimates are only as accurate as the variability in the manufacturing processes of the past products. Further, PCCE focuses on performance and only uses data that are available at the time of RFQ processing. Hence, PCCE adopts resources requirement and cost information retrieved from the most similar product configuration for the new product configuration.

Modification of matching and adoption criteria: Although most manufactured parts require similar sets of resources and resource utilization, several product configurations exist that, though similar, require different sets of resources and resource utilization that vary highly from one another. To deal with the variability in the resources requirement of past products, additional matching criteria and adoption schemes are provided. This allows the user to be able to choose a set of resources requirement that is most appropriate for the new product configuration. For the matching criteria, PCCE provides the following options:

- *Most Recent:* The cost estimation module adopts the most recently used set of resources requirement and cost information.

- *Most Frequently Used:* The cost estimation module adopts the most frequently used set of resources requirement and cost information.

- *Max. Resources:* The cost estimation module adopts the resources requirement and cost information from the past case that requires the highest number of resources. Further, PCCE provides the following adoption options:

- *Most Frequently Used:* The cost estimation module changes the resources' parameters of the retrieved set of resources requirement and cost information to the most frequently used values.

- *Use Max. Rate:* The cost estimation module changes the resources' parameters of the retrieved set of resources requirement and cost information the highest values available.

- *Most Recent:* The cost estimation module changes the resources' parameters of the retrieved set of resources requirement and cost information to the most recently used values.

Database module: The data module lies at the core of PCCE. The module supplies data to the logical BOM processor module and CBR module during their execution. The module consists of two databases, viz., case library database and product structure database.

Product structure database: While the logical BOM and product configuration constraints provide a generic view of the product family and a means to which combinations of component are constrained, product structure database stores attributes of the logical BOM. The product structure database consists of five tables. Each table stores the description of component-products (both logical and physical description) on each level of the logical BOM structure. In this representation, tables are organized to resemble an exploded logical BOM and are linked together by "virtual" goes-into relationships (implemented as entity-relationships in the database. These relationships represent the goes-into relationship in the logical BOM and are used by the product configurator module for product configuration and BOM generation.

11.3.2 Analysis of the experimentation result

For the experiment, a case library was initializing with BOM data, resources requirement data, and cost data from 18 products (that were produced from 1996 through May 1998) as base cases. The experiment simulates the actual product configuration and cost estimation processes using configuration from 10 products (that were produced between May 1998 and April 2000) as test cases. The simulation repeats the following steps for each test case:

1. Enter product configuration of a test case (new product).

2. Modify the price of the purchased components.

3. Estimate the production cost of the new product using most similar search scheme.

4. Compare the estimated production cost with the actual cost data.

5. Update the production processes of the new product accordingly.

6. Save the product configuration and the updated production processes into the case library to be used as old cases.

After the first run, an analysis on the experimentation result was conducted. The analysis reveals that data quality plays an important role in the accuracy of the estimates. Specifically, (1) the shrink factor of test product 4 is found to be 100 times higher than the rest of the test products and (2) one of the parts used in test product 6 and 7 was classified incorrectly. After the problems were found, the entire simulation processes were repeated after correcting each error. The results of each iteration are shown in Table 2. As shown in this table, despite data quality issues, CBR is able to reliably estimate cost of a product (> 90% average accuracy).

Table 2: Experimentation results

Percentage difference between estimated total cost and actual total cost			
Test Product	*Iteration 1*	*Iteration 2*	*Iteration 3*
1	0.38	0.38	0.38
2	14.30	14.30	14.30
3	0.25	0.25	0.25
4	27.60	0.00	27.60
5	0.00	0.00	0.00
6	19.09	19.09	13.44
7	0.00	0.00	0.00
8	1.15	1.15	1.15
9	13.67	13.67	13.67
10	21.35	21.35	7.53
Average % difference	9.88	7.02	7.83
Average % accuracy	90.22	92.98	92.17

11.4 Conclusions

This chapter presented a new parametric approach for rapid cost estimation. Unlike existing parametric cost estimation techniques, which compute an estimate based on the mathematical relationship between product specification and cost, this approach uses CBR to model the cost estimation relationship between product configuration, resources requirement, and costs. The cost estimation relationship is represented in a case library. Specifically, each case in the case library represents the relationship between product configuration, resources requirement, and costs. Hence, the estimation process begins by searching the case library for a similar configuration of a product and component parts for resources required to manufacture a new product variant. The resources requirement and cost from the most similar product and component parts are then retrieved and adopted as the resources requirement and cost for the new product. Further, to deal with variability between the cost estimation relationships, the parameters of the retrieved set of resources and costs may be adapted deterministically or statistically. The advantages of using CBR as a means for estimating cost of a product are as follows:

- Cost estimate can be generated quickly and accurately.

- Both numeric and non-numeric parameters can be used as cost drivers.

- Although it is preferred, this approach does not require significant quantity of past data to work successfully.

- The estimation model can be easily adapted for optimum performance, depending on factors, such as the kind of data available, the drivers, the type of product, and the degree of accuracy of the estimate.

Empirical results show that this approach holds a great promise for rapid and effective response to customers' request for quotation, as they are reliable and able to produce reasonably accurate quote. The results prove that even though the traditional belief that a detailed process plan or routing sheet has to be developed to estimate cost of manufacturing parts and products and to generate a price quotation, the cost of a product can be estimated from the relationship between product specification, resources requirements, and cost, from previous orders.

Although not discussed here, our approach can be used also for manufacturing time estimation by including time factors in the case library for retrieval and adoption in a similar manner as cost factors. Nevertheless, the accuracy of this approach depends on variability associated with the manufacturing processes. For instance, if there is a large amount of variability in resources requirement between cases in the case library, especially when the past products closely resemble each other, the accuracy of the model will be low, and vice versa.

References

[1]   Melin Jr., J.B.: Parametric Estimation, in: Cost Engineering, 36 (1994) 1, pp. 19-24.

[2]   Stockton, D.J.; Middle, J. E.: An approach to improving cost estimating, in: International Journal of Production Research, 20 (1982) 6, pp. 741-751.

[3]   Cochran, E.B.: Using regression techniques in cost analysis, in: International Journal of Production Research, 14 (1976) 4, pp. 489-511.

[4]   Sigurdsen, A.: CERA: An Integrated Cost Estimating Program; in: Cost Engineering, 34 (1992) 6, pp. 25-30.

[5]   Smith, A.E.; Mason, A. K.: Cost Estimation Predictive Modeling: Regression Versus Neural Network, in: The Engineering Economist, 42 (1997) 2, pp. 137-161.

[6]   de la Garza, D. J. M.; Rouhana, K. G.: Neural networks Versus Parameter-Based Application in Cost Estimating, in: Cost Engineering, 37 (1995) 2, pp. 14-18.

[7] Veeramani, D.; Joshi, P. : Methodologies for rapid and effective response to requests for quotation, in: IIE Transactions, 29 (1997), pp. 825-838.

[8] Ketler, K.: Case-Based Reasoning: An Introduction, in: Expert Systems with Applications, 6 (1993), pp. 3-8.

[9] Gonzalez, A.J.; Dankel, D.D.: The Engineering of Knowledge-based System, 1993.

[10] Wongvasu, N.; Kamarthi, S.V.: Inter-Component Compatibility: A New Perspective on How to Represent Relationships Between Items in a Generic Bill-of-Material Structure, in: Proceedings of the Group Technology/Cellular Manufacturing World Symposium, 2000.

[11] Wongvasu, N.; Kamarthi, S.V.: Representing the Relationship Between Items in Logical Bill-of-Material to support Customers' Request for Quotation for Make-to-Order Products, in: Proceedings of SPIE – Photonics East Symposia: Intelligent Systems in Design and Manufacturing III, 2000.

[12] Gupta, S. M.; Veerakamolmal, P.: A Case-based Reasoning Approach for Optimal Planning of Multi-product/Multi-Manufacturer Disassembly Processes. In: International Journal of Environmentally Conscious Design &Manufacturing, 9 (2000) 1, pp. 15-25.

[13] Wongvasu, N.: Methodologies for Rapid and Effective Responses to Requests for Quotation for Mass-Customization Products, Dissertation, Northeastern University, Boston 2001.

[14] Wongvasu, N.; Pittner, S.; Kamarthi, S. V.; Zeid, I.: Trie Representation for Expressing the Compatibility Between Items in Logical Bill-of-Material Structure, Proceedings of SPIE - Photonics East Symposia 2001: Intelligent Systems in Design and Manufacturing, Boston 2001.

[15] Cullinane,T. P.; Chinnaiah, P.S.S.; Wongvasu, N.; Kamarthi, S. V.: A Generic IDFE0 Model of a Production System for Mass Customization, in: Proceedings of Portland International Conference on Management of Engineering and Technology, Portland 1997, pp. 679-684.

Contact:

Professor Ibrahim Zeid
Department of Mechanical, Industrial, and Manufacturing Engineering
Northeastern University, Boston, MA, USA
E-mail: zeid@coe.neu.edu

12 Using TRIZ to Overcome Mass Customization Contradictions

Darrell L. Mann[1] and Ellen Domb[2]
[1] Department of Mechanical Engineering, University of Bath, UK
[2] PQR Group, Upland, CA, USA

The mass customization concept carries with it inherent contradictions between versatility and user benefit versus productivity or cost. Traditional trade-off and compromise based business approaches offer little to help overcome these contradictions. Thus, while many organizations are beginning to recognize the need for mass customization, few actually know how to tackle the issues involved in turning the concept into profit-making reality. The chapter discusses how systematic innovation methods are beginning to be used to successfully to overcome rather than accept the trade-offs and compromises often held to be inherent. Case study examples include design of mass customized bicycle seats, shoe products, novel room lighting solutions and home-customizable food products. The chapter ends with a discussion of these and other disruptive contradiction-breaking technology solutions in general.

12.1 Introduction

"Customers, whether consumers or business, do not want more choices. They want exactly what they want - when, where and how they want it" [Pine/Peppers/Rogers in the HBR 2/1995]

The concept of mass customization – the economically viable creation of products tailored to the specific needs of individual customers – carries with it inherent contradictions between versatility and user benefit versus productivity or cost. Traditional trade-off and compromise based business approaches offer little to help overcome the contradictions. Thus, while many organizations are beginning to recognize the need for mass customization, few actually know how to tackle the issues involved in turning the concept into profit-making reality. The recent introduction of systematic innovation methods into the Design for Mass-Customization (DFMC) environment looks set to change this picture. Central to these new methods is the Theory of Inventive Problem Solving (TRIZ). TRIZ consists of a series of tools and strategies developed through over 1500 person years of research, and the study of over two million of the world's strongest patents.

The key findings of TRIZ research in a mass customization context are:

- that the strongest problem solutions are the ones which successfully challenge the contradictions usually accepted as fundamental;

- that there are only a very small number of inventive principles available to successfully eliminate contradictions for either technical or non-technical problems;

- that technology and business evolution trends are highly predictable, and highly consistent with mass customization.

The chapter builds on pioneering work on application of TRIZ to mass customization problems, and describes how a TRIZ-based systematic innovation methodology has been developed and is now being deployed to generate innovative new solutions across a variety of mass customization problems [1]. The chapter will discuss the methodology through a number of exemplar case studies including the application of TRIZ tools:

- design of inherently customized bicycle seat designs,

- DFMC project in the shoe industry,

- design of novel room lighting systems,

- development 'home-customizable' food product innovation.

12.2   TRIZ basics

We begin, however, with a brief overview of some of the main TRIZ tools and strategies for those unfamiliar with the method. TRIZ provides means for problem solvers to access the good solutions obtained by the world's finest inventive minds [2, 3]. The basic process by which this occurs is illustrated in Figure 1. Essentially, TRIZ researchers have encapsulated the principles of good inventive practice and set them into a generic problem-solving framework. The task of problem definers and problem solvers using the large majority of the TRIZ tools thus becomes one in which they have to map their specific problems and solutions to and from this generic framework. By using the global patent database as the foundation for the method, TRIZ effectively strips away all of the boundaries that exist between different industry sectors. The generic problem solving framework thus allows problem owners working in any one field to access the good practices of everyone working in not just their own, but every other field of science and engineering.

Successful use of the various TRIZ tools requires an approach dissimilar to other creativity methods. Mann discusses the four paradigm shifts – Contradiction, Ideality, Functionality, and Use Of Resources – most commonly observed as important in a TRIZ usage context [4]. We review those elements briefly in the following.

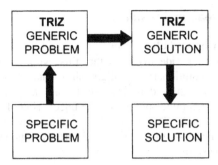

Figure 1:   The basic TRIZ problem solving process

12.2.1 Contradictions / inventive principles

The Contradictions part of TRIZ is constructed on a comprehensive analysis of patents in which the inventor has successfully 'eliminated' design contradictions other design methods assume to be inherent. In using the patent database as a foundation, the analysis has inevitably been dominated by engineering solutions, and so there is an inevitable degree of extrapolation involved in applying the tool to other solutions. Nevertheless, previous research has demonstrated that the exact same solution strategies – or 'Inventive Principles' - apply in non-technical and biological situations [5, 6]. Preliminary work relating the Principles to mass-customization problems seems to further bear out the relevance and scope of the original TRIZ research.

To date, TRIZ researchers have uncovered just 40 Inventive Principles usable by problem solvers to help overcome contradictions. The main message from the TRIZ findings is that there are ways of eliminating contradictions and therefore we should actively seek them out. Without apparently any knowledge of TRIZ, the acute importance of breaking compromises in a mass-customization environment was first recorded by Stalk et al in an article describing the paradigm breaking performance of companies like Schwab, Southwest Airlines and Home Depot after they successfully broke contradictions in their market sector (see Table 1, see also TRIZ analyzes in [7, 8]).

Table 1:   Breakaway growth examples from Schwab, Southwest Airlines and Home Depot [7]

	(%) Industry average growth 1988-1995	*(%) Compromise- breaker growth in same period*
Securities Brokerage	90	520
US Domestic Airlines	80	370
Home Improvement Retailing	40	1500

12.2.2 Ideality / ideal final result

TRIZ founder Genrich Altshuller identified a trend in which systems always evolve towards increasing 'ideality' and that this evolution process takes place through a series of evolutionary paradigm shift characteristics [2, 9]. A key finding of TRIZ is that the steps denoting a shift from one paradigm to the next are predictable and have been under-pinned by the uncovering of a number of generic technology evolution trends. This finding may be expected to play a significant role in helping organizations to predict the future evolutionary potential of any given system or sub-system.

The essential paradigm shift between a conventional design approach and the TRIZ approach is that while traditionally problem solvers start from the knowns of today, the concept of ideality employs a strategy in which the problem solver is asked to envisage the 'ideal final result' situation – in TRIZ terms that situation where the function is performed without any cost or harm – and to then use that as the basis from which to work back to a physically realizable solution. This philosophy is illustrated in Figure 2.

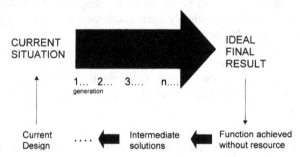

Figure 2: Proposed 'ideality-based' improvement and evolution strategy

As well as offering a successful evolution strategy and real problem solutions, the method also provides a considerable amount of valuable long-term strategy definition data. We concentrate our view here, however, on two simple parts of the tool with particular relevance to the subject of mass customization.

The first comes from the definition of Ideal Final Result (IFR) as 'achieve the function without (additional) resources'. This definition is intended to suggest the concept of the existing system '*solving the problem by itself*'. The key word here is SELF. Self is a very important word in a TRIZ problem-solving context, and the idea of solving problems without complicating the system denotes a powerful step towards increased ideality. Mann discusses the subject in more detail, but of particular interest from a mass customization perspective are solutions that, for example, self-adjust, self-adapt or orient themselves to individual customer desires [10]. The second part emerges from the (usually qualitative) equation used to define the 'ideality' of a system:

$$\text{Ideality} = \Sigma\text{Benefits}/(\Sigma\text{Costs} + \Sigma\text{Harms})$$

For it is in this equation that we see one of the fundamental contradictions of all business problems – the fight between delivering the good things on the top of the equation versus wishing to reduce the cost element at the bottom of the equation. Actually, the fight usually centers on customers – who generally speaking want the benefits – versus the supplier, who generally speaking wants to minimize costs. Of course this fight differs according to a variety of factors – most notably market maturity – but the top-versus-bottom contradiction analogy will be useful to us in this discussion.

12.2.3 Functionality

Although the functionality aspects of TRIZ owe a significant debt to the pioneering work on Value Engineering, the method of defining and using functionality data is markedly different; sufficient at the very least to merit discussion as a distinct paradigm shift in thinking relative to traditional occidental thought processes. Three aspects are worthy of particular note:

1. The idea that a system possesses a Main Useful Function (MUF) and that any system component which does not contribute towards the achievement of this function is ultimately harmful. In a heat exchanger, for example, the MUF is to transfer heat to the working medium; everything else in the system is there solely because we don't yet know how to achieve the MUF without the support of the ancillary components (systems may of course perform several additional useful functions according to the requirements of the customer).

2. In traditional function mapping, the emphasis is very much on the establishment of positive functional relationships between components. TRIZ places considerable emphasis on plotting both the positive and the negative relationships contained in a system, and, more importantly, on using the function analysis as a means of identifying the contradictions, ineffective, excessive and harmful relationships in and around a system. Function analysis thus becomes a very powerful problem definition tool.

3. Functionality is the common thread by which it becomes possible to share knowledge between widely differing industries. A motorcar is a specific solution to the generic function 'move people', just as a washing powder is a specific solution to the generic function 'remove solid object'. By classifying and arranging knowledge by function, it becomes possible for manufacturers of washing powder to examine how other industries have achieved the same basic 'remove solid object' function. '*Solutions change, functions stay the same*' is a message forming a central thread in the TRIZ methodology: People want a hole not a drill.

The emphasis TRIZ places on functionality demands that engineers and scientists adopt a much more flexible approach to the way in which they look for solutions to problems. The age of the specialist is coming to an end; it is no longer sufficient for mechanical engineers to only look for mechanical solutions to their problems when someone from, say, the chemical or software or social or political, etc sectors may already have discovered a better way of achieving the function being sought (Figure 3). This is again important in a mass-customization context.

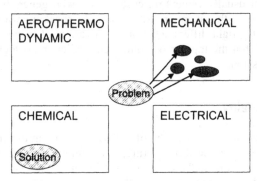

Figure 3: Solution spaces

12.2.4 Use of resources

While the previous three pillars of TRIZ could undoubtedly have been uncovered anywhere where the researchers showed the dedication shown by TRIZ research-ers, the 'resources' pillar is probably uniquely Russian. Resources relates to the unprecedented emphasis placed on the maximization of use of everything contained within a system. In TRIZ terms, a resource is *anything in the system which is not being used to its maximum potential*. TRIZ demands an aggressive and seemingly relentless pursuit of anything – objects, information, time, energy, attributes, etc – in (and around) a system which are not being used to their absolute maximum potential. Discovery of such resources then reveals opportuni-ties through which the design of a system may be improved. In addition to this relentless pursuit of resources, TRIZ demands that the search for resources also take due account of negative as well as the traditionally positive resources in a system. Thus the pressures and forces we typically attempt to fight when we are designing systems, are actually resources. By way of an example of this 'turning lemons into lemonade' concept, Russian engineers often think of resonance as a resource. This is in direct contradiction to most Western practice, where resonance is commonly viewed as something to be avoided at all costs. TRIZ says that somewhere, somehow, resonance in a system can be used to beneficial effect. In effect, resonance is a potent force lever capable of amplifying small inputs into large outputs. Resonance is currently being used to generate beneficial effects in a number of new product developments from vacuum cleaners, paint stripping systems on ships (firing a pulsed jet of water – existing resource! – at the local resonant frequency of the hull), and in helping to empty trucks carrying powder-based substances more quickly.

12.3 Case studies

12.3.1 Bicycle seat

A simple example should serve to demonstrate the key mass customization benefit-versus-cost contradiction in action: Any customer going into any bicycle shop in the world is likely to be surrounded by a plethora of different bicycle saddle designs, all of which are pretty well designed using the same design rules and trade-offs that have been around for over a 100 years. From a mass-customization context, picking up the quote at the beginning of this chapter, there is probably too much choice and too few options. The basic trade-off is one of benefit versus cost. Figure 4 illustrates the basic choices on offer – either cost or benefit via a range of different degrees of compromise (denoted by the hyperbolic 'constant paradigm' curve).

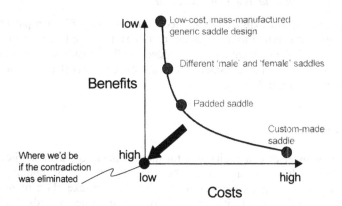

Figure 4: Classic bicycle saddle cost versus customization benefit contradiction

The bottom left hand corner of the picture denotes the point we would be at if the cost versus benefit contradiction did not exist. We might think of this point as our *Ideal Final Result*. The ideality tool, in fact, turns out to be a good start point for designing a better saddle (Mann has previously described a solution derived by solving a technical contradiction within the normal design paradigm [11]). From an ideality perspective, an ideal saddle would 'adapt itself' to the shape of any user without complicating the design relative to current designs. In other words, rather than blasting the customer with too many choices, give them exactly the saddle they want by letting them know that it will work out for itself how best to adapt to their individual shape.

There are a number of TRIZ tools capable of getting us to an inherently cus-tomized comfortable bicycle seat solution. The quickest route in this case takes the 'self-adapting' concept from the ideality definition above and uses it as a prompt to examine a knowledge base to see if anyone has solved a similar problem elsewhere (TRIZ research suggests that it is 99% likely they have). Using this route and the US patent database, quickly identifies the idea of rheopexic gels as a means of providing adaptable *and* stable shapes (the data base search identified patent US4471538: Shock absorbing devices using rheopexic fluid). A rheopexic gel turns out to have highly desirable stress-thickening properties as described in the patent abstract:

"Shock absorbing devices utilize rheopexic fluid contained in a deformable sealed chamber which is subjected to external shock forces. Upon application of the shock forces to the deformable sealed chamber, the rheopexic fluid filled therein is exerted with shear stress which causes the rheopexic fluid to increase its consistency and shock absorbent characteristics as a function of increasing shear stresses applied thereto. Additionally, when used in an application wherein the device is placed against a body part of a user, the rheopexic material "molds" itself to the body portion of the user. When left at rest, the rheopexic material returns to its initial fluid, low-consistency state."

The invention (now expired, albeit cited by a significant 87 other patents since) offers a potential solution to the problem that does not complicate the design of a saddle relative to the conventional gel-padded saddles already on the market; and consequently offers increased benefit (adaptability, comfort) at a fundamental cost no greater than current designs.

12.3.2 Shoes

The largest number of patent citations for the rheopexic gel is for applications as shoe insoles. Such devices again offer the potential for ready adaptability not just to individual shoe wearers, but also to the changes that take place in an individual's foot size and shape at different times of day (evidence records that feet can alter shape by as much as one shoe-size during the course of a typical day). It may be argued that a rheopexic gel insert has complicated the system called 'shoe' relative to current designs, and as such, the benefit of a mass-customized shoe versus cost contradiction has not been completely eliminated in this case. The TRIZ evolution trend known as 'Trimming', however, suggests that ideality can be increased in this case by merging insole and shoe, and incorporating an appropriate gel into the manufacture of the shoe.

It is likely that the reasons we don't all walk around in self-adapting shoes are due to (a) a failure on the part of shoemakers to make a connection between shoe sole design and the integration of a rheopexic gel, and more likely, (b) the probable emergence of other problems when a gel insole is introduced into the shoe design. To take a likely scenario in this second case, it may be that any membrane required to encapsulate the gel would have a negative impact on the breathability of the shoe. If this is the case, then in TRIZ terms, we have found another contradiction, and thus a route into finding ways by which others have successfully overcome similar problems.

12.3.3 Lighting

The concept of 'mass-customized' lighting systems for homes or hotels is already fairly well established through the emergence of systems like that illustrated in Figure 5. Any system that follows the TRIZ-predicted evolution trends away from mechanical systems to 'field'-based systems and towards increased controllability is consistent with an ability to offer customers a highly configurable, adaptable system. The Figure 5 lighting system, for example, is programmable from a PC and allows homeowners to design and pre-set a wide variety of different lighting settings within a room.

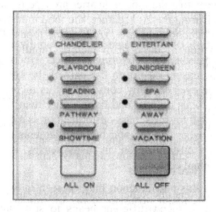

Figure 5: 'Mass customized' programmable light switching

This type of system is able to take ready advantage of existing resources in many homes (ring main to act as communication medium between different switches, dimmer switches, intruder sensors, etc) to give – like most 'field'/software implementable systems – ways of breaking the benefit versus cost contradiction to enable ready mass-customization (albeit with probable 'video – programming' syndrome – 95% of owners understand 5% of the controls on their video player – issues still to be resolved.) Reference [12] discusses this subject further in relation to e-business situations.

Taking this lighting concept a stage further would be a system that allowed the user to modulate not just the level and mix of lighting in a room, but also to modulate the color of those lights. At the present time, the cost versus benefit mass customization contradiction has not been eliminated for such a concept. Any individual user is able to customize the color of lighting in a room only by physically changing light-bulbs or by having multiple bulb holders or by incorporating multiple light filters – each of which necessarily involves cost trade-offs.

Again, TRIZ can help to identify means of breaking the cost-benefit contradictions as they currently stand. From an ideality perspective, the light bulb would change color 'itself'. A search for solutions relevant to this problem would not reveal a viable design at this time. The technology trends uncovered by TRIZ researchers in conjunction with some of the latest design attempts, however, reveals significant evolutionary potential left in the design of lighting systems, and further points towards possible means by which the contradiction breaking, color changing light-bulb might well be a possibility [13].

12.3.4 Home customizable foods

Initial research suggests that the food industries are ripe for mass customization in several sectors [14]. During the 20th century, the food industry necessarily focused on economic mass production of 'safe', ready-made foods. This trend was largely driven by a combination of three main factors; rapidly increasing population, increasing legal responsibility for delivering products which did not cause harm to the consumer, and decreasing desire on the part of the consumer to spend time preparing meals. In the eyes of many consumers, this increase in overall ideality has been achieved by focus on cost and harm rather than 'benefit', and in many senses the emerging trend towards mass customization is a consumer-lead drive to redress the balance between the top and bottom elements of the previously seen ideality equation. Mass customization, in many senses, is about increasing the 'benefit' to each individual customer.

The initial stages of this phenomenon may be seen in in-store coffee blenders that allow customers to blend different beans to suit individual taste, but an increasing number of products are using the same TRIZ inventive principle 1 'Segmentation' to achieve equally attractive customized products. An illustration of a concept for a product to suit households in which different people have different tastes or liking for spicy food could be as follows: A base tomato sauce could be 'customized' by adding spices from an herb mix and/or a chili mix.

12.4   Discussion

History tells us that the successful producers of today's solutions are highly unlikely to create the next generation's solutions [15, 16]. The reasons for this phenomenon are both multitudinous and complex, but probably should include elements such as the failure of organizations to recognize the definitions of success (or 'benefit') registered by customers, and the paradoxical situation in which products are most vulnerable to be superseded at around the time when they are making most profit for the incumbent organization. TRIZ is beginning to change this situation for a number of leading edge companies; it provides powerful insight into the evolution of both technical systems and markets, and it is the only piece of research anywhere which not only recognizes the importance of eliminating contradictions, but also provides systematic solution triggers for actually doing so.

In many senses, 'mass customization' may be seen to be about solving contradictions. TRIZ research has uncovered the fact that the most successful inventions are the ones in which the inventor has not accepted the trade-offs and conflicts most of us take for granted. Furthermore, the research has identified that these contradiction-breaking inventors have used only a small number of inventive strategies in overcoming contradictions. TRIZ thus provides us with potentially very powerful new ways of looking at mass customization issues:

The 'ideality' concept leads us to solutions which 'do things for themselves'. Adaptive systems – like the viscosity changing rheopexic gel solution discussed earlier, or shape-memory alloys, thermo/photo-chromic materials etc – offer much from the perspective of ready incorporation into solutions that inherently 'customize themselves' to the changing desires of individual customers. In most cases they do so in a way that fundamentally breaks the traditional cost versus adaptability benefit contradiction that hampers many mass-customization initiatives. They are thus strategically very important in a design context.

Another pillar of TRIZ, Functionality, similarly offers important messages for mass customization and protection of companies from disruptive technologies. The 'solutions change, functions stay the same' message is vitally important. The fact that each individual customer has subtly different functional requirements (and priorities) is even more so.

Benefits versus costs. It is usually the customer who demands 'benefits' (although not necessarily if they have to pay for them – if cost is equal, they will almost always go for the solution offering greater net benefit), and suppliers who work to reduce cost (it's the only element of the ideality equation they have any really understanding/control of). The overall message is that it is increasingly likely that benefits 'win' over 'cost' – if you don't give them to the customer, it is increasingly likely someone else will.

References

[1]   Mann, D.L.; Domb, E.: Business Contradictions – Mass Customization, on www.triz-journal.com.
[2]   Salamatov, Y.: TRIZ: The Right Solution at the Right Time, in: TRIZ Journal, 1999.
[3]   Mann, D.L.: Hands-On Systematic Innovation, 2002.
[4]   Mann, D.L.: The Four Pillars of TRIZ, in: Engineering Design Conference, Brunel University, 2000.
[5]   Mann, D.L.; Domb, E.: 40 Inventive (Business) Principles With Examples, in: TRIZ Journal, 1999.
[6]   Mann, D.L., Creativity As An Exact (Biomimetic) Science, in: Biomimetics IV, 1999.
[7]   Stalk, G.; Pecaut, D. K.; Burnett, B.: Breaking Compromises, Breakaway Growth, Boston 2000.

[8]   Mann, D.L.; Domb, E.: Using TRIZ to Analyze and Solve Mass Customization Contradictions, in: TRIZ Future 2001, Bath 2001.

[9]   Mann, D.L.: Using S-Curves and Trends of Evolution in R&D Strategy Planning, in: TRIZ Journal, 1999.

[10]   Mann, D.L.: Ideality and Self, in: TRIZ Future 2001, Bath 2001.

[11]   Mann, D.L.: A Comfortable Bicycle Seat, in: TRIZ Journal, 1998.

[12]   Mann, D.L.; Domb, E.: Business Contradictions – Profitable E-Commerce, in: TRIZCON2001, Los Angeles 2001.

[13]   Mann, D.L.: Identifying Evolutionary Potential in Technical Systems, in: Creativity and Innovation Management Journal, 10 (2001).

[14]   Mann, D L.; Winkless, B.: Any Color You Like As Long As It's The One You Want: TRIZ and Customizable Food, in: Food Design Journal, 9 (2001).

[15]   Christensen, C.M.: The Innovator's Dilemma: When New Technologies Cause Great Firms To Fail, Boston 1997.

[16]   Utterback, J.M.: Mastering The Dynamics of Innovation, Boston 1996.

Contacts:

Darrell L. Mann
Department of Mechanical Engineering, University of Bath, UK
E-mail: d.l. mann@bath.ac.uk

Ellen Domb
PQR Group, Upland, CA, USA
E-mail: ellendumb@compuserve.com

Part IV: Interfacing and Integrating the Customer

Getting customers involved and optimally informed

The essence of customization is to provide only and exactly what each customer wants at the right time. The process necessary to reach this objective is the configuration process. Configuration means to transfer customers' wishes into concrete product specifications. While the basic product families, common product platforms and the corresponding manufacturing systems are set up when building a mass customization system, configuration activities take place with every single customer's order. For each order, the individual wishes and needs of a client have to be transformed into a unique product specification. The additional costs arising from the customization process consist largely of information costs in the sales. They are accounted for by the investigation and specification of the customers' wishes, the configuration of corresponding individual products, and the transfer of the specifications to manufacturing. All theses activities are characterized by a high information intensity compared to traditional mass production. An important characteristic of successful customer centric companies is the use of dedicated information systems to capture these additional costs. Called configurator, choice board, design system, tool-kit, or co-design-platform, these systems are responsible for guiding the user through the configuration process. Different variations are represented, visualized, assessed and priced by these systems, enabling the customer to interact closely with the firm's capabilities. While configuration systems theoretically do not have be based on software, all known mass customizers are using a system that is, at least to some extent, IT based. Configuration systems can also include physical measurement tools like 3D-scanners or visualization tools like 3D screens.

Part IV comprises of the discussion of configuration methodologies and modes for customer interaction. The development and implementation of appropriate systems for customer interaction is an important success factor of mass customization – and a field with many open questions (see also Chapter 30 of this book). The papers in this part should help to answer some of these questions. In Chapter 13 *Khalid and Helander* provide an introduction into web-based do-it-yourself product design (DIYD), as installed on many mass customization web sites. DIYD is defined as the selection and configuration of products by customers on their own. However, information about customer needs is usually incomplete, making it difficult to develop a configuration system both in terms of the set of options presented there and the corresponding user interface. Often, customer needs have to be estimated from population preferences or global market diversities. Of particular interest in this context is an investigation into appropriate procedures for customer design conceptualized in different cultures. The objective is to improve the usability of configuration web sites by addressing these differences. How consumers behave in such an environment is discussed by *Kurniawan, Tseng and So* in Chapter 14. Choosing, matching, and swapping components and configuration options and assembling them together to a specific product instead of choosing a ready-made product from a shelf is a totally new way of acquiring products for many consumers. Thus, the authors hypothesize that this may change known patterns of consumer behavior greatly. The authors present a new approach to understanding consumer behavior using the living system theory. While (traditional) marketing models rooted in psychological research assume that information (during the choice process) is evaluated by customers using symbolic processing, the living system theory explains customer behavior as a collection of components. Thus, a new way is offered in this chapter to better understand the essence of being customer centric: the customer.

Bee and Khalid extend this discussion with an empirical study evaluating three DIYD web sites in Chapter 15. Using factor analysis, three generic factors are extracted as important features from a customer's perspective, namely holistic design, navigability, and timeliness. Users seem to prefer a top-down hierarchical approach when designing the sample products (bicycles, watches, and dresses). The study also evaluated success factors for the corresponding design of the configuration system: design procedures, aesthetic preferences, information display, and design pleasure. The design and development of configuration systems corresponding to the needs of (potential) customers is also the topic of Chapter 16 by *Porcar, Such, Alcantara, Garcia and Page*. However, the authors follow quite a different approach and show how consumer expectations can be captured by the Kansei Engineering methodology. The chapter demonstrates how Kansei Engineering can be used to guide (inexperienced) customers in order to quickly find the desired design according to their preferences. This approach may also help manufacturers in cutting down a wide variety of options in manufacturing among which a large percentage does often meet not the preferences of the target group. Focusing production variability on features affecting most users' purchasing decisions may reduce the amount of design options offered, and may thus result in an important contribution to controlling costs and reaching near mass production efficiency in manufacturing.

Hvam and Malis extend the discussion on how to develop and design efficient and effective configuration systems. In Chapter 17 the authors present a documentation tool for configuration processes to foster knowledge based product configuration. Mass customization and similar approaches to create more customer centric product architectures led to an enormous extent of variety and complexity. This calls for an effective documentation system in order to structure this knowledge. Standard configuration systems do not support this kind of documentation. The authors sketch a rather simple application that serves as a knowledge based documentation tool for configuration projects. Their objective is to document complex product models in a way which considers both the development and the maintenance of the products. Part IV concludes with an important plea by *Svensson and Jensen* that the customer should always be at the final frontier of mass customization (Chapter 18). The authors show that despite all technological advances and approaches – such as the ones presented in the previous chapters – the central limiting factor in the expansion of mass customization will be the customer. Often, mass customization has mainly been turned towards product and processes. The authors argue that customers have to come fist. This is an important point that should never be forgotten when designing a customer centric enterprise.

13 Web-Based Do-It-Yourself Product Design

Halimahtun M. Khalid[1] and Martin G. Helander[2]
[1] Institute of Design and Ergonomics Application, Universiti Malaysia Sarawak, Malaysia
[2] School of Mechanical and Production Engineering, Nanyang Technological University, Singapore

Mass customization aims at providing cost-effective products and services to meet individual customer's needs. An implicit assumption of mass customization is that organizations must recognize customers as individuals and understand their needs. A central concept in this regard is the Do-It-Yourself Design (DIYD) approach that is contemplated in this chapter. DIYD is defined as the selection and configuration of product/parts by customers on their own. However, information about customer needs is usually incomplete. To develop and design a catalogue of products/parts (building the base for configuration) and an corresponding user interface for the configurator system, customer needs may be based on estimates of population preferences or global market diversities. In addition, rather general information concerning good taste, new product needs, easy design procedures and rules concerning web usability have to be taken into account. This information may be mapped as functional requirements using a hierarchical approach. Of particular interest in this context is to investigate appropriate procedures for customer design conceptualized in different cultures. Thus, we will investigate in this chapter the top-town hierarchical approach typical for Western engineering in two separate studies. Malaysian and Hong Kong participants supported hierarchical design in a Web-based DIYD process of watches. Additionally, this chapter comments on the design of usable Web sites and human factors issues for DIYD research.

13.1 Towards mass customization

Mass customization is the ability of companies to offer a wide range of products and services using a common production platform to meet the heterogeneous needs of its customers [1, 2]. Flexible work and assembly processes permit the manufacture and distribution of customized goods in high volumes and at a relatively low cost [3].

The objective of mass customization is to efficiently produce highly customized products rapidly and cost-effectively. But it is not about infinite choices. Rather a reasonable number of optional parts may be mixed, which may give the customer the illusion of unlimited choice, while manufacturing remains manageable [4]. This strategy involves more than the manufacturing, logistics system, and marketing. It requires re-engineering of business strategy, from understanding customer needs and translating them into functional requirements, to developing and evaluating usable Web interfaces that are customer-centric. A paradigm shift is needed to transform mass production into global mass customization (Table 1).

Table 1:   Characteristics of contemporary and global manufacturing

Contemporary Manufacturing	Global Manufacturing
Mass production	Mass customization
Standard specifications	Customized specifications
Design for the customers	Design by the customer
Customer-designer-manufacturer link	Customer/designer-manufacturer
Vendor delivery	Direct delivery
Off-line communication	Web-based communication

To transfer the concept of mass customization to the commercial industry, manufacturers have to abandon their mass production techniques and create flexible manufacturing and production systems. One of the biggest changes is in the design and construction of products. Manufacturing companies need to use modular production processes, in which groups of steps can be separated or clipped together like "LEGO blocks" (LEGO is a trademark of the LEGO Group).

There are many examples of mass customization in industry. Cannondale.com can configure over 8 million variations in its bicycles – mainly through variations in color scheme and frame type [5]. Levi Strauss.com offers customized jeans based on individual measurements. Following an order, the jeans arrive two weeks later in the mail, with the customer's own label sewn inside, and a bar code sealed to the pocket lining which stores measurements for simple re-ordering [5]. The production line of BMW can quickly switch from the large 7-series automobiles to the smaller 5 or 3-series. This system may explain why BMW will survive in the future despite their small size [6].

Mass customization requires a different corporate culture, built not around products but around different kinds of customers. Gilmore and Pine [3] identified four distinct approaches to customization: collaborative, adaptive, cosmetic, and transparent. *Collaborative* customizers conduct a dialogue with individual customers through standardized procedures on the internet. The purpose is to help them articulate their needs, and identify the precise offering that fulfills those needs. *Adaptive* customizers offer one standard, but customizable product that is designed so that users can alter it themselves. An adaptive DIYD necessitates a broad understanding of product functions and utilities to enable multiple mappings between utility and customer needs following a hierarchical design approach [7], [8]. *Cosmetic* customizers present a standard product differently to different customers. For example, Planters package peanuts differently for different retail customers. The content is the same but packed in varying sizes and packets to meet individual needs. Cosmetic DIYD requires a dynamic database with varied design options that can be easily customized [9]. *Transparent* customizers provide individual customers with unique goods or services without letting them know explicitly that those products and services have been customized. ChemStation, for example, supplies different detergents to high-volume customers based on their assessed needs.

The interest of this chapter is in collaborative customization. This is clearly the most ambitious and personal alternative, since it is necessary to fully understand the customer and design an interface and options to suit individual customers. The objective of this article is to demonstrate how one can facilitate Web-based design of consumer products for mass customization. There are two objectives: (1) How to predict customer needs based on general research recommendations in human factors engineering, and experimental and consumer psychology; (2) How to present the product/parts catalogue so that it is not overly cumbersome for the customer to design the product. In this context we present the results of two studies. The first investigated customer needs for watch design, while the second highlights usability issues in website design for mass customization. Findings from these studies have implications for designing an e-commerce facility such as www.idtown.com, ewatch-factory.com, and so forth.

13.2 A model of mass customization through DIYD design

A model of mass customization through DIYD design is presented in Figure 1. Human Factors issues of DIYD are addressed in the front-end of Figure 1. There are two main problems in designing for mass customization: First, an e-commerce outfit for customized products, initially, will have a poor appreciation of customer needs. This is because the commodity is new, and mass customization is new. We may improve our customer understanding by applying generic knowledge with support from the research literature in human factors, consumer psychology and marketing. This will supply a flexible and generalized framework, which may be used for mass customization and manufacturing.

Second, it is of great importance to present a user-friendly interface. The customer is only one click away from leaving, and difficult procedures will discourage. We may then identify interface design features that make the Web site more usable, thereby assisting the customer during the various phases of DIYD design. From the estimation of customer needs we may formulate functional requirements and design parameters [8]. The functional requirements identify the goals/purposes of the product at a high level and the design parameters identify the product/parts catalogue. An example is given below.

In the designing of parts there are constraints imposed by design rules for manufacturability. These are particularly important in flexible manufacturing – the

LEGO block based scenario that applies to mass customization. After the design of the product database and the interface, customers may place orders. An order will be sent to the manufacturing/assembly facility with a copy to the supply chain, which will deliver parts just-in-time. After delivery of the product, Marketing may investigate customer satisfaction and customer needs. This information may then be used to upgrade the database of customer needs. The new information will affect the product offerings as well as the interface design.

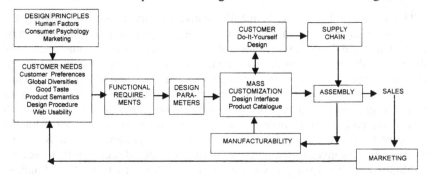

Figure 1:   Estimating customer needs for Do-it-Yourself Design

13.3   Modeling customer needs

In setting up a Web-based interface for design and mass customization, a significant challenge is to help customers formulate their implicit requirements and make them explicit to the manufacturer [10]. In this process one cannot overload the customer with seemingly "trivial" questions. Instead, we can draw from principles concerning "good taste", customer preferences and product variety that have emerged in human factors and marketing research [11]. For example, the psychological principles of good taste suggest that people have universal appreciation of attributes (e.g. aesthetics) that satisfy them [12]. These principles can be implemented in a Web environment, and thereby offer the potential customer a better opportunity to design a satisfactory product.

13.3.1 Population preferences

The purchase of an object is generally preceded by a goal formulation and a decision making process. The greater the amount of money involved the greater amount of time it takes to elaborate goals and make decisions. The purchase of a car is often reflected upon for a year, by evaluating the goal, breaking it down in sub-goals and then elaborating on the sub-goals [13]. There may be primary goals such as: Buy a car to travel, and there may be secondary goals: Buy a car to impress the neighbors. In any of these cases, the decision to buy will be preceded by information collection concerning the price of cars, financing, satisfaction of previous buyers, maintenance record, and so forth. The decision is usually not preceded by probability estimates. Rather the customer will progress in his/her decision using simple heuristics that are learnt response patterns. For example, to evaluate a car, collect information or read the literature. To choose among alternative cars, consult a car expert or car shop.

A manufacturer who can anticipate the customer's needs has a great advantage. Market research has an important function to anticipate customer needs. For new products or for new procedures such as DIYD there is hardly any information available. We will have to resort to generalities about good taste and other findings that may be helpful in setting up DIYD. These are presented below.

13.3.2 Global market

To go global, the Web site must be designed effectively and consistently in multiple foreign languages. This means adapting both content and context for a particular audience [14]. Most customer communication strategies are based on theoretical frameworks developed in the United States [15]. Much research is needed to understand the motivation of customers outside USA. The constructs of individualism and collectivism represent the most broadly used dimensions for cross-cultural comparisons. In many Asian countries individual reasons for purchase are subordinated to the wishes of the group [16]. The concept of mass customization may therefore be less amenable to the Asian population than to the Western world. Attributes of objects such as color may have different meanings to different cultures. For example, Chinese associate red with happiness, while Indians identify red with power and energy. Several studies have shown that people seem to prefer hues of shorter wavelength: blue is preferred over yellow and red [17]. But the preference clearly depends on expectations and the type of situation: blue food would not be preferred over yellow and red.

13.3.3 Customer motivation

In the past, the Do-It-Yourself (DIY) concept implied that a customer assembled a product designed by a company. Here the roles are reversed: A customer "designs" that is, select and configure product parts, while the company assembles. Hence we need to understand why the customer/designer would like to engage in this activity and what the underlying motivation is. Opportunities to express their needs directly via a Web-based system may enhance customer satisfaction [18]. Design ability is possessed by everyone [19]. Through design, individuals have the opportunity to express their personality. The potential benefits can be far reaching, such as maximizing customer values and dealing with actual customer needs.

One of the first Web-based fashion design Websites enables customers to design girl's clothes, inspect the results and order on-line [20]. One customer reported her experience in designing on the Web:

> *"I really enjoyed designing the outfits on your Web site. The dresses turned out beautifully, just how I pictured them. Great fabric and beautiful colors. What a delightful way to buy clothes; I've always wanted to change the color or style of stuff I'm buying and be able to see it instantly. The best use of the Web yet!"*

In order to reflect individuality, people have become increasingly preoccupied with "designer labels"; designed objects are regarded as having status and value. Taste reflects our personality and also reveals part of our personality to others [21, 22]. Through design, individuals have the opportunity to express their personality. Dilnot went even further in claiming that [23]:

> *"Objects (help) make us. Making (and designing) are moments of making (and designing) ourselves ... The object is the substitute for ourselves in the special sense that things work to provide us artificially what nature "neglected" to bestow us."*

Given the opportunity, customers will want to design themselves and purchase individualized artifacts. Customers are no longer passive recipients of goods and services. They tell what they want. They may also be repeat customers, because a learning relationship has developed, which may condition them to return.

Studies on the internet and online buying behavior indicate that e-commerce is indeed a growing trend [24, 25, 26]. The consumer potentials of the Web are driven by several factors, including:

- perceived business advantages (e.g., customization of products and services, shortened time to market for new products and services),

- socio-demographic changes (e.g., a more computer-literate population and increase in families with two working parents, pressure to free up limited leisure time), and

- unique features of a direct marketing channel (e.g., two-way communication between the manufacturer and the consumer, interactive shopping channel that is not bounded by time or geography).

Many studies acknowledge that customers seek variety [27, 28]. Psychological evidence shows that individuals' reactions to a merchandize can be characterized by a "single-peak preference function" – once the optimal level has been reached, then satiation sets in and the customer looks for something different on the next choice occasion [29]. The most common explanation for variety-seeking behavior is that customers have some internal need, or drive, or even an intellectual curiosity that causes them to choose variety over time. Therefore, a DIYD system must be able to support customers with an appropriate product knowledge portal that manages product variety. It must also assist in decision making and commitment to purchase.

13.3.4 Good taste

Psychological research can help in clarifying some of the controversies of what constitutes good taste. Research in Gestalt Psychology emphasizes the equilibrium and over all impression of objects. Objects with visual balance have, for centuries, been said to be particularly pleasing [30]. Objects that are symmetrical and have even weight distribution are regarded as more "balanced" and preferred over other objects [31]. Although these findings seem to support the idea of simple design over complex, Berlyne claimed that people prefer complexity: they look more at complex figures than simple figures [32]. He also mentioned that people prefer objects that are moderately familiar to objects that are over-familiar. Martindale and Uemura [33] suggested that aesthetic preference is linked to the physiological arousal potential of artifacts and that preference diminishes with increased familiarity. Folk psychology is ambivalent in this respect: "old favorites" versus "familiarity breeds contempt".

It is true that objects marketed as "new" can improve sales. On the other hand, Crozier observed that preference is highly correlated with exposure; the more we experience an object, the more we like it [12]. The question of change or newness is essential to the fashion industry; unfashionable designs look wrong. But Berlyne argued that too much novelty is aversive [32]. A gradual, stepwise change in consumer product design features is preferred. Teigen argued that "informativeness" is more important than either novelty or familiarity [34]. He proposed that interest in an object is related to its information value, which is a function of both novelty and familiarity. If objects are overly familiar, then maybe we can at least learn something from using them. This theory seems to agree with Purcell who argued that emotional responses to objects are greater if the objects are different than expected [35].

13.3.5 Product semantics

A suitable starting point for developing product semantics is to observe what objects people surround them with, identify their purpose and what they are used for, how they make sense, and how they are referred to [36]. Thus, Graumann reacted to the research in design psychology that he felt should be directed more towards theories of real world objects rather than abstractions [37]. There is a need to develop categories to describe different meanings associated with consumer products. Accordingly, Bih interviewed a sample of Chinese students after they spent one semester living in New York [38]. Several dimensions of meanings of objects were identified (see also [12, 39]):

- Objects with predominantly functional significance (e.g., radio),
- objects with cultural or personal values or ideals (e.g., Buddhist scriptures),
- objects marking personal achievements (e.g., degree certificate),
- objects to extend memory (e.g., past photographs),

- objects for social exchange (e.g., keepsakes or memorabilia),

- objects to illustrate shared experience (e.g., tea set, as tea-taking is a cherished activity),

- objects to extend the 'core' self and personal values (e.g., soft toys, personal diaries).

As can be expected with young students, objects of social status were not so important. However, for other populations, Dittmar [40] cited furniture, jewelry, and art objects as having prestigious meanings. Crozier [12] extended the classification to include objects of utilitarian nature, such as furniture and kitchen tools. In addition there are objects that reflect symbols of the self, such as mementos of the family, study certificates, sentimental souvenirs, and so forth. Since "meaning is in the mind of the beholder," to render something understandable, one must make use of meanings already familiar to the person [41].

Based on the classification of objects, one can detail several reasons to design and buy various items. We will here take the example of designing/buying a watch. Watches are not only utilitarian. They may be designed to satisfy a variety of purposes: functional, religious expression, memento, fashion, prestige and so forth. This may be achieved by designing the watch face using text, photographs, graphics, expensive materials, and so forth [42]. The quality of a watch depends on a collection of attributes, such as, the type of casing (e.g. metal), precision in time (e.g. chronograph), strength of material (e.g. waterproof), and so forth. It can be concluded that there are several general customer needs and preferences, which make DIYD an interesting and relevant activity. Based on previous research, the items listed in Figure 2 may have generality. They apply to all consumer products and to all design activity.

- Need to acquire/augment one's personality
- Need to express personality through design
- Need to expand collection of important consumer products:
- functional objects, objects of prestige, objects of cultural values, objects marking personal achievements, objects to extend memory, objects for social exchange.
- Prefer objects of moderate complexity and information value
- Prefer objects with good *gestalt* and balance
- Objects must fit customer

Figure 2: General customer needs for Web-based design of consumer products

13.3.6 Human factors principles for web usability

A cursory tour of the WWW shows that the knowledge established in software usability engineering over the last 20 years is generally not applied to Web site design [43]. Typical user interface issues that are poorly addressed pertain to:

- The importance of good conceptual model design to communicate structure, support user models, and facilitate effective navigation.

- The importance of context information to maintain a sense of place, both global (on WWW) and local (in a site).

- The importance of good (simple) screen design.

- The importance of effective feedback.

- The importance of user-centered design and usability testing.

Grose, Forsythe and Ratner studied Web sites and found that the following recommendations being frequently violated [43]:

- Enter data in units that are familiar to users.

- Present only information that is essential to the users at any given time.

- Display information in plain, simple, concise text, whenever possible.

- Make functionality visible.

- Present information in a directly usable format that does not require interpretation or calculation.

- Organize data fields to reflect data that have a naturally occurring order (e.g. sequential).

- Use white space to create balance and symmetry and to lead the eye in the appropriate direction.

- Avoid horizontal scrolling.

- Display functionally related data together.

- Use coding (e.g. color, size, highlighting) to draw attention to critical information, changed items, items to be changed, special areas of the display, and errors of entry.

In addition, graphic features, such as color, location, texture, orientation, shape and size, are important to the design of information sources because of their associative, selective, and ordering capabilities. Poor use of these features can add greatly to the mental effort needed to understand and extract required information from the display [44]. Therefore, design of Web-based DIYD systems must consider issues pertaining to visual performance: how to improve the speed, accuracy, and usability of communicating graphic information to the user. The study by Nel, van Niekerk, Berthon and Davies suggests that for customers to retain interest in a Web site, it must be designed to enhance the "Flow" [45, 46]. Flow represents the extent to which (1) the individual perceives a sense of control over the interaction with the technology, (2) the individual's curiosity is aroused during the interaction, and (3) the individual finds the interaction intrinsically interesting. Web design does not call for a new field of research to support designers. Guidelines for designing usable Web sites can be found in the human factors and Web literature and this information is expanding as the number of Web sites increases [47, 48, 49, 50]. Some crucial guidelines from the IBM Ease of Use group include [51]:

- *Simplicity*: Do not compromise usability for function.
- *Support*: User is in control with proactive assistance.
- *Obviousness*: Make objects and their controls visible and intuitive.
- *Encouragement*: Make actions predictable and reversible.
- *Satisfaction* (feedback): Create a feeling of progress and achievement.
- *Accessibility*: Make all objects accessible at all times.
- *Versatility* (flexibility): Support alternate interaction techniques.
- *Personalization*: Allow users to customize.

Many internet users are discretionary; many have had relatively little computer experience; many are only occasional users; and many may not speak English as their native language. Companies must take into account cultural and conventional differences, as the customer's exit is always just one click away. Thus, usability issues in Web site design seem even more critical than they are for traditional software application design. Customers have many optional web addresses, and they are quick to find. In the following section, we highlight two studies to illustrate functional requirements and the strategies taken in DIYD.

13.4   Functional requirements and design order in mass customization

13.4.1 Identifying customer needs in watch design

Customer needs must be expressed so that one can understand the implications for the design of the product/parts catalogue. Figure 3 illustrates the gradual development of design parameters using a Hierarchical top-down Design procedure. Generalized consumer needs are mapped on to functional requirements that are mapped on to design parameters.

One customer may require a "sporty" watch, and through a thoughtful design one can enhance the 'sporty' self-concept. In Figure 4 there is a progression from customer needs to functional requirements to design parameters. This implies that functional requirements are derived from customer needs, and design parameters are derived from functional requirements. For each of these there are four levels of abstraction, which are used to break down customer needs, functional require-ments and design parameters from abstract goals to concrete design. Figure 3 does not detail all the information necessary to design a watch. In particular, the lower levels of abstraction would need to be better specified in a real design case. None-the-less, this exercise demonstrates how goals can be broken down into sub-goals at lower levels of abstraction, and how design parameters may be derived from functional requirements, which in turn are derived from customer needs.

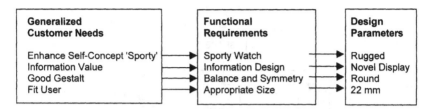

Figure 3: Mapping of generalized customer needs onto functional requirements and design parameters at four levels of abstraction

13.4.2 Modeling the design task for DIYD

Designing on the Web demands specific considerations of the design task. In the Western engineering tradition, design is usually undertaken top-down [52], [8], as illustrated in Figure 3. This is however not the natural order of design. Many studies illustrate that in real life design, designers are driven by their associations and may jump back and forth between abstraction levels [53].

In this study we were therefore interested in how items should be displayed and what the implied selection order should be – top-down, bottom-up or random. Since our goal was to design a global interface (for *idtown.com* in Hong Kong), we were particularly interested in the design order of Asian customers. In a top-down design the type of watch may be considered first, since the choice of watch will determine design parameters. To an engineer it would make sense to present design items following a hierarchical order: type of watch shape face, size, strap, time format, dial, and numeral (Table 2). After each design selection it should be possible to view the design, and change the design elements with a minimum of cursor control actuations (Figure 4).

Table 2: Hypothesized hierarchical order of design parameters (Note: Several tied ranks which are obvious from Figure 4)

Rang	Design Parameter
1	Type : sports, work, fashion, pop, party, cyber, alarm, company, chronograph, etc.
2	Shape : round, square, rectangle, hexagon, etc.
4	Face : color, pattern, text, graphics, photo, etc.
4	Size : length, width, diameter
4	Casing: metal, plastic, rubber
6.5	Strap : metal, leather, cloth, plastic, etc.
6.5	Time format : digital, analog
8.5	Dial : hour, minute, seconds-hand, date, etc.
8.5	Numeral : none, 4 or 12 numbers, Roman, Arabic, Chinese, modern, etc.

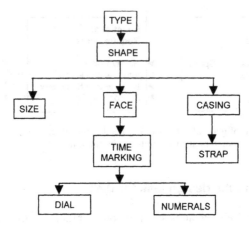

Figure 4: Hypothesized design hierarchy

To model customer needs and to confirm the design hierarchies as hypothesized in Figure 4, a study was conducted among 48 Cognitive Science students at The Universiti Malaysia Sarawak, aged between 20-41 years (mean=27.8), and 91 Industrial Engineering students at The Hong Kong University of Science and Technology, aged between 18-33 years (mean=20.7). Subjects completed a 15-item questionnaire that measured, among others, their reasons for buying a watch, their rankings of design parameters, and their interest in participating in Web-based DIYD. Their needs for having or buying a watch are summarized in Table 3.

Table 3: Common customer needs for a watch

Customer Needs	Malaysian (n=48) %	Hong Kong (n=91) %
Show time	81.25	51.60
Manage time	31.25	3.30
Style, fashion	14.58	33.00
Esthetics, attractive design	14.58	27.50
Useful, need	16.67	8.80
Accessory	10.42	4.40

Subjects were then asked to anticipate that they were designing a watch on the Web. There were nine design items (see Table 2). Subjects were asked to rank order items that would be selected first, second, third, and so forth in watch design. Subjects were not told that they could provide the same ranks to different items. The obtained ranks from this study were compared with the hypothesized ranks in Table 2. Spearman rank correlations were calculated for the ordinal data.

The results showed highly significant relationships between the hypothesized and measured ranks, $r_s = 0.77$, $p<0.05$ for Malaysian, and $r_s = 0.98$, $p<0.001$ for Hong Kong subjects, suggesting a tendency for top-down and hierarchical approach in design. About 54.2% of Malaysians would participate in DIYD on the Web, while only 27.5% of Hong Kong students would do so; 62.6% of Hong Kong subjects were uncertain compared to 35.4% Malaysians. In a follow-up study by Oon and Khalid [54] 48 subjects, aged between 21 to 36 years (mean=25), were asked to design products on three different DIYD websites (*see also Chapter 15 of this book*):

- www.squash-blossom.com for designing girls' dresses,

- www.idtown.com for designing watches, and

- www.voodoo-cycles.com for designing bicycles.

The subjects completed a questionnaire for each website which required them to rank their preferred design strategy in designing the different products. The configurator system for each website was typically set up following a top-down sequential order, but subjects could choose the design options in a non-sequential random order if so desired. Spearman rank correlation was performed on the ordinal data between hypothesized and measured ranks. The results were significant at the 5% probability level, confirming that subjects applied the top-down procedure despite the opportunity to design flexibly. Another aspect of the study was to evaluate the usability of the DIYD websites. Good design of websites for mass customization is crucial in capturing and sustaining online customers. This helps in identifying design features that should be avoided in future development of such websites. Usability is measured in terms of user satisfaction and site effectiveness in supporting design activity. The ANOVA was performed on the usability data. The results showed significant effects of website features and e-product catalogue on user satisfaction. Details of the study are reported in the chapter by Oon and Khalid in this book.

13.5 Discussion

Web-based DIYD is a new phenomenon, and there is limited experience in how to design an attractive web environment, which makes it easy to mass customize. We showed in the study that a hierarchical design structure is preferred. In this case contextual design elements - type of watch, shape, face - are selected prior to detailed design elements - dial, numerals and so forth. The subjects anticipated this approach. The study showed a greater enthusiasm amongst Malaysians than Hong Kong subjects. The interest of the former may have been affected by the Multimedia Super Corridor project – a highly visible government initiative to help catalyze national development in the information technology field [55].

13.5.1 Customer needs

Figure 1 provides a flow model of mass customization. Assessments of customer needs are used to derive functional requirements and design parameters. These distinctions have the effect of formalizing product design, which according to Finke, Ward and Smith is beneficial for design creativity [56]. In particular, by focusing on the higher levels of abstraction, that is, goal and purpose of the product, designers arrive at more creative design solutions. Only after a thorough exploration of the design objectives can one then seek solutions. Figure 2 illustrates this approach for design of watches.

Customer needs are difficult to assess, particularly for new products, since there is not yet any feedback from marketing. With mass-customization the assessment of customer needs is further complicated since we need to understand not only "average needs" but also the "variability" in needs. Measures of variability are then the basis for selecting what options should be included in mass customiza-tion. On the other hand mass customization provides a built-in flexibility in design and manufacturing, which may compensate for the lack of precision in estimating customer needs. It is not necessary to get the estimates 100% right, since the customization process could steer sales in the appropriate direction.

Using the internet and one-to-one marketing, companies can elicit information about individual customers. Previously such information could be collected only through a direct sales force, often at high costs. With online data gathering, cost has fallen sharply. Some customer needs may be predicted by incorporating research findings from human factors and consumer psychology. The chapter discussed several findings concerning "good taste" in product design. These can be used as guideline to design options for mass customization. For example: a new design of an old product should not look different from the old. Moderate changes are usually preferred over radical changes.

Since products are used for a variety of purposes in a variety of tasks, it is helpful to establish a list of product usage - so called *product semantics.* Several potential uses of a watch were suggested in Figure 3, and many of these were verified in our study, see Table 1. By exploring product semantics, or the natural usage of products, one can expand the market for single products considerably. In our case we could suggest many creative uses of a watch – not only for keeping track of time. But customer needs go beyond product features. A future e-business should try to build customer loyalty through relationships. The relationship is not only with the company but also with fellow customers [47]. Customers like to help other customers, and real value is added by enabling customers to interact with one another [57]. Therefore facilities to contact other consumers should be available in the interface.

13.5.2 Web design options

There are several limitations in the use of mass customization. First there are limits set by the options displayed by the manufacturer. Customers will therefore trade-off their requirements considering available options and product price and decide if the opportunities for DIYD are satisfactory. The design of an attractive Web environment is crucial. Reasonable options must be presented – in an order that allows logical and user-friendly procedures for assembly of the various design elements. There is also a limit due to lack of user knowledge of the product to be designed. Detailed design with many design elements will be difficult since they require specific know-how. At least the cursory customer will not be interested in spending much time on activities with unknown outcomes. We may therefore be limited to fairly simple design, which may not fully realize customer needs.

13.5.3 Research directions

Although Web interfaces are an established part of the HCI repertoire, there is still a lack of theoretical framework to support design of such interfaces. Given the cognitive complexity of Web interaction, it is unlikely that craft style approaches will produce effective interfaces. Through a description and analysis of DIY non-expert customer behavior, it may be possible to design Perceptual User Interfaces that integrate perceptive, multimodal and multimedia. This is already an emerging area of research [58].

Virtual collaborative design: A number of issues have emerged from recent developments in Web-based collaborative design. DIYD requires a link between the customer and manufacturer who may design the product collaboratively in real-time [59]. The core element of a typical collaborative system is a shared space, through which customers, manufacturers and suppliers coordinate their cooperative design activities. Intelligent engines and agents, which reside in this shared space, support the activities of the participants, including detecting design changes and scheduling meetings for synchronous problem solving sessions. In this collaborative design system there are several human factors issues. For example,

- how to format and present information that enable designers and/or customers to utilize it efficiently via verbal, graphical and written communication;

- how to provide a preset procedure to support decision making in team design. Distinctions between "Functional Requirements", "Design Parameters" and "Process Variables" will structure and clarify the communication. In addition, design problem solving may be organized top-down, using abstraction hierarchies, such as in IDEF0, see Klein [60]. This procedure will also improve creativity in design;

- how to be intuitive to use by the DIY designers and/or customers. Usability of DIYD system is an important issue besides cultural and individual differences that may impede communication and collaboration.

Kansei engineering: Research in Kansei Engineering has the purpose to uncover customer's needs, by measuring customer affect or feelings (or *Kansei*), as they are exposed to different product design features. It is a 30 years old methodology for translating the customer's psychological feeling about product features into the design domain [9, 61, 62]. Customer needs may be documented by combining Kansei engineering with methods from ethnographic studies. The results can then be evaluated using statistical tools such as Hayashi's Quantification Theory Type 1, factor analysis and cluster analysis.

Unobtrusive research: Customer behavior on the web offers interesting opportunities for unobtrusive research. Presently event-recording software is used in marketing. This can be set up to record user behavior as an e-commerce user enters a web store. Reaction times or dwell times may be recorded as well as errors in navigation. Given that there are opportunities to record the behavior of many users who enter a web site, these types of analyzes can be extremely powerful. Event recordings can be used as unobstrusive, and very realistic reaction time experiments. The design of web pages may be altered (according to some theory), and web users' reaction times evaluated. It would also be possible to evaluate customers search behavior and quantify the trade-off function. There is, for example, a potential to study customers learning behaviour. Johnson, Lohse, and Mandel showed that there is a strong learning effect of visiting an e-store [63]. Twenty visits to a facility will half the search time (from about 400 seconds to 200 seconds). Future studies could address what types of store environments and web design features reduce user reaction time and are easy to learn.

13.5.4 Conclusion

In mass customization, it is up to the customers to formulate functional requirements and to choose among available design parameters to satisfy their needs. It is therefore essential that design elements chosen by the manufacturer are relevant to customers and that they are easy to select. Poor usability will discourage many customers, and web pages must therefore be designed so that navigation and selection of design elements are easy.

Mass customization has many of the required ingredients to make it a successful enterprise strategy: compelling content, interactivity, personalized experience, and customer control over the design [57, 64]. As mass customization becomes more important in the third millennium, design of e-commerce web sites have to meet the requirements of the customer, the manufacturer and the suppliers in a dynamic global customize-supply-manufacture chain. This provides many opportunities for interdisciplinary research from related fields of psychology, communications, management, economics, engineering and human factors.

References

[1]   Helander, M.; Khalid, H.M.; Tseng, M.M.: Mapping customer needs in Web-based DIY design of consumer products, in: Tseng, M. M. (Ed.): Applications and Practice, Proceedings of the 3rd International Conference on Industrial Engineering Theories, Hong Kong 1998.

[2]   Tseng, M.M.; Du, X.: Design by customers for mass customization products, in: CIRP Annals, 48 (1998) 1, pp. 103-106.

[3]   Gilmore, J. H., Pine II, B. J.: The four faces of mass customization, in: Gilmore, J. H.; Pine II, B. J. (Eds.): Markets of One, Boston 2000, pp. 115-132.

[4]   Schonfeld, E.: The customized, digitized, have-it-you-way economy, in: Furtune, September 28th 1998, pp. 115-124.

[5]   Wind, J.; Rangaswamy, A.: Customerization: the second revolution in mass customization, in: eBusiness Research Center Working Paper, Penn State Smeal College of Business Administration, 1999.

[6]   The Economist: The car industry, May 6th 2000, p. 65.

[7]   Helander, M. G.; Li, L.: Optimal sequence in product design, in: Proceedings of the 42nd Annual Meeting of Human Factors and Ergonomics Society, Santa Monica 1998.

[8]   Suh, N.: Principles of Design, Oxford 1990.

[9]   Nishino, T.; Nagamachi, M.; Ishihara, S.; Ishihara, K.; Ichitsubo, M.; Komatsu, K.:Intelligent linkage between Internet Web browser and Kansei database, in: Karwowski, W.; Goonetilleke, R. (Eds.): Proceedings of HAAMAHA '98, Manufactruing Agility and Hybrid Automation-II, Hong Kong 1998, pp. 55-58.

[10]   Bailetti, A. J.; Litva, P. F.: Integrating customer requirements into product designs, in: Journal of Product Innovation Management, 12 (1995), pp. 3-15.

[11]   Khalid, H M.: 2001, Can customer needs express affective design, in: Helander, M. G.; Khalid, H. M.; Tham, M. P. (eds.): Proceedings of International Conference on Affective Human Factors Design, London 2001, pp. 190-198.

[12]   Crozier, R.: Manufactured Pleasure: Psychological Responses to Design, Manchester 1994.

[13]   Blythe, J.: The Essence of Consumer Behavior. New York 1997.

[14]   Cheng, S.: Globalize your Web site, in: E-Business Advisor, October 1999, pp. 22-26.

[15]   Maheswaran, D.; Shavitt, S.: Issues and new directions in Global Consumer Psychology, in: Journal of Consumer Psychology, 9 (2000), 59-66.

[16]   Hofstede, G.: Culture's consequences: International differences in work-related values, Beverly Hills 1980.

[17]   McManus, I. C.: The aesthetics of color, in: Perception, 10 (1981), pp. 651-666.

[18]   Oon, Y. B.; Khalid, H. M.: Usability evaluation of design by customers websites for mass customization, in: Proceedings of 2001 World Congress on Mass Customization and Personalization, Hong Kong 2001.

[19]   Cross, N.: Discovering design ability, in: Buchanan, R.; Margolin, V. (Eds): Discovering Design, Chicago 1995.

[20]   Middaugh, A.: Squash Blossom, on www.squash-blossom.com/design/design.html.

[21]   Khalid, H. M.: The ability of selected personality variables to discriminate and predict brand preference, in: Journal of Man and Society, 6 (1985), pp. 104-115.

[22]   Abelson, R. P.: Beliefs are like possessions, in: Journal for the Theory of Social Behavior, 16 (1986), pp. 223-250.

[23]   Dilnot, C.: The gift, in: Margolin, V.; Buchanan, R. (Eds.): The Idea of Design, Cambridge 1997.

[24]   Jarvenpaa, S. I.; Todd, P. A.: Customer reactions to electronic shopping on the World Wide Web, in: International Journal of Electronic Commerce, 1 (1997) 2, pp. 59-88.

[25]   Keeney, R. L.: The value of Internet commerce to the customer, in: Management Science, 45 (1999) 4, pp. 533-542.

[26]   Lohse, G. L.; Spiller, P.: Quantifying the effect of user interface design features on cyberstore traffic and sales, in: Proceedings of CHI, New York 1998, pp. 211-218

[27]   Kahn, B.; Huffman, C.: Variety for sale: Mass customization or mass confusion?, in: Journal of Retailing, 74 (1998) 4, pp. 491-513.

[28]   Ulrich, K.; Randall, T.; Fischer, M.; Reibstein, D.: Managing product variety: a study of the bicycle industry, in: Ho, T.; Tang, C. (Eds.): Product Variety Management: Research Advances, New York 1998.

[29]   Coombs, C.; Avrunin, G. S.: Single peaked preference functions and theory of preference, in: Psychological Review, 84920 (1977), pp. 216-230.

[30]   Zee, A.: Fearful Symmetry: The Search for Beauty in Modern Physics, New York 1986.

[31]   Locher, P.; Smets, G.; Overbeeke, K.: The contribution of stimulus attributes, viewer expertise and task requirements on visual and haptic perception of balance, in: Proceedings of the 12th International Congress of the International Association for Empirical Esthetics, Berlin 1992.

[32]   Berlyne, D. E.: Aesthetics and Psychobiology, New York 1971.

[33]   Martindale, C.; Uemura, A.: Stylistic change in European music, in: Leonardo, 16 (1983), pp. 225-228.

[34]   Teigen, K. H.: Intrinsic interest and the novelty-familiarity interaction, in: Scandinavian Journal of Psychology, 28 (1987), pp. 199-210.

[35]   Purcell, A. T.: Environmental perception and affect: a schema discrepancy model, in: Environment and Behavior, 18 (1986), pp. 3-30.

[36]   Krippendorff, K.: On the essential contexts of artifacts or on the proposition that "Design is Making Sense (of things)", in: Margolin V.; Buchanan, R. (Eds.), The Idea of Design, Cambridge 1995, pp. 156-184

[37]   Graumann, C. F.: Psychology and the world of things, in: Journal of Phenomenological Psychology, 4 (1974), pp. 389-404.

[38]   Bih, H. D.: The meaning of objects in environmental transition: experiences of Chinese students in United States, in: Journal of Environmental Psychology, 12 1992, pp. 135-147.

[39]   Csikszentmihalyi, M.; Rochberg-Halton, E.: The Meaning of Things: Domestic Symbols and the Self, Cambridge 1981.

[40]   Dittmar, H.: The Social Psychology of Material Possessions. Hemel Hempstead 1992.

[41]   Margolin, V.; Buchanan, R.: The Idea of Design, Cambridge 1995.

[42]   Khalid, H. M.: Uncovering customer needs in DIY product design, Proceedings of QERGO International Conference on TQM and Human Factors, Linköping 1999, pp. 343-348.

[43]   Grose, E., Forsythe, C., Ratner, J.: Using Web and traditional style guides to design Web interfaces, in: Forsythe, C.; Grose, E.; Ratner, J. (Eds.): Human Factors and Web Development, Mahwah 1998, pp. 121-136.

[44]   Wiebe, E. N., Howe, J. E.: Graphics design on the Web, in: Forsythe, C.; Grose, E.; Ratner, J. (Eds.): Human Factors and Web Development, Mahwah 1998, pp. 225-239.

[45]   Nel, D., van Niekerk, R.; Berthon, J. P.; Davies, T.: Going with the flow: Web sites and customer involvement, in: Internet Research: Electronic Networking Applications and Policy, 9 (1999) 2, pp. 109-116.

[46]   Csikzentmihalyi, M.: Flow: The Psychology of Optimal Experience, New York 1990.

[47]   Vora, P.: Human factors of designing usable Web sites: E-commerce and its Impact on Future HCI Work, in: Proceedings of IEA/HFES 2000 Congress, Santa Monica 2000.

[48]   Nielsen, J.: A web based collection of papers on Web usability, on www.useit.com/papers/.

[49]   Nielsen, J.: Designing Web Usability, Indianapolis 2000.

[50]   Helander, M. G.; Khalid, H. M.: Modeling the customer in electronic commerce, in: Applied Ergonomics, 31 (2000), pp. 609-619.

[52]   Tilson, R.; Dong, J.; Martin, S.; Kieche, E.: Factors and Principles Affecting the Usability of Four E-Commerce Sites, in: Proceedings of Human Factors and the Web, on www.research.microsoft.com/users/marycz/hfweb98/tilson/index.htm

[52]   Hubka, V.; Eder, W. E.: Design Science: Introduction to the Needs, Scope and Organization of Engineering Design Knowledge, New York 1995.

[53]   Guindon, R.: Designing the Design Process: Exploiting Opportunistic Thoughts, in: Human-Computer Interaction, 5 (1990), pp. 305-344.

[54]   Oon, Y. B.; Khalid, H. M.: How to do it yourself? – Evaluation of configurator systems in Design by Customer websites, in: Proceedings of the 3rd Malaysian Ergonomics Conference, Johor Bahru 2002.

[55]   Khalid, H. M.: Human factors of IT-based solutions for Worldwide Manufacturing Web, in: Human Factors and Ergonomics in Manufacturing, 10 (2000) 1, pp. 99-113.

[56]   Finke, R. A., Ward, T. B.; Smith, S. M.: Creative Cognition, Cambridge 1992.

[57]   Windham, L.: Creating compelling content, in: E-Business Advisor, 18 (1999) 2, pp. 24-29.

[58]   Turk, M.; Robertson, G.: Perceptual user interfaces, in: Communications of the ACM, 43 (2000) 3, pp. 33-34.

[59]   Khalid, H. M.: Human factors of virtual collaboration in product design, in: Lim, K. Y. (Ed.): Proceedings of APCHI/ASEAN Ergonomics 2000 Conference, Amsterdam 2000, pp. 25-38.

[60]   Klein, G.: Sources of power: How people make decisions, Cambridge 1998,

[61]   Nagamachi, M.: Kansei Engineering; a powerful ergonomic technology for product development, in: Helander, M. G.; Khalid, H. M.; Tham, M. P. (Eds.): Proceedings of the International Conference on Affective Human Factors Design, London 2001, pp. 9-14.

[62]   Tsuchiya, T.; Matsubara, Y.; Ishihara, S., Nagamachi, M.: Detecting non-linear relations between variables of Kansei evaluation data, in: Proceedings of the IEA 2000/HFES 2000 Congress, San Diego, 6 (2000), pp. 376-379.

[63]   Johnson, E. J.; Lohse, G. L.; Mandel, N.: Designing Marketplaces of the Artificial: Four Approaches to Understanding Consumer Behavior in Electronic Environments, Working Paper, Colombia School of Business, New York 1999.

[64]   The Economist: Mass customization: A long march, July 14th 2001.

Contacts:

Professor Halimahtun M. Khalid
Institute of Design and Ergonomics Application,
Universiti Malaysia Sarawak, Malaysia
E-mail: mkmahtun@idea.unimas.my

Professor Martin G. Helander
School of Mechanical and Production Engineering,
Nanyang Technological University, Singapore
E-mail: mahel@ntu.edu.sg

14 Modeling Consumer Behavior in the Customization Process

Sri Hartati Kurniawan, Mitchell M. Tseng and Richard H. Y. So
Department of Industrial Engineering and Engineering Management
The Hong Kong University of Science and Technology, Hong Kong

The essence of being customer centric is to provide only and exactly what each customer wants at the right time. The process to reach this objective is the configuration process. During configuration a consumer can choose different components and assemble them together to a specific product. With this new way of acquiring products, it is predicted that consumer behavior will change as well. This chapter presents a new approach to understanding consumer behavior using the living system theory. The paper follows the living system theory to explain consumer behavior by specifying its components, relations, and organization. The (traditional) marketing models rooted in psychological research have made many assumptions about human behavior, for example, the human brain is assumed to be similar to the computer brain and the processing of information is assumed to be symbolic processing. Living system theory is seen as an alternative solution. It has the capability of explaining customer behavior as a collection of components and their organization. Thus, it may become possible to emulate the consumer buying process by explaining this process as a collection of components and their organization. We will discuss this approach for the configuration process of a mass customization system. A case study is presented in order to illustrate the concept.

14.1 Introduction

14.1.1 Consumer behavior and choice

Today customers have their rights to vote and quote for their preferences in a product, mainly because of the increase in market competition. This forces the product providers to react by enhancing their capabilities to serve individual customers and to suit each individual need in order to retain customers. Many companies react by providing product customization to meet such needs. Following this, companies come up with the concept of doing customization in a more economical way, namely mass customization. Mass customization aims at providing goods and services that best meet individual customers' requirements with near mass production efficiency [1].

The essence of customization is to provide only and exactly what each customer wants at the right time [2]. Von Hippel [3] stressed the importance of the customer's involvement in designing products in mass customization, since the customer has the very best understanding of his / her own needs, and can relay the information to the manufacturer. Peppers and Rogers [4] also emphasize the understanding and categorization of customers as necessity requires in order for the product / service providers to be able to customize their offerings. The key issues are the need to fulfill what customers want and provide a method for the customers to interact with the product provider in a better way. One possible solution is to better understand the customers such that the product or service provider will be able to provide the right product at the right place and the right time to avoid customers sacrifice and maintain high customers' satisfaction. There is a growing interest in uncovering consumer behavior in order to develop better support for the business model of customization and personalization.

The traditional mass-producer follows a take-it-or-leave-it approach towards the consumer. However, customized product providers offer unique products to suit individual consumer preferences. The most widely known way to customize a product is by configuration process. In this process, the consumer can choose different components and assemble them together to form a product. With this new way of acquiring products, it is predicted that consumer behavior will change as well.

Research on understanding the customer, known as consumer behavior study, has been the subject of investigation in the marketing and psychology area for decades. However, research on traditional consumer behavior was geared toward explaining how consumers behave with the intention to predict aggregate trends. In the case of customization and personalization, the focus of modeling has shifted from passive to proactive. Now, there is a need to understand and model individual consumers as well as provide guidance for them, in order to determine exactly what they want. The traditional models of consumer behavior are no longer suitable for modeling the consumer while customizing products. Therefore, the way of modeling consumer behavior should be changed to suit the customization scenario.

14.1.2   Research on consumer behavior in product customization

Research on consumer behavior during customization is scarce. Moreover, there is no formal model specified yet due to the complexity of modeling the consumer in the buying process. Marketing research has summarized three different aspects that make the modeling of consumer behavior difficult- the product aspect, the consumer aspect, and the environment aspect. The product aspect mainly refers to the context-dependent choices; there are many situational aspects that could change product preferences over time or in different situations [5, 6]. Consumer aspect refers to different consumer characteristics which could change preferences, for example mood, emotion or impulsive feeling [7]. Environmental aspect refers to the aspects outside the consumer, such as influence of family or social environment. [8]. When the three aspects are applied to customization and personalization, the recording of consumer behavior becomes complicated since the modeler has to be able to capture individual data for each consumer.

In answering the challenge to model consumer behavior in customization and personalization, there are two streams of disciplines receiving a great deal of attention these days. The first one involves capturing consumer preference in customization and personalization through data mining and profiling methods (e.g. [9]). Data mining and profiling enable the researcher to collect demographic and consumer preference data, and extract consumer behavior from their product preferences. With the support of advanced Web technology, the product provider is able to collect a large sample of consumer data.

Another way to understand consumer behavior in customization and personalization is by extending marketing theories to this area. This effort is mainly done using empirical research. Huffman and Kahn investigated consumer behavior with a variety of selection tasks. They found out that the process could be enjoyable when consumers were informed about product variety and were able to learn while gathering information. However, it could also be frustrating when consumers were confused by the number of product variants [10].

The first type of research is interested in the results of customization and personalization. The data are extracted from consumer preferences and the consumer demographic data. The second one is interested in the process of customization and personalization, such as the decision-making process when customizing products. It would be an ideal case if the modeler could account for these two measurements at the same time. Thus, we take up the challenge of proposing a new alternative to investigate consumer behavior in customizing products. The new approach should take the process and results of customization into account and integrate these two aspects to create a more complete model.

The objective of this chapter is to provide an alternative way to model consumer behavior using system theory; in particular living (biological) system theory. The marketing models rooted in psychological research have made many assumptions about human behavior, for example, the human brain is assumed to be similar to the computer brain, and the processing of information is assumed to be symbolic processing. Living system theory is seen as an alternative solution. It has the capability of explaining the system as a collection of components and their organization; therefore it is possible to emulate the consumer buying process by explaining this process as a collection of components and their organization. Advances in neuroscience have enabled human behavior to be explained by using a lower level of explanation, which is the mechanism of neurons and their relationships. Therefore, it will be more appropriate to explain human behavior by revising the assumptions and proposing a new methodology.

14.2 Literature review

14.2.1 General system theory

The General System Theory was first coined by von Bertalanffy in 1968, and has been widely applied since then. Von Bertalanffy's minimal definition of a system is still accepted by most researchers in System Theory. He defined a system as elements – relations – wholes [11]. A system consists of elements, but elements will behave differently when observed in isolation compared to elements in their relationships with other elements. Therefore investigation of a system has to take into consideration elements as well as their relationships. General system theory is a theory of wholeness, which means that the whole is more than the sum of its parts. By investigating elements in their relationships with other elements, there will be distinct characteristics and phenomena emerging from the system which form wholeness with the systems' elements and the relationships between them.

14.2.2 Living system

The investigation of living systems is also in line with the General System Theory. Living system theory, too, recognizes the system as a collection of interrelated elements which form a whole. In the living system theory, the elements, relations, and wholes are expressed with different notions, which are stated in Table 1. Hutardo and Landeck present arguments that applying Living System Theory in consumer behavior research could explain many different phenomena that previously had been taken as assumptions [12]. The theory of living systems explains the living system as a cognitive system, which is an attempt to explain how cognition can be explained by biological phenomena. Cognition is now being seen as actual behavior realized through organization of components in their domain of interactions in order to maintain the existence of the organization. Cognition is interacting, which means that it is a behavior of a living system. Through structural coupling, cognition interacts with its environment to find a viable, not necessarily optimal, way to continue the process of maintaining its identity.

From a biological point of view, cognition is seen as an embodied action. It means that cognition depends upon the kinds of experiences that come from having a body with various sensorimotor capacities and embedded in a more encompassing biological, psychological, and cultural context [15]. So, in order to study a system, one must take the world as it is, as perceived by different individuals, and observe regularities showed up from the enactment of sensorimotor relationship. From the observers' point of view, there will be a coherent pattern emerging, which can describe the world [16, 17]. In Living System Theory, the explanation of behavior can be represented in relations between components. The relations between components strengthen when the behavior is performed repeatedly. A person is capable of modifying and elaborating on a collection of components to construct new behavior [18].

Table 1: Components, relations, and wholes of living system

General System Theory	Living System
Elements	Components: any subunit participating in the network of relations and processes, which define a living system in the space in which its organization is manifested [13]
Relations	Domain of interactions: the set of all interactions into which an entity can enter [14, p.8].
Wholes	Organization: the relations that define a machine as a unity, and determine the dynamics of interactions and transformations which it may undergo as such a unity, constitute the organization of the machine [14, p.77].

14.3 Modeling consumer behavior using living system theory

14.3.1 Assumptions of the model

In modeling consumer behavior using living systems, there are certain assumptions made in order to simplify the model. Assumptions regarding the consumer behavior modeling are as follows:

- Consumers can choose to select components from product family architecture to be assembled into a product (mass customization) or may prefer to modify components (customization). The customization process is with respect to the company capability.

- Consumer behavior is composed of a series of activities. These activities are defined as the processing unit, which is the smallest undividable unit of activities. These activities are called *Basic Cognitive Activities*.

- In the beginning, there are no pre-specified relationships between basic cognitive activities. The relationships between basic cognitive activities are formed gradually when the consumers perform activities. When the same activities are performed repeatedly, stronger relationships between basic cognitive activities will accumulate and become known as experience.

Assumptions regarding the relationship between consumer and product provider are:

- No impulse or sales incentives buying in the purchasing situation.

- Product provider understands the general consumers' needs at this moment, and presents the range of product components that can satisfy general consumers. This study focuses on the structure and mechanism of consumers' behavior in customizing products.

- The understanding of the relationship between consumers and product provider focuses on the understanding of how consumers take information from product provider, select, or choose to modify product components, as long as the product provider has the capability to produce the modified components (e.g. modify length or width).

- The information influencing consumers is regarded as perturbation to the system.

14.3.2 Components of the model

(1) Basic cognitive activities are the elements or components of the system. It describes the smallest processing unit of activities performed by consumers when performing a customization process. There are two distinct groups of basic cognitive activities included here. The first group is basic cognitive activities in product component selection. This process is regarded as the customization process in which a consumer selects different product components from available variants, to be assembled together to form a complete product. The second group is basic cognitive activities in product configuration process.

The basic cognitive activities of product selection are adapted from Bettman, Johnson, and Payne [19]. In their research of consumer decision-making process, they defined Elementary Information Processes (EIPs), which is actually a way to decompose the decision-making process into smaller elements of activity. The basic cognitive activities defined are:

- *Read*: read a value of information - normally 'information' here refers to the attribute value of a product.

- *Compare*: compare two or more attribute values.

- *Difference*: difference of two attribute values.

- *Add*: addition of two attribute values.

- *Product*: put a weight into an attribute value by multiplying weight and the attribute value.

- *Eliminate*: eliminate an alternative / attribute from consideration.

- *Move*: go to next information.

- *Select*: announce preference.

The basic cognitive activities of component modification for customization are adapted from Du [20]. Du explained the product variety generation as a function of three activities: attaching, swapping, and scaling. The authors extend the concept into customization as a function of these activities by further decomposing swapping activities into attaching and detaching:

- *Attach*: attach two components together.

- *Detach*: detach one component from another component(s).

- *Scaling*: change the dimension of the component.

At this stage of modeling, the activities in the customization process are limited to the selection of available variants to be assembled together, attaching components together, swapping, and scaling.

(2) Perturbation: A living system is a closed system. There is no input or output to the environment, but the system can be perturbed by outside events. With this notion, the components of a cognitive system are actually able to regenerate themselves and maintain the identity of the system. The outside events perturb the system causing the components of the system to interact with each other. In the modeling of consumer behavior, product information and preference of the product components and attributes are regarded as perturbation. The perturbation, which is product information or product preference, will cause the consumer to respond by performing a basic cognitive activity. This activity will trigger the activation of another basic cognitive activity, and so on.

(3) Behavior as collection of activities: Behavior can be regarded as a series of basic cognitive activities performed sequentially. When a series of basic cognitive activities is performed, it will form a pattern of behavior. For example, in product selection, if a consumer always employs elimination starting from most to least important attribute, it is known that the consumer employs Elimination-by-Aspects decision-making strategy [21].

14.3.3 Mechanism of the model

The mechanism of the model can be explained by the concept of behavior-based artificial intelligence [22, 23]. The system to be modeled can be seen as collection of activities that compete with each other. An activity can be executed if it fulfills certain conditions. Consumers can be seen as such a system, with the basic cognitive activities as the elements of the system. The relationship between basic cognitive activities is a predecessor – successor relationship. At the initialization stage, there are no relations specified. These relationships are formed once a basic cognitive activity is performed, followed by another basic cognitive activity. This will form a predecessor – successor relationship as illustrated in Figure 1.

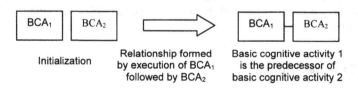

Figure 1: Predecessor-successor relationship between components

When a consumer performs a basic cognitive activity followed by another basic cognitive activity, the sequencing will describe the relations between them. It will serve as descriptive tool of the model, describing how the consumer behaves based on what (s)he has done in the past. However, there is a need for the model to have predictive powers. The prediction of consumer behavior will be explained by the activation of basic cognitive activities.

Every basic cognitive activity has an energy level at a particular time t. The energy is propagated from perturbation to a basic cognitive activity and from one basic cognitive activity to another basic cognitive activity. If the energy level in a basic cognitive activity has passed the threshold, then the basic cognitive activity is activated. The activated basic cognitive activity then passes its energy to other basic cognitive activities, and so on.

The energy activation rules are adapted from Maes [22]:

- Activation by perturbation: the outside perturbation will trigger the components of the system to react with it, for example: product information will trigger the activation of basic cognitive activity 'read'.

- Activation of successors: when a basic cognitive activity is activated, it will transfer its activation energy to its successors.

- Activation of predecessors: if a basic cognitive activity is not activated, it will transfer its activation energy to its predecessors in order to push itself to be activated. When a predecessor is finally activated, it will activate its successor eventually.

By activation rules specified above, at each particular time t, there are activities competing with each other to be activated, depending on their activation energy levels.

14.4   Case study

To explain consumer behavior for product customization, a study case will be presented. In order to simplify the explanation, the case study will be limited to the component selection only. In other words, the case study does not include the process of designing components by a consumer. In the subsequent sections, the task will be illustrated, then the cognitive activities related to the task will be presented, and the mechanism of basic cognitive activities activation will be further explained.

14.4.1   Task description

Suppose a consumer is customizing a shirt. The supplier provides different variants, representing different levels of attribute values. E.g., for the attribute 'pattern' three variants exist: 'vertical stripes', 'horizontal stripes', and 'no pattern'. The consumer also has preferences representing constraints or attribute values that (s)he likes, such as no black color wanted. The variants and preferences are stated in the Table 2; with the values marked with bold representing the values to be selected by the consumer. The task would be to select from the different variants available by comparing the variants with the preferences. The consumer would select variants that match with his / her preferences, and if there are more than one variant matching his / her preferences, then the consumer would further compare them and select one variant.

Table 2: Case example of component selection for customizing a shirt

	Pattern (P)	*Material (M)*	*Color (C)*
A	Vertical strips	**Cotton**	**Blue**
B	**Horizontal stripes**	Rayon	White
C	No pattern	Silk	Black
Preference	Want some pattern	No pref.	No black

14.4.2 Basic cognitive activities involved in variant selection

The sequence of the selection process can be observed as a series of basic cognitive activities performed one after another to form a behavior. In relation to the variant selection task presented in the last section, the sequence of basic cognitive activities when customizing a shirt is illustrated in Table 3. In this particular case, the consumer compared pattern variants, and decided to keep pattern A and B, and eliminate pattern C. Then the consumer compared pattern A and B, and decided to select pattern B. The consumer then compared three materials and decided to select material A. For color selection, the consumer compared color variants, decided to keep color A and B while eliminating color C, and after the second comparison, decided to select color A. The detailed step-by-step sequence is described in Table 3 by reading the first column downwards, shifting to the second column, reading it downwards, shifting to the third column, and reading it downwards.

The sequence of basic cognitive activities can be described in a graphical representation. The basic cognitive activities are represented in boxes. The arrows with numbers represent the sequence of basic cognitive activities movement. Movement here means that a basic cognitive activity is activated after another basic cognitive activity is activated, meaning that the consumer performs a particular basic cognitive activity. For example: the arrow numbered 1 from basic cognitive activity 'read' to basic cognitive activity 'move' means that the consumer reads a piece of information (in this case study, the consumer reads the information on pattern preference) then moves / shifts his / her attention to another piece of information. When the sequence from Table 3 is translated into graphical representation, the result is shown in Figure 2. The frequency of the movement can be translated into the weight of relationship between two basic cognitive activities. The higher the number of interactions between two basic cognitive activities, the higher the weight will be. In this case study, the consumer activated basic cognitive activity 'read' after 'select' three times, resulting in a weight equal to three. The translations of movements into weights are illustrated in Figure 3.

Table 3: Sequence of basic cognitive activities in customizing a shirt

Basic cognitive Activities (read column wise)		
Read Pattern Preference (PT)	Select PB	Read Color A (CA)
Move	Move	Compare CT and CA
Read Pattern A (PA)	Read Material A (MA)	Select CA
Compare PT and PA	Select MA	Move
Select PA	Move	Read Color B (CB)
Move	Read Material B (MB)	Compare CT and CB
Read Pattern B (PB)	Select MB	Select CB
Compare PT and PB	Move	Move
Select PB	Read Material C (MC)	Read Color C (CC)
Move	Select MC	Compare CT and CC
Read Pattern C (PC)	Move	Eliminate CC
Compare PT and PC	Compare MA, MB, and MC	Move
Eliminate PC	Select MA	Compare CA and CB
Move	Read Color Preference (CT)	Select CA
Compare PA and PB	Move	

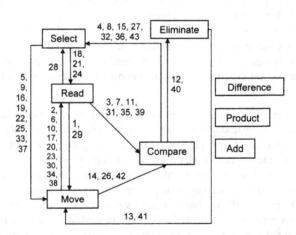

Figure 2: Graphical representation of basic cognitive activities sequence

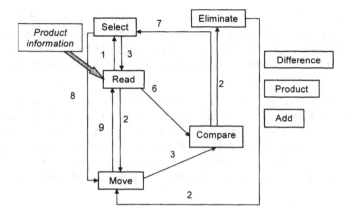

Figure 3: Relation weights in customizing a shirt (the value of the weight is the frequency of movement as illustrated in Figure 2) and activation by perturbation

14.4.3 Activation of individual basic cognitive activities

The two previous sections described how consumers behave when customizing a product, particularly in relation to the selection of available variants. The next step is be to predict the consumer behavior given the data of the movement of basic cognitive activities and their corresponding weights. The prediction of behavior will be explained by the activation of individual basic cognitive activities. This part will explain the activation of basic cognitive activities, given certain conditions and driven factors. There are three different mechanisms to explain activation of basic cognitive activities which will be explained in the following three sections.

Activation by perturbation: Product information can be seen as perturbation to the system. It will trigger an activation of basic cognitive activity to react with the perturbation. A table is used to record the energy levels of each basic cognitive activity. The values in the table will be updated once there is a change in the energy level value. In our case study, the product information will bring energy to be transferred to basic cognitive activity 'read'. When the energy level of 'read' exceeds the threshold (in this case, assume all thresholds = 100), the basic cognitive activity 'read' will be activated. Once activated, the energy level will be reset to zero. The activation by perturbation is described in Figure 3 (above) by the arrow "product information", the corresponding values are given in Figure 4.

Activity	Energy level		Activity	Energy level		Activity	Energy level
Read	70	Information Energy = 50	Read	120	Activate 'read'	Read	0
Compare	45		Compare	45		Compare	45
...	...		...	...		..	...

Figure 4: Energy transfer in the activation by perturbation recorded in table form

Activation of a successor: When the basic cognitive activity 'read' is activated (indicated by highlighted box), it will transfer its energy to its successors. In this case, the successors will receive a portion of energy belonging to basic cognitive activity 'read'. The amount of energy received depends on the weight of relations between 'read' and its successors.

If the energy level of a basic cognitive activity increases and passes the threshold, then it will be activated. In this case, the basic cognitive activity 'read' will transfer its energy to basic cognitive activities 'select', 'move' and 'compare' with the amount of energy transferred proportionate to the weight of relations between them. As a result, basic cognitive activity 'compare' is activated. Once activated, its energy level is preset to zero. The activation process is illustrated in Figures 5 and 6.

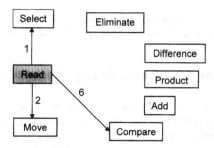

Figure 5: Activation of a successor

Activity	Energy Level		Activity	Energy Level		Activity	Energy Level
Compare	30	Energy transfer from 'read'	Compare	30+(6/9)*120=110	Activate 'compare'	Compare	0
Select	35		Select	35+(1/9)*120=48		Select	48
Move	40		Move	40+(2/9)*120=67		Move	67

Figure 6: Energy transfer in the activation of a successor recorded in table form

Activation of a predecessor: Basic cognitive activities that are not activated will spread their energy backwards. The energy spread to their predecessors will increase the predecessors' energy level or even activate them. This mechanism explains the notion of competing activity in the living system.

In this case, the basic cognitive activity 'compare' is activated (denoted by highlighted box); therefore basic cognitive activity 'compare' will not spread its activation backward. But other basic cognitive activities that are not activated will spread their activation energies backward. As the result, basic cognitive activity 'move' will be activated. The activation process of predecessors is described in Figures 7 and 8.

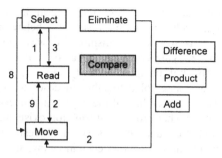

Figure 7: Activation of a predecessor

Not activated	Predecessor	Weight	Energy transfer	Energy before transfer	Energy after transfer
Select	Read	1	4	0	4
Read	Move	9	40	67	**107**
Read	Select	3	13	48	61
Move	Select	8	35	61	96
Move	Eliminate	2	9	25	34
Move	Read	2	9	4	13

Activity	Energy after transfer
Read	13
Move	**107**
Select	96
Eliminate	34

Activate 'move'

Activity	Energy after transfer
Read	13
Move	0
Select	96
Eliminate	34

Figure 8: Energy transfer in the activation of predecessor recorded in table form

14.4.4 Recording and prediction of the preferences

In predicting consumer's preferences, the model takes previous experiences into account. The values of the basic cognitive activity 'select' need to be recorded for the purpose of prediction. In the case study presented above (Table 2) the variants 'vertical stripes', 'silk', 'rayon', and 'white' were the selected ones, while the variants 'horizontal stripes', 'cotton', and 'blue' were selected twice, Consumer's preferences were 'horizontal stripes', 'cotton', and 'blue'. The more a consumer performs a selection process, the stronger the historical data record will become, and the higher the probability of being able to predict the consumer's preferences.

14.5 Consequences of using living system theory to model consumer behavior in customization

As a result of living system theory implementation in the area of consumer behavior, it is expected that there will be grow in interest in the investigation of this theory. There will be a great deal of effort spent in studying how the living system works. Research in neuroscience will provide inspiration for the explanation of neurons and relations of the neuron in order to explain the translation of a particular relationship into observable behavior. In another words, the way a living system works will inspire the way consumer behavior is described. The way of studying consumer behavior will change. With this new approach, plenty of effort will be spent on studying the 'processes', and how the activities are selected and assembled together to describe behavior. In studying consumer behavior, the understanding of the buying and decision-making process should no longer focus on what product information exists and what information needed to make certain buying decision. Instead, the investigation should focus on understanding consumers' historical experience in facing similar situations.

In the academic area this new approach to modeling consumer behavior will provide a new understanding of the latter. The immediate effect is the investigation of the topic with the elimination of certain assumptions used in marketing and psychological research. The results drawn from the study can be regarded as a more valid finding by arguing that the consumer is a living system, and every explanation of consumer behavior should be explained from the living system behavior point of view. Industry can also take advantage of this research. There are various possible implementations of this research in the area of consumer profiling, consumer segmentation, and understanding consumer behavior when customizing products. The ways of offering customization are moving towards a more automated means, via the Web. The Web has to perform many functions of the traditional shop, including the presence of sales-persons that have always been a company's best source in understanding consumer behavior. With the Web, the role of sales-person must be replaced with an intelligent way of recording, extracting, and predicting consumers' behavior. The model provides an alternative way to build the capabilities required of a virtual sales-person.

In order to understand consumer behavior at a particular time in a particular situation, it is impossible to measure all influencing variables determining one's behavior. Instead, in order to understand the behavior, the effort must focus on the investigation of how a person built his / her behavior repertoire throughout his/her development in his/her natural environment. An individual's history and development builds his/her capability, character, experience and habits. Capabilities and internal factors such as the way a person perceives information, emotional aspects, and problem solving capabilities, are built throughout his/her development and interaction with his / her particular environment, such as cultural background and education.

Autopoiesis gives a very different view in understanding consumer behavior. It focuses more on the general explanation of how biological setup, historical background and experience of a person can explain his / her cognitive ability and processes, which in turn determine behavior. The components of the model are small, undividable components, as in human cells or brain neurons. The relations between components will give rise to what the Cognitivism approach regards as 'memory'. A stronger relationship between components implies that the person has accumulated certain knowledge about certain things. Behavior is no longer described as a fixed routing of a sequential model. Rather, behavior is something dynamic and adaptive to the environment. Understanding comes by doing something repeatedly. So, knowledge is not something that can be created by implementation of production rules. Instead, knowledge is a successful historical coupling between a person and his/her environment. Knowledge is embedded in the process and it will drive behavior.

This chapter presented a perspective of new consumer behavior modeling in customizing products using living system theory. Living system theory is seen as an appropriate philosophical approach to understand and model consumer behavior because of its ability to explain the behavior of human beings as a biological system. We also presented some preliminary ideas on the model by explaining the components of the model and their interrelationships, and how those relations form a behavior. A behavior was explained as a collection of basic cognition activities. The activity selections could be explained using three different rules. There is still much work to be done in relation to this model. So far, the model only accounts for the selection and modification of components, but extension of activities is needed, and may include browsing activities, design activities, etc. Implementation, testing, and validation of this model are also required, in order to refine it.

References

[1]   Tseng, M. M.; Jiao, J.: Design for Mass Customization, in: Annals of the CIRP, 45 (1996) 1, pp. 153-156.

[2]   Pine II, B. J.; Gilmore, J. H.: The Experience Economy: Work is Theatre and Every Business a Stage, Boston 1999.

[3]   Von Hippel, E.: Economics of Product Development by Users: The Impact of "Sticky" Local Information, in: Management Science, 44 (1998) 5, pp. 629-644.

[4]   Peppers, D.; Rogers, M.; Dorf, R.: The One to One Fieldbook: The Complete Toolkit for Implementing a 1 to 1 Marketing Program, New York 1999.

[5]   Simonson, I.; Tversky, A.: Choice in Context: Tradeoff Contrast and Extremeness Aversion, in: Journal of Marketing Research, 29 (1992), pp. 281-295.

[6]   Tversky, A., Simonson, I.: Context dependent preferences, in: Management Science, 39 (1993), pp. 1179-1189.

[7]   Luce, M. F., Bettman, J. R.; Payne, J. W.: Choice processing in emotionally difficult decisions, in: Journal of Experimental Psychology: Learning, Memory, and Cognition, 23 (1997), pp. 384-405.

[8]   Blackwell, R. D.; Miniard, P. W.; Engel, J. F.: Consumer Behavior, Fort Worth 2001.

[9]   Hamuro, Y., Katoh, N.; Ip, E. H.; Cheung, S. L.; Yada, K.: Discovery of Interesting Rules for Purchase Behavior using String Pattern Analysis, in: Tseng, M. M.; Piller, F. T. (Eds.): Proceedings of the First World Congress on Mass Customization and Personalization, Hong Kong 2001.

[10]   Huffman, C., Kahn, B.E.: Variety for Sale: Mass customization or Mass Confusion, in: Journal of Retailing, 74 (1998) 4, pp. 491-513.

[11]   Von Bertalanffy, L.: General System Theory: Foundations, Development, Applications, New York 1968.

[12]   Hutardo, P. S.; Landeck, M.: Exploratory Research on the Implications of the Theory of Autopoiesis in International Consumer Behavior, in: Proceedings of Southwest Review International Business Research, 1999.

[13]   Gaia E.; Whitaker, R.: The Observer Web: Autopoiesis and Enaction, on www.enolagaia.com/AT.html.

[14]   Maturana, H. R.; Varela, F. J.: Autopoiesis and Cognition: The Realization of the Living, Boston 1980.

[15]   Varela, F. J.; Thompson, E.; Rosch, E.: The Embodied Mind: Cognitive Science and Human Experience, Cambridge 1991.

[16]   Brooks, R. A.: Intelligence without Representation, in: Artificial Intelligence Journal, 47 (1991), pp. 139-159.

[17]   Varela, F. J.: Autopoiesis and a Biology of Intentionality, in: McMullin, B.; Murphy, N. (Eds.): Autopoiesis and Perception: A Workshop with ESPRIT BRA 3352 (Addendum), Dublin City University, Dublin 1992.

[18]   Ford, M. E.; Ford, D. H.: Humans as Self-constructing Living Systems: Putting the Framework to Work, Hillsdale 1987.

[19]   Bettman, J. R.; Johnson, E. J.; Payne, J. W.: A Componential Analysis of Cognitive Effort in Choice, in: Organizational Behavior and Human Decision Processes, 45 (1990), pp. 111-139.

[20]   Du, X.: Architecture of Product Family for Mass Customization, Ph.D. Thesis, Hong Kong University of Science and Technology, Hong Kong 2000.

[21]   Tversky, A.: Elimination by Aspects: A Theory of Choice, in: Psychological Review, 79 (1972), pp. 281–299.

[22]   Maes, P.: A Bottom-Up Mechanism for Action Selection in an Artificial Creature, in: Wilson, S.; Arcady-Meyer, J. (Eds.): From Animals to Animats: Proceedings of the Adaptive Behavior Conference 1991, 1991.

[23]   Maes, P.: Learning Behavior Networks from Experience, in: Varela, F. J.; Bourgine, P. (Eds.): Towards a Practice of Autonomous System, Cambridge 1991.

Acknowledgements: This research is funded by Research Grants Council of the Hong Kong SAR Government through the grant (HKUST 6164/00E).

Contact:

Sri Hartati Kurniawan and Professor Mitch M. Tseng
Department of Industrial Engineering and Engineering Management,
The Hong Kong University of Science and Technology, Hong Kong
E-mail: ling@ust.hk and tseng@ust.hk

15 Usability of Design by Customer Websites

Oon Yin Bee and Halimahtun M. Khalid
Institute of Design and Ergonomics Application, Universiti Malaysia
Sarawak, Malaysia

The Design by Customer (DBC) approach is aimed at enabling companies to be more sensitive to what the customer really wants. The concept implies that users design a product using options offered by the company in a configuration system, while the latter assembles the product. There are constraints in providing the configuration system online, such as the types of product to offer, attributes of product for customer to design, and so forth. Therefore, design of a catalogue (representing the customization options) and of a corresponding configuration system is critical so that customers can be supported effectively in the design process. More important, what is designed is what the customer gets in the final product. This chapter reports the results of an experimental study that evaluates three DBC Web sites on user preferences of Web site features and e-catalogue-cum-configuration system. Using factor analysis, three generic factors were extracted for the DBC Web site features, namely: holistic design, navigability, and timeliness, while for the configuration system itself, the factors extracted represent design procedure, aesthetic preferences, information display, and design pleasure. The results also showed that users preferred top-down hierarchical approach for designing bicycles, watches and dresses. The Spearman rank correlation performed on the ordinal preference data showed significant relationships between the hypothesized and measured ranks for these Web sites. On the basis of this study, we derived specifications for an online configuration system of future Web sites.

15.1 Introduction: DIYD and understanding customer needs

Rapid development in new technologies drives customers to demand high-quality and low-priced products that meet their needs [1]. Manufacturers have to compete to overcome factors such as product variety and the speed to market. To be agile in product development, manufacturers need to understand customer needs, and to seize quickly changing market opportunities, as well as build products speedily in response to those needs [2]. In the light of this development, the customer profile has changed. Customers today are more liberated, choosey, informative and demanding. Keeping the customers satisfied requires manufacturers to mass customize their products. Mass customization is the selling of highly individual products on a mass scale. Products are built to order, and manufacturing takes place only as and when there is an order from a customer [3]. The advantage of mass customization over mass production is that products are designed to customer specifics through direct communication with the manufacturer.

Although many companies tried to move their business onto the Web, not all of them are successful due to concerns on profitability and user satisfaction [4, 5]. Customization of a product is essentially a feature of e-commerce. The customization process enables companies to provide customers with exactly what they need, with the goal of satisfying them. Offering electronic product catalogues within a configuration system on the Web allows customers to design their own desired product based on options provided by the company. The term catalogue refers in this context to a set of possible customization options (degree of variety) and their representation in a configuration structure. The structure is represented to the customers by the configuration tool. To date, only a few companies have provided online configuration system services to their customers.

The Design by Customer (DBC) approach via a configuration system is aimed at enabling companies interact directly with their customers. By doing so, manufacturers can gather customer requirements, besides giving them opportunities to express their choices from what they can offer [3]. However there are limits to mass customization for the options offered may not be afforded by the technology or the type of product to be designed. For example, a product with detailed engineering features may be too complex to be designed online by customers with no technical knowledge [6]. Despite the opportunities provided by new technologies, companies are still designing products in the traditional way. Through market surveys, they try to predict potential customer needs and designs for the customers. Understanding customer needs in this rapidly evolving world of technologies can be difficult.

15.1.1 Understanding customer needs

Understanding customer needs and the desires of the potential customers gives added advantages to the company in terms of securing a wider market. There are many approaches that may influence consumers in perceiving and choosing their desired options in the configuration system. The Gestalt approach suggests that people perceive some forms by grouping and segregating them from the surroundings [7]. Visual attention, according to Treisman [8], is then allocated to features that popped out, and this can influence the way people define the object. The features that popped out may include color, line curvature, target contrast, number, and proximity. Theories such as Gestalt can influence human preferences indirectly when choosing their desired product options. A product part which has popped-out features tends to get more attention. This could mean that customers either prefer or dislike the features.

Understanding customer needs is the first step in the product development life cycle [9]. Customer needs can vary from person to person. Jordan [10] claims that user needs tend to shift from usability-based to pleasure-based requirements; a feeling of pleasure towards a product has become an important factor. He identified customer needs at three levels: the lowest level as functionality, the next level as usability and the highest level as pleasure. To produce a product, user's requirements of product functionality must be satisfied first. Then users would seek ease of use for the product, as well as its emotional benefits. Pleasureability of a product is a new challenge for manufacturers, not only in terms of understanding user's emotion or cognitive processes, but also in capturing the physical processes that could induce pleasure and fun feelings to users [10].

Evaluation of a product (e.g. aesthetic preference) is linked to the users' physiological arousal [11]. Product attributes give different meanings to different individuals, due to differences in anthropometrics, functional and cognitive characteristics [12]. Additionally, environmental factors influence product choice. These together affect customer requirements [13]. Detailed knowledge of customer satisfaction and attitudes can influence important operational decisions that a company makes. Research on consumer attitudes helps to uncover what product and service features that are desired by the customer. To increase the sales of a product, the first step is to understand customer needs toward the product. Customer needs vary, depending on many factors, such as culture in terms of ethnic preferences (e.g. colors); product semantics in terms of its meaning to the customer (e.g. photographs); holistic design in terms of its overall shape or form (e.g. symmetry); past personal experiences in terms of familiarity of the product [14].

15.1.2 The Do-It-Yourself Design concept

Helander, Khalid and Tseng [6] coined the concept of Do-It-Yourself Design (DIYD) as used in e-design. It implies that the user designs a product using options offered by the company while the latter assembles the product. This gives customers an opportunity to express their needs through the Web while enhancing their satisfaction. The DBC term is used synonymously with DIYD [6]. In the context of our study, 'design' involves selecting product parts based on options that are considered to meet certain functionality and aesthetics criteria. In design theory, the design task requires creativity and an interdisciplinary approach [15]. These design options are determined first by the company, based on customer needs [16]. Clearly, users have specific needs that must be satisfied in the design process. With a Web-based DBC system, users are able to view the changes made to the product online prior to the purchase.

Additionally, the design procedure in DBC approach must be simple and easy to do as complex tasks can only deter customers. Khalid [17] confirmed a top-down hierarchical approach of mapping customer needs to functional requirements and to design parameters in a study on watch design among Malaysian and Hong Kong university students (this study is described above in Chapter 13 of this book).

15.1.3 e-Catalogue and configuration system

Manufacturers with a wide range of products have to develop their products instantly for the market. Product catalogues are one mode for presenting these products. Paper-based product catalogues are commonplace. They usually feature a variety of products both in pictorial and textual form. Based on the products offered, customers can order the product by phone or mail. The disadvantages of paper catalogue include [18]:

- Long processing time,

- repetitive correcting,

- high costs and time, and

- much effort in updating the catalogue.

Given the short span of a product lifecycle, the printed catalogue has to be published rapidly. This can be overcome by publishing an electronic catalogue on the Internet. An electronic catalogue has the advantages of

- reducing production and distribution costs,

- quicker data exchange between manufacturer and customer,

- quicker access by customers to select and read relevant information,

- enhanced presentation of product parts in graphical and text format,

- easier expansion of a product family based on the point product platform.

Helander and Khalid [5] discussed the benefits of online store (and e-catalogue) relative to paper catalogue and retail store. In the online market, the e-catalogue is intended to support online configuration of products [19]. The term catalogue reefers in this context to a set of possible customization options (degree of variety) and their representation in a configuration structure. The structure is represented to the customers through the configuration tool. Nielsen [20] claimed that the current standard user interface for user-driven product design is the configurator that consists of a long list of pull-down menus of every specified component. The configuration process implies accomplishing the goal of getting a specific product from a product family according to a customer's order [21]. The process starts by understanding customers' needs and transferring them into the input of the configurator. The output of the specific product configuration should approximate customers' requirements, depending on the options provided in the configurator [21]. Usually the options are presented as 2-dimensional or 3-dimensional images. These images give a better visualization to the customers compared to text based options. Given the limitations of human information processing, the design procedure must afford easy steps without overloading the customers. Ideally the items should not exceed ten and that good choices must be provided for customers to select [6].

15.1.4 Web usability

Web usability is a growing field of research due to the increasing use of the Web in e-commerce. Many Web sites have failed to realize their potential because they disregard users' needs and requirements [22]. Vora identified some of the usability-compromising situations such as: slow response time, difficulty in finding information, opaqueness of technology, poor design and implementation [23]. He proposed a methodology for designing Web sites, including personalizing the Web content. Design of a Web user interface must take into account human factors issues for the user is only a click away from leaving the Web site [5].

The International Standards Organization (ISO 9241-11) defines usability as the extent to which particular users can attain particular goals with efficiency and satisfaction in a particular environment [24]. Efficiency considers the amount of effort needed to achieve a goal measured by the time on task or errors in task performance. User satisfaction indicates a specific level of cognitive comfort or pleasure that users experience in using the product and their acceptance of the product in accomplishing their task goals [25]. A person's feeling of pleasure or disappointment can be estimated by comparing the perceived performance of a product with that expected. Subjective evaluations can provide estimates of user satisfaction, thereby usability.

Kim [26] identified four dimensions in designing a usable Web site: content, structure, navigation and graphic design. The content involves decisions concerning the type and scope of information to be included in the Web site. Users should be able to understand the structure of the Web site easily, including the functions provided in the DBC system. A good navigational design helps the user to explore the Web site smoothly, besides making the system easy to use. A good graphical display can influence customer's feelings toward the Web site, especially if the information is presented explicitly on the screen.

Nielsen found ten common mistakes in Web site design, namely [27]:

- Using many frames that confuse the users when browsing the Web pages.

- Employing the latest Web technology that may not be supported by many users' computer.

- Scrolling text and running animations that have an overpowering effect on human peripheral vision.

- Using complex URL that confuses the users on the structure of the Web site, besides being difficult to remember if the URL path is too long.

- Orphan pages that do not link to other Web pages or the main page.

- Long scrolling pages that are tedious to scroll or browse.

- Non-standard link color that confuses the users about which link to click or unaware that a link exists.

- Lack of navigational support for finding information they need in the Web site.

- Outdated information due to poor updates of the Website.

- Long downloading time that causes users to abandon the site.

Usability of the DBC system is therefore a prime concern in the design stage by customers. Website developers should consider the ease of use of the system rather than increase its functionality that may only lead to increased complexity of the system. Users in fact have become more aware of the ease-of-use of a system and are readily intolerant to poor interface design [28].

15.2   The experiment

Our study is aimed at evaluating the usability of DBC Web sites with the goal of identifying features for building future Web sites for mass customization. The research questions to be addressed include: Who will use the Web site? Is there a gender bias towards specific Web site? Do users enjoy designing their own product? Will they buy the end product they designed online? and so forth. This study also identified generic factors that influence user satisfaction towards the Web sites, the e-catalogue–cum-configuration system, and design procedure preferred by users.

15.2.1   Hypotheses and experimental design

The main hypotheses predicted significant effects of certain Web site features, such as downloading, scrolling, information display on user satisfaction, and that the e-catalogue-configurator impacts differently to users in terms of aesthetic preferences, design procedure and ease of design. In addition, this study predicted that the design sequence of top-down hierarchical approach is preferred in the three Web sites that were evaluated.

A within-subjects 3 (Web sites) x 2 (gender) factorial design was used. All subjects performed the tasks for each Web site. The assignment of subjects to experimental conditions was counterbalanced across tasks and gender in order to neutralize the effects of task order.

A cursory survey of the internet reveals a few Web sites that enable users to design products on their own in line with the DBC concept. Most Web sites allow users to choose the product parts that are displayed in textual or form-filling format (see www.designapc.com/design.cfm). In our study, the e-catalogue comprised a variety of 2D images of the product's parts as options for customers to choose. Users could also view changes to the design instantly unlike other Web sites. Three Web sites, meeting the DBC criteria were used in the study. They are documented in Table 1 and had the following generic features:

- The product to be designed is a simple consumer product that can be performed by users on their own.

- The design task was well structured in easy-to-follow sequences.

- The product parts were displayed in 2D graphics with product description in text.

- The use of color and product variety to differentiate design options.

- The Web sites appeal to both gender groups.

Table 1: Sample web sites of the empirical study

Web Site 1	Web Site 2	Web Site 3
Squash-Blossom	Idtown	VooDoo-Cycles
www.squash-blossom.com	www.idtown.com	www.voodoo-cycles.com
designing small girls' dresses	designing a watch	designing a bicycle

15.2.2 Dependent measures

The following dependent measures were obtained in the study:

(1) User satisfaction on Web sites features: A questionnaire consisting of items rated on a five-point scale was used to assess subjects' satisfaction towards the Web sites features. These include ease of access to the DBC Web page, downloading time and ease of page scrolling, use of graphics, colors, font size and font type, navigation and links, layout of content, quality of the images, and language used.

(2) User satisfaction on e-catalogue configurator features: Another set of questionnaire using a five-point scale was used to evaluate subjects' satisfaction towards the e-catalogue configuration system features. The measures include design procedure and sequence, availability of guidelines, product variety and choices, ease of making changes, task completion, design pleasure, look of the graphics and colors, image of the product material, reliability of the final product with that designed, product pricing, purchase intent, and adequacy of product information.

(3) Design procedure: For the design procedure, subjects were asked to indicate the steps they would take in designing each of the product in the three Web sites. They were given a list of steps to rank order. At the end of each design task, subjects were asked to rank their preferred sequence of design steps. Our hypothesized rank orders for each configuration system are as follow (Figure 1):

- *Squash-blossom:* There are six items for subjects to design, namely, type of dress, pattern or color of dress, leggings, hat, dress decoration and hat decoration. Our hypothesized sequence of design is: Type of dress > Pattern or Color of dress > Leggings or Hat or Dress Decoration > Hat Decoration.

- *Idtown:* There are also six items for subjects to design: type of watch, face, case, strap, hour hand, and minute and second hands. The hypothesized sequence of design: Type of watch > Face > Case or Strap > Hour or Minute Hand and Second Hands.

- *Voodoo-cycles:* There are five items for subjects to design, namely: type of bicycle, frame, fork, type of saddle and pedals and other measurements such as size, length, and so forth. The hypothesized sequence of design is: Type of bicycle > Frame > Fork > Type of Saddle and pedals > measurements (e.g. size, length, etc.).

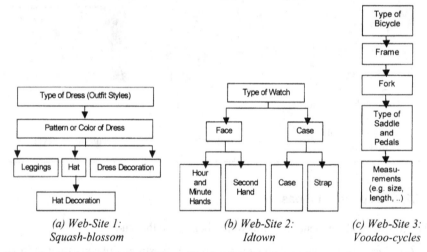

| *(a) Web-Site 1:* | *(b) Web-Site 2:* | *(c) Web-Site 3:* |
| *Squash-blossom* | *Idtown* | *Voodoo-cycles* |

Figure 1: Hypothesized design hierarchies for the Web-sites in the experiment

15.2.3 Procedure of the experiment

Forty-eight subjects, aged between 21 to 36 years (mean = 25), participated in the study. There was equal number of female and male subjects assigned randomly to each Web site. The subjects comprised mainly students and staff of the Universiti Malaysia Sarawak, with a maximum income of about US$1, 500.00. All subjects have used the internet and can surf on their own. About 87.5% of the subjects have never visited a DBC Web site, while 6.3% have bought a product online. The study was conducted in the Ergonomics Laboratory at the Institute of Design and Ergonomics Application.

Table 2: ANOVA results for effect of Web site features on user satisfaction (* p<0.05 ** p<0.01 *** p<0.001)

Measure	Web site	Mean	Std Dev	F value $F(2, 47)$
Can you get into the Web page which allows you to design the product? *(1=No, difficult; 5=yes, easy)*	1	4.63	0.73	4.39 *
	2	4.33	0.93	
	3	4.23	0.97	
How did the Web page load up? *(1=slow; 5=fast)*	1	4.06	0.89	6.67 **
	2	3.33	1.33	
	3	3.69	1.13	
What do you think of the overall graphics on the Web page? *(1=not nice at all; 5= very nice)*	1	3.29	0.92	2.87
	2	3.69	0.83	
	3	3.56	1.01	
What do you think of the colors of this Web site? *(1=not appealing at all; 5= very appealing)*	1	3.19	0.96	2.36
	2	3.58	0.99	
	3	3.48	1.09	
How is the size of the text font? *(1=too small; 5= too big)*	1	3.25	0.84	1.29
	2	3.06	0.70	
	3	3.10	0.52	
What do you think of the font type used in this Web site? *(1=difficult to read; 5=easy to read)*	1	3.92	1.01	1.19
	2	4.17	0.88	
	3	4.10	0.99	
Can you navigate through the Web site? *(1=No, difficult; 5=Yes, easy)*	1	4.10	1.02	.025
	2	4.13	0.98	
	3	4.15	1.01	
What do you think of the layout of this Web site? *(1=disorganized; 5=well-organized)*	1	3.35	1.06	1.98
	2	3.65	1.12	
	3	3.67	1.00	
Do you find some broken links in the Web site? *(1=Yes, often; 5=no, not at all)*	1	4.13	1.28	2.62
	2	4.23	1.13	
	3	3.83	1.31	
How is the resolution/quality of the images on the Web site? *(1=poor; 5=good)*	1	3.33	1.19	2.63
	2	3.77	0.88	
	3	3.63	1.08	
How is the language used in this Web site? *(1=difficult to understand, 5= easy to understand)*	1	4.31	0.85	4.61 *
	2	4.27	0.96	
	3	3.88	1.16	
Do you need to scroll down to have a full view of the page content? *(1=Yes, often; 5=No, not at all)*	1	1.98	1.25	8.53***
	2	2.60	1.25	
	3	2.98	1.45	

A standard computer system, Pentium III 450MHz with 64Mb RAM, was used for each subject. The Internet Explorer 5.0 was chosen as the Web browser for access to the given Web sites via the university's local area network. Subjects had to complete four sets of questionnaires in either English or Bahasa Melayu (Malay language). The first questionnaire is on subject demographics; the remaining questionnaires contained questions relating to usability and sequence of design for the Web sites. The order of questions in these questionnaires was randomized for each Web site as a measure against test-taking attitude. This is seen as important as the usability questionnaire was the same for each Web site. The tasks involved designing the product according to the design procedures given on the respective Web site. Subjects were first briefed on the tasks. They then completed the demographics questionnaire. Each subject performed the task according to the prescribed order of Web sites. They visited the first Web site and accessed the Web page that allowed them to design the product. Having done so, they answered the respective questionnaire. The procedure was repeated for the remaining sites. The questionnaire was collected before performing the next condition.

15.3 Results

The data were analyzed using one-way repeated measures ANOVA as summarized in Table 2 for Website features and in Table 3 for e-product catalogue. Factor analysis was performed on the subjective ratings data. Principal component method with varimax rotation was used to extract features that are generic to the Website and e-product catalogue design, as summarized in Tables 2 and 3, respectively, across gender. Spearman rank correlation was used to test the subjects' ranks with that hypothesized. The results showed significant differences for certain features of the Web sites and product catalogues between for the three Websites. Only significant results are discussed here, although it is also important to note the remaining findings.

15.3.1 Effect on user satisfaction

Website features (Table 2): The results showed that the design page for Web site 1 (Squash-blossom.com) was the easiest to access relative to Web sites 2 (Id-town.com) and 3 (Voodoo-cycles.com), $F(2, 47) = 4.39$, $p<0.05$. The time taken for uploading the Web sites was significantly faster for Web site 1 than the other Web sites, $F(2, 47) = 6.67$, $p<0.01$. But subjects had to scroll more frequently the Web pages of Web site 1 in order to view the page contents than Web sites 2 and 3, $F(2, 47) = 8.53$, $p<0.001$. In terms of the terminology used in the design process, Web site 3 could not be easily understood by novice subjects compared to Web sites 1 and 2, $F(2, 47) = 4.61$, $p<0.05$.

E-Catalogue / configurator features (Table 3): The e-catalogue of Web sites 3 and 2 were found to be significantly more colorful and attractive than Web site 1,

F(2, 47) = 4.68, p<0.05). In terms of the material options displayed for subjects to choose, Web site 2 was found to be less impressive than the other Web sites, F(2,47)=4.97, p<0.01. Also, the information given on the product was significantly minimal for Web site 2, F(2, 47) = 22.12, p<0.001 compared to the other two product catalogues. The design procedure was found to be significantly flexible for Web sites 1 and 2 than 3, F(2,47) = 5.39, p<0.01. Subjects found it difficult to follow the design sequences in Web site 3. It was significantly easier to complete the design task for Web site 2 than 1, F(2,47)=8.89, p<0.001. The pricing of the product 'bicycle' for Web site 3 was viewed to be significantly expensive than the products 'dress' and 'watch', F(2, 47) = 36.29, p<0.001. As a result, subjects would buy these products relative to the bicycle, F(2,47)=6.37, p<0.01.

Further, *effects of the gender on user preferences* of Web site features and e-catalogue were tested. The ANOVA results revealed that female subjects found the colors of Web site 1 significantly more appealing than their male counterparts, F(2, 47) = 9.91, p<0.001. They also thought that the layout was well-organized, (F(2, 47) = 5.97, p<0.01), and that the 2D images of the product parts were good (F(2, 47) = 8.32, p<0.001). This suggests that females preferred Web site 1 better than the other Web sites in terms of these features. In terms of the e-catalogue, male subjects found the choices in Web site 3 met their requirements (F(2,47)=4.00, p<0.05), the colors were more attractive (F(2,47)=4.93, p<0.01), and the graphics were more appealing (F(2,47)=3.94, p<0.05). In terms of the configuration process, female subjects found the design sequence relatively more flexible for Web site 1 (F(2,47) = 4.04, p<0.05), compared to their male counterparts, and they enjoyed designing small girl dresses compared to male subjects (F(2, 47) = 13.83, p<0.001). On the other hand, male subjects enjoyed designing the bicycle in Web site 3 and that they would purchase it compared to female subjects, F(2, 47) = 3.35, p<0.05.

Table 3: ANOVA results of effect of e-catalogue (configurator) features on user satisfaction (* p<0.05 ** p<0.01 *** p<0.001)

Measure	Cata-logue	Mean	Std. Dev.	F values (2, 47)
Is it easy to design the product online? *(1=No, difficult; 5=yes, easy)*	1	4.00	1.15	3.33
	2	4.40	0.79	
	3	3.96	1.29	
How were the instructions/ guidelines for you to design the product? *(1=difficult to understand; 5=easy to understand*	1	3.77	1.12	1.62
	2	4.10	0.90	
	3	3.83	1.26	
Do the choices given meet your requirement? *(1=No, not at all; 5=Yes. definitely)*	1	3.02	1.21	0.93
	2	2.98	1.08	
	3	3.23	1.06	
Is it easy for you to make any changes after you have selected the options you want? *(1=No, difficult; 5=Yes, easy)*	1	3.92	1.33	1.91
	2	4.29	0.92	
	3	3.85	1.25	*contd.*

Table 3 *(contd.)*

Measure	Cata-logue	Mean	Std. Dev.	F values (2, 47)
How is the sequence for designing? (1=not flexible; 5=flexible)	1	3.75	1.00	5.39 **
	2	3.75	1.19	
	3	3.17	1.29	
Is it easy for you to complete the task of designing your own product? (1=No, difficult; 5=Yes, easy)	1	4.06	1.12	8.89 ***
	2	4.65	0.60	
	3	4.25	0.89	
Do you enjoy designing the product? (1=No, I dislike; 5=Yes, I like)	1	3.54	1.54	2.41
	2	3.98	1.06	
	3	3.54	1.35	
Are the graphics appealing? (1=Not at all; 5= Yes, definitely)	1	3.27	1.23	2.77
	2	3.48	1.07	
	3	3.75	1.02	
Is the product material displayed? (1=No; 5=Yes)	1	3.94	1.33	4.97 **
	2	3.08	1.60	
	3	3.83	1.36	
Do you think that the final product will look the same as what will be delivered to you? (1=No, it will be different; 5=Yes, it will look the same)	1	3.31	1.08	2.07
	2	3.63	1.21	
	3	3.29	1.03	
How is the price? (1=expensive; 5=cheap)	1	2.83	0.95	36.29 ***
	2	2.94	0.89	
	3	1.60	0.82	
Will you buy the product if they allow you to design your own? (1=No; 5=Yes)	1	3.58	1.41	6.37 **
	2	4.04	1.09	
	3	3.23	1.40	
Is there information to describe the product? (1=No; 5=Yes)	1	3.58	1.23	22.12 ***
	2	2.63	1.20	
	3	4.19	1.10	

15.3.2 Factor analysis of Website features and e-catalogue / configurator

Table 4 presents the results of a factor analysis for Web site design across gender groups. Three factors were extracted after Varimax rotation: Factor 1 represents the first cluster of attributes for holistic design with defining attributes "graphics", "color", "image quality", "language," and "text". Factor 2 for the second cluster navigability with attributes "site access", "font readability", "broken links", and "page layout". Factor 3 for the third cluster timeliness with attributes "page scrolling" and "page loading". These three factors accounted for 70% of the total variance. Factors 1 and 2 contributed about 30% and 28% respectively, while Factor 3 about 12% only.

Table 4: Mean, standard deviation and factor loadings of Website features after varimax rotation

			Factors		
Measure	*Mean*	*Std. Dev.*	*1.* *holitsic design*	*2.* *navi-gability*	*3.* *timeli-ness*
What do you think of the overall graphics on the Web page?	3.51	0.62	0.86	0.13	
What do you think of the colors of this Web site?	3.42	0.59	0.81		
How is the resolution/quality of the images on the Web site?	3.58	0.65	0.68		0.54
How is the language used in this Web site?	4.15	0.77	0.63	0.59	
How is the size of the text font?	3.14	0.50	0.57	0.36	-0.10
Can you navigate through the Web site?	4.13	0.66	0.17	0.82	0.32
Can you get into the Web page which allows you to design your own product?	4.40	0.69	0.21	0.75	-0.23
What do you think of the font type used in this Web site?	4.06	0.69	0.49	0.69	0.19
Do you find some broken links in the Web site?	4.06	1.01	-0.13	0.60	0.39
What do you think of the layout of this Web site?	3.56	0.76	0.42	0.55	0.33
Do you need to scroll down to have a full view of the page content?	2.52	0.88			0.83
How did the Web page load up?	3.69	0.80		0.45	0.63

There were four factors extracted for e-catalogue-configurator features (Table 5). Factor 1 design procedure is defined by attributes "instructions", "task completion", "ease of design", "ease of modification", and "sequence flexibility". Factor 2 aesthetic preferences represent "design choices", "graphics appeal", "product pricing", "final product look", and "overall color". Factor 3 information display with defining attributes "information availability", and "display of product material", while Factor 4 design pleasure represents "product purchase" and "design enjoyment" attributes. Together these four factors contributed 72% of the total variance. Of this, Factors 1 and 2 contributed 26% and 24% respectively, while Factors 3 and 4 about 11% each.

15.3.3 Spearman rank correlation on design procedure

The Spearman rank correlation was performed on the ordinal preference data for the hypothesized sequence of design and subjects' rank order. The results showed significant relationships between the hypothesized and measured ranks for all three Web sites. For the design procedure in the Squash-blossom.com Web site, the results (r_s = 0.94, p< 0.01,) showed that users tend to follow the design sequence from 'Type of Dress' to 'Pattern or Color of Dress', then 'Leggings or Hat or Dress Decoration' and finally 'Hat Decoration'. Users also followed a top-down hierarchical approach in Idtown.com from 'Type of Watch' to 'Face', then 'Case or Strap' and lastly 'Hour and Minute Hand or Second hand'. The results of Spearman rank correlation, r_s = 0.88, p<0.05, too showed significant relationship between the hypothesized and users' ranks. Similarly with the Voodoo-cycles.com Web site, subjects preferred a top-down approach. The results, r_s = 0.90, p< 0.05, confirmed that subjects would design from 'Type of Bicycle' to 'Frame or Fork', then 'Type of Saddle and Pedals', and finally 'Measurements'.

Table 5: Mean, standard deviation and factor loadings of e-catalogue features after varimax rotation

			Factor			
Measure	*Mean*	*Std. Dev.*	*1. design procedure*	*2. aesthetic preferences*	*3. information display*	*4. design pleasure*
How were the instructions/ guidelines for you to design the product?	3.90	0.76	0.81	0.37		
Is it easy for you to complete the task of designing your own product?	4.32	0.68	0.80	0.13		0.18
Is it easy to design the product online?	4.12	0.79	0.71	0.33	-0.12	0.32
Is it easy for you to make any changes after you have selected the options you want?	4.02	0.67	0.67	-0.29	0.46	
How is the sequence for designing?	3.56	0.81	0.58	0.11	0.29	0.11
Do the choices given meet your requirement?	3.08	0.77	0.28	0.82		
Are the graphics appealing?	3.50	0.73	0.21	0.73		0.25
How is the price?	2.46	0.53	-0.11	0.63	0.33	-0.22
Do you think that the final product will look the same as what will be delivered to you?	3.41	0.83		0.61	0.27	0.41
How about the overall colors of the catalogue?	3.36	0.68	0.36	0.59	0.25	0.24
Is there information to describe the product?	3.47	0.70	0.20	0.31	0.79	
Is the product material displayed?	3.62	0.81		0.14	0.79	0.30
Will you buy the product if they allow you to design your own?	3.62	0.92		0.24	0.23	0.78
Do you enjoy designing your own product on the site?	3.69	0.84	0.44		0.10	0.71

15.4 Discussion

(1) Design of Web site: Our study has shown that design of Websites must take into consideration human factors issues as documented by Forsythe, Grose and Ratner [29]. Holistic design features such as graphic, color, image resolution, language and text are important attributes to consider in Website design. Equally important is the navigational support to users. It was easier to access the Web page for Web site 1 than Web site 3. This is because ambiguous words were used to describe a link. Inconsistencies in specifying link anchors and inappropriate labeling of links can create navigation problems [30]. Therefore language plays an important role in site design. Web sites with easy-to-understand terms will be preferred. Web site 2, for example, was easier to understand than Web site 3. Borges claimed that links with problematic words or names that did not provide a good hint to users could cause navigation problems and waste user's time in searching for the right link to click [31]. Web sites that are slower to load up can cause users to click away. GVU's Ninth WWW User Survey found most users experienced this problem [32]. About 53% of respondents in the survey would simply abandon the Web page as they dislike waiting. In our study, Web site 2 was significantly slower than Web site 1. Clearly, tedious scrolling of Web pages can put customers off. This was particularly true for Web site 1. In fact, only 10% of the users would scroll beyond the information that is shown on the screen [33]. This means that users have a tendency to ignore information placed at the bottom of the page. If long scrolling pages are necessary, the important links or contents should be placed at the top in order to attract the user's attention.

(2) Design of the e-catalogue and configurator: The design procedure and aesthetic preferences both play an important role in e-catalogue-configurator development. From the findings, Websites 2 and 1 have better design sequence compared to Web site 3. Khalid confirmed the top-down strategy in design [16]. Flexible design sequence is also preferred as it enables customers to take shortcuts in the design process. As such Web site 2 that structured the task hierarchically allows users to complete the design task better than the other Web sites.

Customers have different preferences for color. The Asian value differs considerably from Western [9]. In this study, Web site 3 was seen as more appealing in terms of color representation than Web sites 1 and 2. This site belongs to a company operating in Europe compared to Web site 2 that operates from Asia. Other variables that influence user satisfaction include the pricing of products, the variety of product material and information on the respective product parts. These contributed toward attracting customers to the DBC Web site and committing them to purchase.

(3) Gender impact: Designing a watch seems to induce the same feeling in both female and male users. However, designing dresses tends to be more female biased relative to male, while the latter preferred designing the bicycle. It was not surprising then that female subjects rated Web site 1 better in providing design options. They also enjoyed designing the dress, appreciated the color and graphics of the product material, and would purchase the dress designed online.

(4) Design procedure: The way to design a product may vary between simple consumer products and complex engineering products. In our study, the results showed that users preferred a top-down hierarchical approach for designing bicycles, watches and dresses. To a certain extent, our study confirmed the findings by Khalid for watch design [17].

15.5   Conclusion and future work

We have shown that usability of DBC Web site is critical in attracting captive customers. Throughout the Do-It-Yourself Design process, subjects were indeed concerned about the ease of completing the task, the flexibility of the design sequence, the display of product options in color, and so forth. To support browsing, users need an easy-to-find link, clear language, fast downloading time, and minimal scrolling to view the contents. In short, many of the human factors concerns in poor Web design were confirmed in this study. However, the insignificant findings do imply that the selected DBC Web sites have common features which met user needs, such as: ease of design, product choices, good images, availability of guidelines, ease of modification to design, pleasurable to design. Together these have implications for the design of future DBC sites:

- The Web site should provide links for users to get into the design page and they should be clear and easy to find.
- The site must be able to be loaded up quickly.
- The language used must be simple and easy to understand.
- The site must avoid long scrolling pages.
- The design procedure for the configurator should be flexible and easy to complete.
- The e-product catalogue should be colorful and well-organized.
- The product material and product information must displayed clearly.
- The pricing of products should be cheaper than the market price in order to attract users to design and buy online.

In short, Web site features such as holistic design and navigational support are important design features, while the e-catalogue must be aesthetically appealing, and the design procedure in the configuration system must be easy to follow as well as flexible. The design procedure should follow a top-down approach to make the design task easier to perform. This could be done by first identifying customer needs, then functional requirements of the product, and lastly mapping these to design parameters provided in the configuration system. Understanding customer needs can help to increase user satisfaction during designing. The selected Web sites in this study used 2D images for their e-catalogues. About 60% of the users had difficulty in imaging the real product. About 62% preferred to rotate the product if allowed. Further research to improve the e-catalogue seems the next logical step. One possibility is to use virtual prototypes that can give better visualization of the product to users. Mohageg, Myers, Marrin, Kent, Mott, and Isaacs claimed that 3D content is often best understood when the data can be viewed interactively from different points of view [34]. The user can rotate the object to view different sides. This could help the users to be clearer of the full view of the product they wish to purchase.

References

[1]   Tseng, M. M.; Jiao, J.: Su, C. J.: Virtual Prototyping for Customized Product Development, in: Integrated Manufacturing Systems, 9 (1998) 6, pp. 334-343.

[2]   Anderson, D. M.: Agile Product Development for Mass Customization, Niche Markets, JIT, Built-To-Order, and Flexible Manufacturing, New York 1997.

[3]   The Economist: Mass Customization: A long march, July 2001, pp. 63-65.

[4]   Kopalle, P. K.; Kannan, P. K.: Dynamic Pricing on the Internet: Importance and Implications for Consumer Behavior, in: International Journal of Electronic Commerce, 5 (2001) 3, pp. 63-83.

[5]   Helander, M. G.; Khalid, H.M.: Modeling the customer in electronic commerce, in: Applied Ergonomics, 31 (2000), pp. 609-619.

[6]   Helander, M. G.; Khalid, H.M.; Tseng, M. M.: Mapping customer needs in Web-based DIY design of consumer products, in: Proceedings of the 3rd Annual International Conference on 'Industrial Engineering Theories, Applications and Practice, Hong Kong 1998.

[7]   Morgan, T. C.; King, R. A.; Weisz, J. R; Schopler, J.: Introduction to Psychology, New York 1986.

[8]   Treisman, A.: Features and Objects in Visual Processing, in: Scienctific American, 255 (1986), pp. 114-125.

[9]   Khalid, H. M.: Can customer needs express affective design?, in: Helander, M. G.; Khalid, H. M.; Tham, M. P. (Eds.): Proceedings of the International Conference on Affective Human Factors Design, London 2001, pp. 190-198.

[10]   Jordan, P.: Creating Pleasurable Products, in: International Encyclopedia of Ergonomics and Human Factors, 1 (2001), pp. 1095-1097.

[11]   Martindale, C.; Uemura, A.: Stylistic Change in European Music, in: Leonardo, 16 (1983), pp. 225-228

[12]   Buti, B. L. et al.: Ergonomics Industrial Design: Principles for Product Ergonomics, in: Proceedings of IEA World Conference, 3rd Latin American Congress, 7th Brazilian Ergonomics Congress, 1995, pp. 59-64.

[13]   Tseng, M. M.: 2000, Design by Customers – Expanding the Global Manufacturing Enterprise by Including the Customers, in: Proceedings of Professional Congress 'The Future of Work', World Engineers' Convention, Dusseldorf 2000, pp. 47-56.

[14]   Khalid, H. M.; Helander, M. G.: A Framework for Customer Needs in Product Design: Theoretical Issues in Ergonomics Science, 2002.

[15]   Schon, D.: The Reflective Practitioner, London 1983.

[16]   Khalid, H. M.: Mass customization and Web-based Do-It-Yourself product Design, in: Proceedings of the 14th Triennial Congress of the IEA and 44th Annual Meeting of the HFES, 2 (2000), San Diego 2000, pp. 766-769.

[17]   Khalid, H. M.: Uncovering Customer Needs for Web-Based DIY Product Design, in: Proceedings on the International Conference on TQM and Human Factors, Linköping 1999, pp. 343-348.

[18]   Hitzges, A.; Krieger, M.: The Way to the Electronic Catalogue, on www.swt.iao.fhg.de/swt/Publikationen/ecARH.pdf.

[19]   Jiao, J.; Tseng, M. M.: Web-Based Sales Support to Online Mass Customization by Developing Electronic Catalogs, in: Proceedings of the Congress on Mass Customization and Personalization, Hong Kong 2001.

[20]   Nielsen, J.: Customers as Designers, in: Jakob Nielsen's Alertbox, on www.useit.com/alertbox/20000611.html

[21]   Dong, J.; Xiao, T.; Qiao, G.: Study on Configuration Method Based on Constraint and Fuzzy Decision, in: Proceedings of the World Congress on Mass Customization and Personalization, Hong Kong 2001.

[22]   Sano, D.: Designing Large Scale Websites: A Visual Design Methodology, New York 1996.

[23]   Vora, P.: Human factors methodology for designing Web sites, in: Forsythe, C. et al. (Eds.): Human Factors and Web Development, Mahwah 1998, pp. 153-172.

[24]   Jordan, P.: Usability and Product Design, in: International Encyclopedia of Ergonomics and Human Factors, 2 (2001), pp. 1426-1428.

[25]   Ahonen, P.: Customer's satisfaction, on www.cerco.org/OUR WORKINGGROUPS /WGQ/Ahonen.html.

[26]   Kim, J.: Toward the Construction of Customer Interfaces for Cyber Shopping Malls, in: Focus Theme, 7 (1997) 2, pp. 12-15.

[27]   Nielsen, J.: Top Ten Mistakes in Web Design, in: International Encyclopedia of Ergonomics and Human Factors, 1 (2001), pp. 738-740.

[28]   Thomas, D. B.; Dzimbor, G.; Bohr-Bruckmayr, E.: Ergonomics in Product Testing, in: Proceedings of the International Conference on 'Marketing Ergonomics, 1989.

[29]   Forsythe, C.; Grose, E.; Ratner, J.: Human factors and Web development, Mahwah 1998.

[30]   Goonetilleke, R. S.; Duffy, C.; Jacques, D.: Usability and beyond: a macro view of usability, in: Proceedings of HCI International Conference 2001, 1 (2001), Mahwah 2001, pp. 365-369.

[31]   Borges, J. A.: Page Design Guidelines Developed Through Usability Testing, Human Factors for World Wide Web Development, 1997, pp. 137-152.

[32]   Graphic, Visualization, and Usability Center (GVU): GVU's 9th WWW User Survey, Georgia Institute of Technology, www.gvu.fatech.edu/user_surveys/

[33]   Vora, P. R.: Internet and the World Wide Web, in: International Encyclopedia of Ergonomics and Human Factors, 1 (2001), pp. 701-704.

[34]   Mohageg, M.; Myers, R.; Marrin, C.; Kent, J.; Mott D.; Isaacs, P.: A User Interface for Accessing 3D Content on the World Wide Web, in: Proceeding of CHI'96, on www.acm.org/sigchi/chi96 /proceedings/desbrief/Mohageg/mfm.txt.htm.

Contact:

Professor Halimahtun M. Khalid
Institute of Design and Ergonomics Application,
Universiti Malaysia Sarawak, Malaysia
E-mail: mkmahtun@idea.unimas.my

16 Applications of Kansei Engineering to Personalization

Practical ways to include consumer expectations into personalization and customization concepts

Rosa Porcar, María-José Such, Enrique Alcántara, Ana-Cruz García and Alvaro Page
Biomechanics Institute of Valencia (IBV), Valenicia, Spain

This chapter proposes two practical ways to include user preferences in a personalization system aimed at psychological perception and based on a Kansei system. In many mass customization systems the consumer as an inexperienced designer can get lost and will become frustrated about the huge amount of offered possibilities. We will discuss how Kansei engineering can be used to guide customers in order to quickly find the desired design according to their preferences. Secondly, the number of possible design options and combinations in a modular personalization system can be higher than stocking, logistic and manufacturing capabilities. Focusing production variability on features affecting most users' preferences and purchase decisions may reduce this amount of design options. We will use case studies form the office furniture and footwear industry to support these ideas.

16.1 Introduction

As previous innovations in production and design systems, mass customization offers important challenges. Focusing the problem only on its aspects more related to the user (but keeping in mind the importance of production and logistics), two principal key points are to be solved. These two points are the consequence of the two kind of values a product has: functional and symbolic [1]. The functional values, in a mass customization approach, must be revised and new methodologies to generate criteria for best personal fit must be built up. The symbolic values can be afforded using the information provided by Kansei Engineering studies. In a simplistic approach, mass customization can be described as the possibility offered to the user of building up a product as a sum of different parts of his choice. For each of these parts (i.e. design elements), the user can choose among different options (categories).

For example, in constructing an office table, one design element may be the supporting frame, and corresponding categories: L shape, four legs, boards, etc. The goal of the manufacturer dealing with mass customized products will be having the range of possibilities as broad as possible. The symmetric goal of the customer will be choosing his own product from a really wide and rich range of offered possibilities.

But this variability, desired by both companies and customers, holds a problem. From the manufacturer's point of view, each category that is offered in each design element represents an extra cost in stocked references, logistics and production organization management. How to reduce this cost to the minimum without lacking richness choice appearance to the consumer? From the customers' point of view, the huge number of combinations offered, that would be a richness to an experimented designer, can result into time expending, frustrating and discouraging to a layperson. Therefore, it seems important to find an equilibrium between the variability offered and the amount of available choices. The aim is to keep production and logistic costs as well as the customers options into reasonable and maneuverable limits.

16.2   Kansei Engineering

'Kansei engineering is an ergonomic technology of consumer oriented product development. It focuses not on the manufacturer's intention but rather on the customer's feelings and needs (Kansei)' [2]. Kansei type II is a computerized expert system containing the rules to transfer the consumers' feeling and image of a product into design details. Constructing a type II Kansei engineering system requires a lot of information regarding [3]:

Semantic space of the user-product relationship: Aimed at describing, in the customer own words, all the subjective perceptions regarding product appearance. This collection of words or expressions are organized into an space of independent axis which can be obtained using Differential Semantics [4].

Level of fuzziness of each axis in user perception: Not all the axis obtained have equal clarity (level of consensus) in judging a product. This information, added to the next, provides useful keys in modeling the user perception.

Importance of each axis into purchase decision: As a complementary action of building a Kansei System, it is possible to obtain the contribution of each axis to the purchase decision [5, 6].

Contribution of each design element into each semantic axis: Using diverse techniques (Hayashi Quantification Theory [6], GLM, fuzzy logic), it is possible to calculate the contribution (or the relevance) of each design element to each axis.

Similarities or differences in the contribution of each category: In addition, it is derived information about how similar or dissimilar are category influences within a design element into each axis. The axis independence plays an important role in building up the perception modeling as it simplifies the statistical methods used to obtain the relationships. But the weaknesses of this is obvious: categories and design elements may have different (see contradictory) contribution to each axis. For example, in an imaginary product, to obtain a valuation of 'warm' it must be green, and to obtain a valuation of 'innovation' it must be red; how to obtain a product warm and innovator?. Here relies the importance of knowing the contribution of each axis to the purchase decision as well as its clarity or consistence.

Kansei engineering has multiple applications in the conceptual and applied design. It is a direct help in design innovation. The rules obtained in a Kansei system construction can be used from two complementary perspectives. On the one hand, having a proposed design, the expected customers response could be evaluated in terms of semantic axis. On the other hand, deriving from a desired consumers' answer, several designs could be produced (as combinations of design elements) which will fit (with a certain level of probability) the users' expectations. These two ways are used in this chapter with other complementary tools. All results presented in this chapter derive from an R&D project on the application of Type II Kansei Engineering to *office tables* and *casual footwear*. The scenario and target users to the former were electronic catalogues and purchasers, being to the latter display stands and final users. The complete results were implemented into software named Kn6/IBV.

16.3 The consumer as a designer

In this section, an application of Kansei resulting into mass customization is presented. It is based on the idea of guiding the user to achieve earlier and easier his own customized products. The majority of people are able to say if they like a product (or a service). They are capable of choosing between houses, or watches or car insurance. But being able to construct each of those things merely adding parts is not so easy. Despite the amount of available money, some people look for help in styling their house or choosing their suits (wardrobe). The ability needed to visualize the final result of a combined choice is not a common value and it increases as the product complexity and/or the number of offered possibilities do.

In many personalization systems, an image of the each choice step resulting product in some perspectives is offered. So, the user is able to check the effect of all categories of the actual design element with the previous selections done. But, to obtain a real exploration of the effect of the actual design elements possibilities with all the previous, the user should be able to go backwards and visualize each possibility. This means dozens (or hundreds) of combinations, no matter how restricted the choice possibilities are. Part of this problem is overcome if there exists a rigid hierarchy in the design element selection. This hierarchy may come from natural nested design (one choice is only available if a fixed previous one has been selected), or from induced selection, if there is available information about the relative importance of each design item. In the majority of cases, no nested multiple combinations are really available.

This huge number of paths offered to the user, (a freedom considered highly added value to design professionals), is in fact not very practical, and after some intents, the customer use to get lost. The results to the manufacturer are obvious: failed sell, hurried sell with no satisfaction or returned product (as many companies offered this guaranty if product does no fulfill customer expectations). There is also another possible threat to companies offering customized products: loosing their brand image. Being one of the most important values of a company, the brand image can be put into serious damage if each user is able to freely construct their product without limitations. As the customized service offers broader possibilities, the risks of lacking brand image increase. Kansei engineering could help in the concerns presented above.

As stated above, a Kansei system provides the relationships between the image of a product and their visible design elements. Its contribution to solving the problem of the inexpert designer is immediate. The system asks the user about his desired image of the product, in terms of punctuation in each of the semantic axis (Kansei words). This punctuation can be a point (in a scale of several points, named Semantic Differential Scales) or a range of possible values. It is also possible, if desired, to fix any of the categories of one or more design elements. The latter will restrict the field of valuable solutions and makes the search quicker.

Once established, using the inverse rules, the system provides a list of possible products expressed as multiple combinations of design elements that fit the aimed perception. The user freedom is not restricted but guided. From these offered solutions, the user can filter the solutions containing one or more category of a design element, so narrowing the list of solutions and making easier the final choice. A Kansei system can also contribute to prevent brand image from excessive changes. As this image is constructed as a combination of possible values in the semantic axis, it is possible to restrict the range of offered solutions avoiding the user to choice certain combinations in terms of 'concepts' defined by axis. The advantage of this system is obvious: the company concentrates its efforts in the 'image' and not in the total possible combinations of design elements, which is easier due to two factors. On the one hand, the number of semantic axis is usually much lower than design elements, if the product is not extremely simple; so, this makes the problem solution faster. On the other hand, the limitation of the solutions is based on the user feelings and perceptions (as far as the Kansei system is constructed modeling the user thinking). In addition, on a previous exploitation, it can be stated what the image of the company is to his mind, which could be far away from what the company staff suppose.

16.4   Narrowing choice possibilities – manufacturer's point of view

Personalizing a product in a mass customization approach means letting the user choice among many possibilities in each design element (internal or external). In the external (apparent) design elements, the production possibilities can be extended. But, each category that is offered in each design element represents an extra cost in references and production management. So, reduction of the cost of the offered possibilities will be a goal of companies. At the same time, increasing the number of choices implies richness in a mass customization system. So the question arise: where companies should put their effort into design variability to obtain the maximum richness appearance keeping the cost under reasonable limits? Kansei engineering offers valuable information to this point.

As stated above, constructing a Kansei engineering system requires a lot of information regarding:

- The words (Kansei axis) related to the customers' perception of a product.

- A decomposition of the product into customer perceptible design elements. For example in a chair: armrest, backrest, seat, legs, upholstery, etc.

- Each design element has many categories available. These categories are the only selections the user can do. For example a design element can be office table surface color, and the corresponding categories: white, wood and grey.

- Each design element has a specific and different contribution to each axis.

- Each category has different contribution to each design element in each axis.

Having a Kansei system constructed in which you have considered as many categories as a previously defined market segment offers, some rules, based on the information described above and aimed at reducing variability, are proposed:

- Focus on the design elements affecting more to the axis perception. Sort them by their total contribution (i.e. as a sum) to the evaluation. Decisions regarding where to offer more variability must be focused on these design elements.

- Explore, for those design elements, the categories having similar weight and fuse them together. If a consistent pattern does not exist, decide taking into account both: the importance into purchase decision and the fuzziness of the axis.

- Keep to a minimum the range of categories pertaining to design elements with no clear contribution to axis perception. Decide this variability taking into account only production facility or marketing strategy (i.e. to reach an 'impressive' figure of theoretical possible combinations available).

16.5 Case study results

Some of the results presented in this section derive from the software Kn6IBV® that has been developed in Spanish. In these cases, screens on original language are presented and explanations are offered in captions.

16.5.1 Customer personalized design selection

A first application of Kansei engineering into mass customization is presented by means of a personalized design selection example for *office furniture* (office table). The job sequence a consumer would carry out with the developed software Kn6/IBV® is presented.

First of all, a 'translator' is available. The user could focus the design selection with an assistant expert system able to gather the desired psychological perception. The format of the information requested for that aim is shown in Figure 1. It contains a list of the 8 principal axis extracted (which are able to explain a 62% of the observed variance in the psychological perception) for the identification of the semantic space of the psychological perception of Spanish office tables purchasers. Figure 1 shows an example where a consumer describes his needs. Extreme positive punctuation are asked on axis of innovation feeling (1), psychological comfort (2), space availability feeling (4), privacy image (5) and physical strength perception (8), such an extreme negative punctuation in terms of identification with the nature (3) and hierarchical image (7). Finally, a positive non-extreme punctuation is entered for the idea of pure forms perception (6).

The customers' choice is represented by selecting a scale punctuation of each of the 8 axis used to describe an office table. Responses can range from 'total disagreement' (left) to 'total agreement' (right).

Figure 1: Evaluation of customer demands using the software Kn6/IBV®

E.g., if a consumer would like to select the category no cabinet frame (the less frequent), an output of nine possible products would be extracted (from the previous 2343) in coherence with the original perception entry (Figure 2a). A view to the histograms extracted after this second step shows the categories included in the final solution set. Figure 2b shows, for the supporting frame design item, the only two possible categories: trestle or arch supporting frames. An output of 2343 solutions (as combinations of design elements) out of 4.756.340.736 theoretical possible combinations is extracted according to the psychological perception described by the consumer. A summary of the frequency of each design category in the solution set is available by means of histograms. Figure 2a shows an example of those histograms for the cabinet design item.

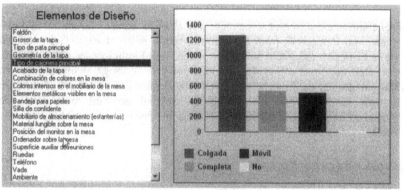

(a) After definition of the solution, set, for each design element the number of proposed solutions in each category is displayed. In the example, frequencies for type of cabinet are shown (in columns from left to right: pendant, complete, movable and no cabinet frame).

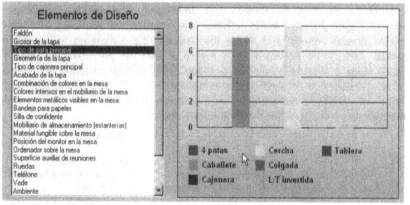

(b) In the example, frequencies for type of supporting frame after filtering initial solutions set are shown (in columns from left to right: four legs, trestle, cabinet, arch, screen suspended, L/T shape and panel).

Figure 2: Display of possible solutions according to customer's requirements using the software Kn6/IBV®

Figure 2b shows that the main category that leads to the asked perception is the pendant frame cabinet. After this result, the consumer may refine his search according with personal requirements from any of the categories included in the solution set. With no more than three or four steps, a consumer may obtain a clear set of product design that fit properly his expectations. Time needed to achieve a convenient solution is reduced in several orders of magnitude. The possibility of user dropping the process is highly reduced. Therefore, customer satisfaction is promoted.

16.5.2 Manufacturer point of view – focusing on production variability

Kansei engineering provides some more key clues to manage variability in a mass customization framework from the manufacturers' point of view. Going on with previous example, the implementation of the initial solution set (2343 different product configurations comprising combinations of 24 design elements each with several categories) needs to be narrowed into a more comprised set of combinations with feasible production variability. Several assistant tools have been developed to help the manufacturer fulfilling this task to get an optimized set of products.

Importance-frequency diagram

Although the customer perception is modeled by means of punctuation in semantic axis, not all have the same importance in taking a purchase decision. To gather that information, an indirect study of the influence of the perception axis in the purchase decision in market was developed (dimensions according to Figure 1). Figure 3 shows in the ordinate axis the observed frequency of positive response in each concept and in the abscissa axis the correlation with purchase decision. The results extracted from Figure 3 give the principal axis that the product image (office table) should stimulate if the average customer is intended. That is, a criterion to be used by the manufacturer, in the task of optimizing the solution set, should be the exclusion of the design combinations with low impact in the concepts more correlated with purchase decision.

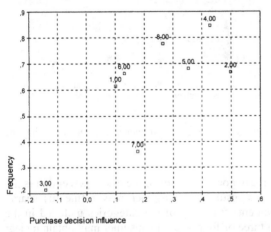

Dimensions

Meaning of the axis:

1 Innovation feeling (bold and eccentric);

2 Psychological comfort (warm and pleasant);

3 Identification with the nature (professional or domestic);

4 Space availability feeling (practical, roomy);

5 Privacy image (communication skills: provide intimacy, favor the concentration);

6 Pure forms perception (simple);

7 Hierarchical image (serious);

8 Physical strength perception (tough).

Figure 3: Importance-Frequency diagram

Brand image

Another criterion to be used by a manufacturer in the task of reducing variability in the solution set may be to define a psychological perception as a brand image and act consequently, limiting the possible solutions to a set that guarantees that desired image. Most companies have an idea about their brand image. It can be defined ideally, as a combination of semantic axis and can be checked using another Kansei graphic tool: the 'profiler'. This is the representation of the market evaluation of one or several products in terms of semantic axis. In this diagram, the mean value for one (or more) product is represented taking the extreme percentiles of products' valuations in the market as a frame reference.

As an example, in Figure 4 two samples of representative models of *casual footwear* are shown. The graphs represent the profiles of both products. The axis take the similarities between both products into account, being on the left with the most similar. The brand image can be defined as the average customer perceived aspects in company's representative products. This communality can be expressed in terms of the ten first axis and the market differentiation relies on the ideas of thermal comfort, smartness and innovation.

If brand image must be created or shifted to a more desirable position, some clues to drive successfully this may be extracted for the footwear company corresponding to results reflected in Figure 3 above. Looking at the concepts with a low punctuation in the ordinate axis and with not so low punctuation in abscissa axis, and acting consequently in terms of design elements leading to that.

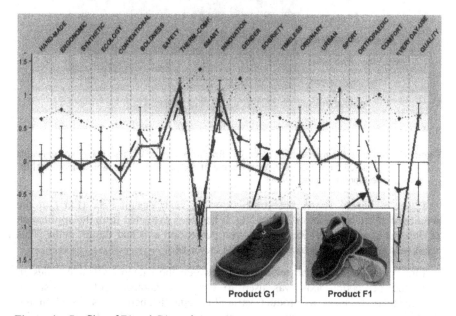

Figure 4: Profiles of F1 and G1 products

Selection of design items

Two coefficients (C and D) have been developed with the aim of defining rules to assign a coefficient representative the contribution to the perception required for each design item. The first one, C, is a combination of axis fuzziness and importance of design element.

(1) $C = \sum_{i,j}(LF_{1,i} * LF_{2,ji})$ where:

$LF_{1\ i}$: load factor indicator of the level of fuzziness of the axis of perception i, if aimed in the psychological perception.

$LF_{2\ ji}$: load factor indicator of the contribution of design element j in the prediction of the axis i.

The level of fuzziness of the principal concepts extracted from the semantic universe of the *office table case* has been studied. It is defined as the portion of the total variability inside each concept able to be predicted from the design items input source. As shown in Figure 5, relevant differences drive into the definition of a level of fuzziness loading factor for each concept (LF_1). The more the R^2, the less the level of fuzziness.

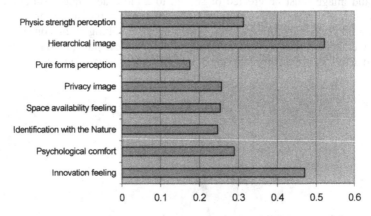

Figure 5: Inverse estimation of level of fuzziness (LF_1) of the principal concepts extracted from the semantic universe of the product office table (theoretical regression maximum R^2)

Figure 6 finally shows the contribution of each design item (Eta^2) in the prediction of the axis innovation feeling. It has been extracted from the application of General Lineal Model Regression with perception axis and design items. This result was implemented into a second load factor LF_2.

Finally, the study of the relationship between the categories inside a design item and perception axis was used to configure another tool to assist the manufacturer decision in product variability establishment. It is based in the definition of a coefficient D where a combination of three loading factors is used.

(2) $D = \sum_{i,j} (LF_{1,i} * LF_{2,ji} * LF_{3,kji})$ where:

LF_1 and LF_2 are already defined, and

$LF_{3,kji}$: is the load factor indicator of the relative contribution of category k inside the design item j in the prediction of the axis i.

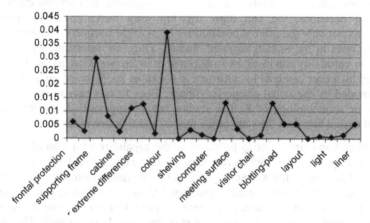

Figure 6: Contribution (LF_2) of each design item in the concept of innovation feeling in the product office table (Eta^2 in a GLM)

Figure 7 shows an example of the output (coefficient D) of this tool for the initial definition of defined categories for the supporting frame design item of the office table, taking two perception axis into account (innovation feeling and hierarchical image). The output encourages the definition of office tables where only three categories of supporting frame design item may be defined, with no sensible reduction in user perception variability. The definition of them, according to the three clusters inferred based on coefficient D will depend on multiple factors. An example of application could be the definition of three new categories as a mixture of each one with similar D punctuation.

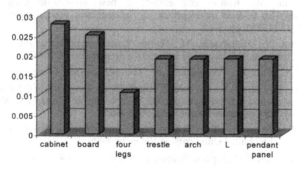

Figure 7: Coefficient D for the supporting frame design item of the office table (two perception axis: innovation feeling and hierarchical image)

16.6 Conclusions

We have proposed two possible uses of Kansei for mass customization. Both approaches apply the results obtained from a 'standard' study in which a broad spectrum of the market product is considered. Concerning the question which design elements and categories affect more the semantic axis and the consequent purchase decisions, we can draw the following conclusion: The approach represents not only the offered possibilities by a single company but nearly all the existing ones in the market segment studied. Thus, the information provided to focus the variability can result in a suggested offer broader than actual. The image selected by a user can lead to constructive solutions inside brand image that were not provided by the actual categories offered. Both factors have to be seen not as threats or shortcomings of Kansei studies but as opportunities to improve the offered design based on the state of the market and user preferences.

Another important point to be stated is the capability of a Kansei system to model and predict the user perception. In the work presented, the rules are calculated to predict the average opinion. On this assumption, the rules lead for certain axis to a seventy-percent reliability. Of course high response dispersion exists, and many personal user characteristics not taken into account in this work (like age, gender, socio-economic level etc) can be included as well. This would lead, at least in theory, to a better prediction of perception. The application of Kansei system affects only to image aspects and does not take important aspects like functionality and usability into account. Thus, we can conclude that our proposed approach is just the first link in the chain of user oriented design processes.

References

[1]   Rosenblad-Wallin, E.: User-oriented product development applied to functional clothing desing, in: Applied Ergonomics, 16 (1985), pp.279-287.

[2]   Nagamachi, M.: Kansei engineering: a new ergonomic consumer-oriented technology for product development, in: International Journal of Industrial Ergonomics, 15 (1985), pp. 3-11.

[3]   Matsubara, Y.; Nagamachi, M.: Hybrid Kansei engineering system and design support, International Journal of Industrial Ergonomics, 19 (1997), pp. 81-92.

[4]   Osgood, C.; Suci, G.; Tannembaum, P.: The measurement of meaning, 1957.

[5]   Faulkner, T.; Caplan, S.: The role of human factor specialists in the development of consumer/commercial products, in: 4th Symposium of Human Factors and Industrial Design in Consumer Products, St. Paul 1985.

[6]   Porcar, R: Application of multivariate analysis to obtaining design criteria for office furniture, Politechnical University of Valencia, 1999.

[6]   Hayashi, C.: Method of quantification, in: Toyokeizai, Tokyo 1976.

Contact:

Rosa Porcar
Biomechanics Institute of Valencia (IBV), Valencia, Spain
E-mail: rporcar@ibv.upv.es

17 Knowledge Based Product Configuration

A documentation tool for configuration projects

Lars Hvam and Martin Malis
Centre for Product Modeling, Department of Manufacturing Engineering
and Management, Technical University of Denmark, Lyngby, Denmark

How can complex product models be documented in formalised way that consider both development and maintenance? The need for an effective documentation tool has emerged in order to document the development of product models. The product models have become more and more complex and comprehensive. A lot of knowledge is put into these systems and many domain experts are involved. This calls for an effective documentation system in order to structure this knowledge in a way that fits to the systems. Standard configuration systems do not support this kind of documentation. The chapter deals with the development of a Lotus Notes application that serves as a knowledge based documentation tool for configuration projects. A prototype has been developed and tested empirically in an industrial case-company. It has proved to be a success.

17.1 Introduction

The need for documenting product models has emerged in several commercial configuration projects followed closely by the Technical University of Denmark. Documentation is essential in relation to the construction, maintainability and further development of product models. Without a proper documentation system it has proved impossible to overview the entire model as soon as the model gets more complex.

The technological advances, increased global competition and customized product demands have changed the traditional way of doing business. Mass-producing companies are changing their business processes to serve the market with personalized products and services. The move from mass production towards mass customization, as described by Pine, has led to new interesting opportunities within many areas [1]. One of these new opportunities is to use information technology in order to build product models which contains knowledge about the product. Not all companies moving towards mass customization are coming from the traditional field of mass production. In this chapter an engineering company is looked upon. How can benefits from mass customization be transferred to this kind of company? Many engineering activities could be conceived as a mass production activity itself and advantages from mass customization obtained here as well. Improving these activities could be called industrialization of the engineering activities.

Danish industry is characterized by many small companies, which traditionally have been very flexible and adaptive to the customers needs. Product related knowledge within these types of companies is typically located as tacit knowledge in the mind of the engineers. In larger companies the picture is the same. Although knowledge has been looked at as a critical asset not much has been done to store it and make it useful to the rest of the company. In larger organizations the task of storing knowledge has become more obvious in the past decade with extensive use of databases as a result. At the Technical University of Denmark research has been done in order to combine product models and knowledge acquisition. Knowledge integrated product- and product related models are defined by Hvam et al. as: "A knowledge base which contains part of or all of the knowledge and information associated with the product in different phases of the product's life cycle, e.g. sales, design, production, assembly, service and reuse" [2]. Furthermore Krause uses the following definition [3]: *"Product related models contain knowledge and information about the systems related to the product's life cycle, while the product model contains knowledge and information about the product's structure and functional properties"*.

In the following section the product model definitions are put into perspective by describing the use and development of knowledge based product models.

17.1.1 Procedure for building product models

The term "specification process" is used to describe the activities carried out within the areas of order acquisition and order fulfillment, i.e. activities such as making quotations, BOM's, drawings, routings etc. The activities of the specification process can be supported by product models as illustrated in Figure 1.

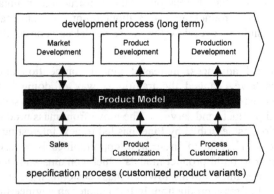

Figure 1: The engineering system [2]

A procedure for building product models to support the specification processes has been developed by the Centre for Product Modeling at the Technical University of Denmark [2]. The procedure is based upon an analysis and redesign of the business processes in order to support the processes with product models. It includes a product analysis by using a product variant master technique and CRC cards. Finally, the product model is designed by using object-oriented modeling. The procedure is shown in Table 1. A short description of the contents of the phases is given below.

Table 1: A procedure for building product models [2]

Phase	Description
1	*Process Analysis.* Analysis of the existing process. Statement of the functional requirements to the process. Design of the future process.
2	*Product Analysis.* Products and eventually life cycle systems are analyzed. Structuring and formalizing knowledge about the products and related life cycles in a product variant master.
3	*Object-oriented Analysis.* Creation of object classes and structures. Description of object classes on CRC-cards. Definition of user interfaces. Requirements to the IT solution.
4	*Object-oriented Design.* Defining and development of the model for a specific programming tool.
5	*Programming.* Own development or use of standard software. Programming the system.
6	*Implementation.* Implementation of the product model in the organization. Training users.
7	*Maintenance.* Maintenance and further development of the product model.

Phases 1 & 2: An AS-IS analysis of the current processes is made. Based on this a conceptual ideal process is designed. Similar to BPR methodology, described by Hammer [4], the future process design (TO-BE) is made. Product variant masters are used to support the product analysis (phase 2). The technique is described by Mortensen et al. [5]. The objective is to get an overview and describe the product assortment. The product variant master consists of two main elements: a generic part-of structure and a generic kind-of structure, as shown in Figure 2.

The product variant master is a complete description of the product assortment. "The generic part-of structure describes the modules, assemblies and parts that exist in all the products within the assortment or the product family. Each element is described by attributes that are determined during configuration. The Generic kind-of structure describes the modules, assemblies and parts which are changeable in the product assortment" [5].

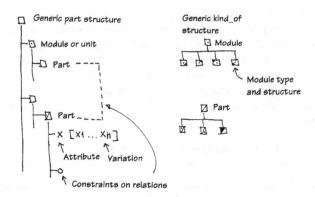

Figure 2: Contents of a product variant master [5]

Phases 3 & 4: The object-oriented Analysis is carried out by using a class diagram as described by Booch et al. [6]. The Unified Modeling Language (UML) is used as notation. An example of the three relations (generalization, aggregation and association) is shown in Figure 3. CRC cards are being used as a supporting modeling tool. Changes have been made to the traditional CRC cards described by Bellin & Simone in order to use them for configuration purposes [7]. They are used as a communication tool between the engineers and system maintainers. The structure of the revised CRC cards is shown in Figure 4.

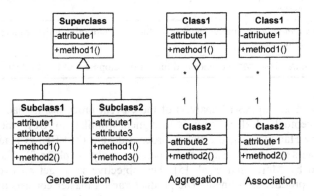

Figure 3: Notation for class diagram [6]

Phase 5: The actual development of the configurator can either be carried out by using a standard configuration system or by own development. The actual amount of programming depends of course on this selection. Own development sets up extensive requirements to programming skills within the company. The use of standard configuration systems gives more freedom to allocate resources to other areas of the development of the entire model, since the actual demands to programming skills are relatively small.

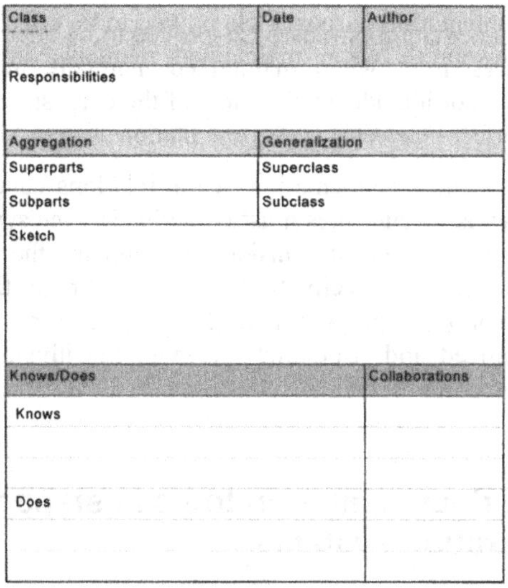

Class		Date	Author
Responsibilities			
Aggregation		**Generalization**	
Superparts		Superclass	
Subparts		Subclass	
Sketch			
Knows/Does			**Collaborations**
Knows			
Does			

Figure 4: CRC card for object-oriented modeling [8]

Phases 6 & 7: Implementation, user acceptance, maintenance and follow-up are very critical factors. User acceptance is completely crucial if the system is to be a success; if the users are not satisfied, the system will not survive long. The users' acceptance of the system can be obtained by involving the users already in the analysis phase by developing an early prototype of the system, which the users can comment on. Also training and current information of the users will facilitate the users' acceptance of the system. Product and product related models can be viewed as "living modeling." The models will soon loose their value if they are not further developed and maintained. By using the structure from object-oriented analysis and documentation (class diagrams, CRC cards etc.) it becomes considerably easier to maintain and further develop the models:

"Application of product modeling introduces a new way of doing business. New tasks are introduced into the organization. E.g. a salesman will have to use the product and product related models in order to configure a product, and a product designer will have to build up and maintain the information and rules describing the products. This calls for commitment and ongoing motivation from the top management. Besides this, both users and model managers need education and training in using and maintaining the product and product related models" [2].

17.1.2 Need for a documentation tool

The documentation of product models is not a new issue. Until now documentation has been made by using Word documents and drawing tools (e.g. Visio) to sketch the product variant master and class structures. The technique of using

CRC cards for documentation purposes has proved to be effective, but the models require a lot of resources when maintained manually in Word documents. Furthermore they do not include an overview of the class structures. This calls for a better and more effective electronic documentation system.

Most important is the documentation of the relations concerning the product assortment. Valid variants and rules must be defined in the system. In order to do this, and to preserve the knowledge inside the company, the development of the documentation system must facilitate the tasks of both the domain experts (knowledge possessors) and programmers (knowledge users). The system should be based on formalized and structured information without limiting the users' creativity.

17.2   Lack of documentation tools in standard configuration systems

A survey has been carried out in spring 2001. Five standard configuration tool providers have been tested in order to find out how the documentation of product models was integrated in their standard configuration system. The configuration tool providers were SAP, Cincom, Siebel, Invensys-CRM and Oracle. It was found that none of the above suppliers could support the product model with tools for documenting the product variant master, class diagrams and CRC cards. The programming was the documentation itself. None of the suppliers provided complete object-oriented configurators even though some claimed to be using object-oriented principles. But functions such as encapsulation and inheritance were not included. Data and rules/constraints are not collected in common object classes. This leads to more difficult maintenance procedures, as soon as the product models become more complex. Product data and rules were separated in all the mentioned configuration tools and could only be related to each other by naming folders in the same name.

The standard configuration systems demanded different levels of programming skills when developing the models. Users without any programming skills could handle some of the systems (e.g. Cincom and Oracle). If the system should be integrated with ERP or other programs, extensive programming would be necessary. All the suppliers offered different helping tools when working with product attributes and constraints/rules but the general picture was that more complex models would be difficult to manage without a "paper model" to begin with. The survey showed a lack of documentation tools that were able to support the procedure for building product models, described in above. It would be necessary to develop this system outside the standard configuration system.

17.3 General requirements to the documentation system

Since the standard configuration systems did not offer any sufficient internal nor external documentation a study has been made to state the requirements such a system must comply with in order to build the documentation system. Based on the experience gained from several configuration projects the following requirements can be stated to a new electronic documentation tool:

1. *Easy to maintain:* Easy access to knowledge and logical structure of classes and rules. Furthermore automatic updating procedures addressed to the responsible employees/functional departments.

2. *The system must facilitate a structured design of knowledge* following the proposed structure described above. The knowledge could be represented in tables, syntax rules or plain language (product variant master, class diagrams and CRC cards).

3. *Central storage of data:* A central storage place (server) minimizes multiple versions of the model.

4. *Distribution of data using networks* (e.g. Intra-, Extra- or Internet): It should be possible for users to interact with the server independently of location.

5. *Multiple user access:* Multiple employees are often involved in the development of complex product models. These employees represent different functions e.g. R&D, purchase, construction and sales. The knowledge put into the product model is therefore on several levels. It is necessary to structure this knowledge in a way that can be understood by fellow colleagues and developers of the configuration tool.

6. *Automatic updating* from product variant master to CRC cards and class diagram: To ease maintainability it would be useful to link the product variant master to the CRC cards and class diagram so updating in one area would affect the two others.

7. Version control: Numbering and status of different parts of the model and the entire model itself have to be documented.

8. *Integration with configuration tool:* The survey described above showed insufficient documentation possibilities inside the standard configuration tools hence the documentation system should be separated from the configuration tool. Close integration between the configuration system and documentation is advantageous to ensure that the content of the two systems is the same. The ideal is to link the two systems closely together so that when a rule in the documentation system is changed it is automatically corrected in the configurator.

17.4   Description of a Lotus Notes based system

In this section a principal solution of the documentation system is sketched, based on a list of requirements focusing on Lotus Notes. The documentation system is more than a documentation collection. If the system should work properly searching and linking functions must be available. This calls for a relation-based application. The Lotus Notes application has been chosen due to its wide acceptance as a standard system throughout the world. Lotus Notes is suited as a document database. It is not a relation-based database. Traditionally, knowledge has been stored as documents in companies but features from the relation-based databases would be useful when creating a documentation system. A relation database complies to the following two criteria:

- The data structure is based on tables. A flat structure where the data inside the tables is in focus.

- Defining the database is independent from the physical storing structure.

Earlier problems with the DataBase Management System (DBMS) meant slow optimization and searching processes. With the development of DBMS this has been improved. Relation-based databases are now much more flexible. Notes can be programmed to act as a relation-based application using special Notes applications.

17.4.1   Requirements to the Lotus Notes application

The general requirements to the system were listed in the last section. In extension to these requirements a set of new requirements are listed, specifying the demands to the Lotus Notes application.

Structure of the product variant master: The application must show different levels of the product assortment as follows:

1.	Level 1			2.	Level 2	
	1.1	Level 2			2.1	Level 2	
	1.2	Level 2			2.2	Level 2	
		1.2.1	Level 3			2.2.1	Level 3
		1.2.2	Level 3				

The numbering system must be generated automatically and changes/insertions handled as they arise. Changes in the specific product related data should affect the structure of both the class diagrams and CRC cards. This should work both ways meaning that if the structure is changed anywhere in the documentation system, it is automatically being updated in the corresponding systems. The *graphical appearance* of the *class diagram* in the documentation system is shown in Figure 5.

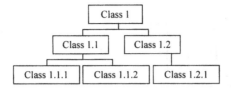

Figure 5: The principal structure of the class diagram

Structure of the CRC cards: The application must incorporate the following issues concerning the formalizing of knowledge on CRC cards:

- The CRC cards should have a structure as described in the first section of this chapter.

- Creation of specified fields.

- Search possible for specific fields.

- Tabbing between fields.

- Actual status of the CRC card e.g. development, implementation or in use.

- Linking to other documents e.g. drawings, tables or word documents.

User control: A set of procedures must be built into the documentation system in order to control the access and editing possibilities of a specific CRC card:

- Author's name.

- Editing names.

- Responsible/document owner.

- Automatic generation of emails when an update is needed.

17.4.2 Activities during the development and implementation of the configurator

In general terms the activities during the development of the system can be sketched as shown in Figure 6. The process is initiated by knowledge contributions from domain experts. Together with the model developers the knowledge is documented by using a product variant master, CRC cards and class diagrams. The result is stored in a central documentation system. Programmers or engineers use the knowledge database to create the configurator. Once the system has been implemented it is necessary to maintain the model. Maintenance routines should be built into the model during the development phase. The operating procedure of the system is shown in Figure 7.

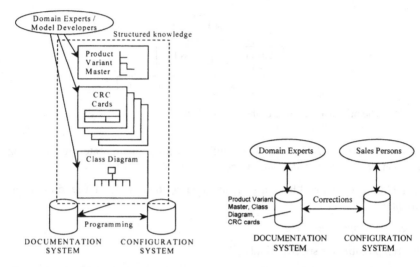

Figure 6: Activities during development   **Figure 7:** The documentation system

17.5 Empirical case study at Niro A/S

A case study research was initiated at Niro A/S ("Application of Product Models in Extended Enterprises"). The aim of this project is to develop a configurator to support the sales process and investigate how the implementation of the configurator changes the business processes, and how relations to suppliers/customers are affected. The configurator should give the sales staff sufficient technical information about the components to give a detailed quote for a complete spray drying plant.

Niro A/S is an international engineering company that has a market leading position within the area of design and supply of spray drying plants. World-wide the Niro Group of companies consists of 36 companies and 2050 employees creating approx. 340 Mio. US\$ in turnover a year. The products are characterized as highly individualized for each project and services that are customized. During the development of the product model the need for an effective documentation system has emerged. The aim was to develop a prototype of a documentation tool as specified above. The prototype should later be transferred into an actual documentation system.

17.5.1 System structure

Early in the project it was decided to separate the documentation system from the configuration software. Lotus Notes Release 5 is implemented throughout the company as a standard application, and all the involved people in the configurator project have the necessary skills to operate this application. The documentation tool is therefore based on the Lotus Notes application. The prototype of the documentation tool should give a good base for validation of the functions needed for a full-scale documentation system.

A Notes-template has been developed and used for many different databases in the company. The design template figures as a master to other databases, which refer to the design template. When changes are made to the design template the changes are transferred to the referring applications. The documentation system was also based on the design template instead of a special application. A special application would give more features and the opportunity to custom design for this specific use but the standard functions from the design template were sufficient in this case. The design template provides functions such as searching for specific words, linking between documents and different sorting procedures (e.g. by date, title, category, author or document owners. The system is built as a hierarchical response system. The user interface (UI) shows the whole-part structure of the product variant master as described in section 1 and the corresponding CRC cards (Figure 8).

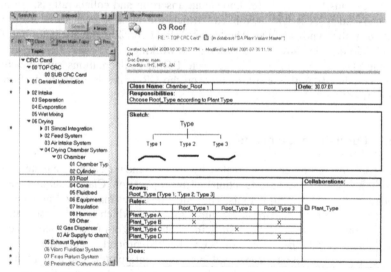

Figure 8: User interface displaying product variant master (left) and corresponding CRC card (right)

17.5.2 Product variant master and CRC cards

The User Interface in Figure 9 is divided into two main parts, the product variant master and the CRC cards. The *structure of the product variant master* is the same as shown in Figure 3. Only the whole-part structure of the product model is transferred to the documentation tool. Main documents and responses construct the tree. The response design needs two master CRC cards to create the structure. One top card for creation of main classes and a sub card to create responses.

The left side of the interface contains the *CRC cards* and a header. The CRC cards describe the knowledge and the header shows information used for the Lotus Notes application such as title, author, document owner, editors and readers. Since the standard configuration software does not provide full object-orientation the CRC cards described above has been changed to fit the partly object-oriented system. The fields for Modeling and aggregation have been erased. The aggregation relations can be seen from the product variant master and generalization is not supported. To ease the domain experts overview an extra field to describe rules has been added. In this way methods (does) have been divided into two fields, does and rules. "Does" could be e.g. print BOM while "Rules" could be a table stating valid variants.

The CRC Card is divided into three sections. The first section contains a unique class name, date of creation and a plain language text explanation of the responsibility of the card. The second section is a field for sketches. This is very useful when different engineers needs quick information about details. The sketch describes, in a precise manner, the attributes, which apply to the class. The last section contains three fields for knowledge insertion and collaborations. Various parameters such as height, width etc. which the class knows about itself are specified in the "knows" field. The "knows" field contains the attributes and the "rules" field describe how the constraints are working. The "does" field describes what the class does, e.g. print or generate BOM. Collaborations specify which classes collaborate with the information on the CRC card in order to perform a given action.

17.5.3 Future improvements

First of all the prototype was created by using standard Notes templates. This gives limitations according to functions of the application. Class diagrams are not included in the documentation tools. They must be drawn manually. Implementing class diagrams would demand a graphical interface, which is not present in the standard Lotus Notes application. Since the prototype is not fully implemented yet the maintenance routines have not been checked properly. The first feedback indicates a positive effect when developing the configurator, and a more simple maintenance is forecasted even though the generation of a reminding email to document owners after defined intervals is yet to be tested.

17.6   Conclusion and perspectives

The need for a structured documentation system for knowledge based product models has emerged. It has proved necessary to formalize the knowledge on which the systems are based in a way that consider development and maintenance of the configurator. A survey has been carried out in order to study whether standard configuration systems include the needed documentation system. Five software suppliers (SAP, Cincom, Siebel, Invensys-CRM and Oracle) were involved in this survey. It was found that the documentation of the configurators was not sufficient. A separate documentation tool had to be created.

According to a procedure of building product models a prototype of this documentation system has been developed. It is based on a set of requirements that originate from several industrial configuration projects in Danish industry. The requirements have not been accomplished fully in the prototype but the main structures have proved to be working successfully. The documentation system is created in a Lotus Notes application. It consists of a product variant master with related CRC cards. It is based on a whole-part structure of the product assortment. CRC cards are used as a tool to translate knowledge from domain experts into conventional inputs to the programmers of the configurator. The CRC cards are an element from object-oriented analysis. Since the standard configuration systems available on the market are not fully object oriented the CRC cards have been revised. Other elements from object-oriented analysis such as class diagrams are not implemented in the documentation system. They must be created outside the system.

Work is currently being carried out at the Centre for Product Modeling at the Technical University of Denmark in order to set up requirements for an object-oriented documentation system. The system should be based on a single object-oriented model. One of the most powerful advantages of the tool will be to allow the user to view and edit information through different views (e.g. product variant master, CRC cards, class diagrams etc.). Today the product master and CRC-cards are developed and maintained side by side as two different elements of the documentation, regardless that some information is hereby duplicated. By using a single model as a core of the tool, information is kept in a single repository and the product master and CRC-cards will then serve as different views on the model.

The results from the first testing of the prototype show that the development of the actual configurator has become more simple, since the knowledge is formalized in a way that facilitates the programmer. Furthermore the use of an electronic documentation system has proved to be easier to maintain than the former paper models.

References

[1]   Pine II, B. J.: Mass Customization, Boston, 1993.

[2]   Hvam, L., Riis, J., Malis, M.; Hansen, B.: A procedure for building product models, in: Product Models 2000-SIG PM, Linköping 2000.

[3]   Krause, F.: Knowledge integrated product modeling for design and manufacture, in: The Second Toyota Conference, Aichi 1988.

[4]   Hammer, M.: Reengineering Work: Don't Automate, Obliterate, in: Harvard Business Review, 1991.

[5]   Mortensen, N. H.; Yu, B.; Skovgaard, H. J.; Harlou, U.: Conceptual Modeling of Product Families in Configuration Projects, in: Product Models 2000-SIG PM, Linköping 2000.

[6]   Booch, G.; Rumbaugh, J.; Jacobson, I.: The Unified Modeling Language User Guide, 1999.

[7]   Bellin D.; Simone S. S.: The CRC Card Book, 1997.

[8]   Hvam L,; Riis J.: CRC Cards for Product Modeling, in: The 4th annual international conference on industrial engineering theory applications and practice, San Antonio 1999.

[9]   Vang, S.: Relation databases and SQL (Danish), Technical Publishing, Copenhagen 1996.

Contact:

Professor Lars Hvam
Centre for Product Modeling, Department of Manufacturing Engineering and Management, Technical University of Denmark, Lyngby, Denmark
E-mail: lhv@ipl.dtu.dk

18 The Customer at the Final Frontier of Mass Customization

Carsten Svensson[1] and Thomas Jensen[2]
[1] Department of Manufacturing Engineering and Management, Technical University of Denmark, Lyngby, Denmark
[2] Danish Technological Institute, Center for Production, Taastrup, Denmark

Mass customization is no longer new. A decade of industrial experience have shown how this business paradigm has been used – and abused. Some companies report on a successful implementation leading to a radically improved business while others have not managed to fully exploit the promised potential. At this point in the evolution of mass customization one may look back and examine these cases with the purpose to empirically determine the factors influencing a successful application of mass customization. However, one may also look into the future and speculate how mass customization may be further exploited. With this in mind, we will discuss in this chapter which factors may limit the further expansion of mass customization. We will argue that the customer is the major limiting factor at the final frontier of mass customization. Until now mass customization has mainly focused on the product. In this chapter we show that there is a need for an increased focus on the fulfillment of customer needs. As a result manufacturers have to balance new trade-offs if the paradigm of mass customization becomes a commodity. This chapter's objective is to open a discussion within research communities working with mass customization. Thus, more questions are raised than answered.

18.1 The paradigm of mass customization

Mass customization was first identified in "Future Shock" by Toffler [1] and later described in "Future Perfect" by Davis [2]. Davis describes mass customization as the contradiction of mass production. The idea of mass customization is based on the observation that there is a customer interest in products that are adapted since the adaptation will increase perceived performance. As the standard of living has increased during the last 50 years individualization has received increased focus. Customization is coming within the reach of many consumer. At the same time there have been massive developments of technologies, e.g. information processing, production control, flexible manufacturing systems, logistics and mechatronics – enabling efficient customization.

18.1.1 Why customization?

In the natural cycle of industrial evolution *order-winners* will over time turn into *order-qualifiers* [3]. In the 1970's quality and function where order winners, but during the 1980's and 1990's products have become increasingly uniform regarding attributes such as quality and performance, and thereby they are turned into order qualifiers. Faced with this challenge manufactures are forced into the search for new parameters of differentiation. Some manufactures are using mass customization as a tool to escape the enervating price competition between uniform products. Customization is one mean by which manufactures strive to differentiate their products in a world of similarity. Since customers are individual and have different preferences there is a fertile soil for the development towards personalization and adaptation. Customization is intended to add increased customer perceived value to a product, since a customized product – compared to a standard product – increasingly fulfills the need of its customer. Historically companies have made a strategic decision on craft production, build-to-order, or mass production. Within these traditional paradigms, a customer may select between a customized product with a high price or standardized product at low costs. With the emergence of mass customization the added value of customization should now be manufactured with no or only little extra cost. The differentiation is, however, only effective (i.e. the product is sold) if the customer considers customization to be of value. Recognizing the customer as the major (limiting) factor in the successful exploitation of mass customization is the central theme of this chapter. We will focus on products for consumer markets, however, many arguments are also valid for other customized products.

In the context of manufacturing, some major drivers for mass customization can be observed:

(1) Globalization: Markets are becoming increasingly global. Competition in many markets is fear. Low cost manufactures in developing countries possess a challenge that many western manufactures can not match. Therefore, western manufactures must turn to other means of differentiation than price. Adding to this spreading of mass customization are market expansions. Products are no longer targeted at one geographic market only, but moreover towards a global market. Thus, product variations must be made efficiently possible in order to adapt to local legislation, taxation, climate, and taste.

(2) Market turbulence: Pine [4] made an investigation regarding the development in speed and dynamics of the markets. He showed that many manufactures experienced a more turbulent and unpredictable market. This turbulence created a need for agility. Mass customization can be seen as a tool to achieve flexibility towards a dynamic market. Through the use of modularization mass customization can reduce the risk in product development as a large number of variants can be launched to insure a wide market coverage. Product development can be performed in modules. Time to market is short compared to traditional product / variant development. This is particular relevant for innovative products, since first mover benefits are of large importance in a turbulent markets.

(3) Outsourcing of technologies: A major source of the increasing similarity in product performance are outsourced technologies. If suppliers invent a break-through technology, these suppliers sell the technology often to all markets leaders. Gore-Tex breathable fabric, for example, is sold to virtually all manufac-tures of high-end sports garment. The background for this development is the increased focus on core competencies as products are becoming increasingly complex and often hold a wide range of technologies. The manufacturers of consumer products do not have the resources to invest in the development of all the competencies used in their products. In the following, manufacturers need other means of differentiation towards their (end-)customers than the technology of their products.

(4) Driver for conversation: Another factor driving customization is the social experience a customer gets when buying a product through a sales person. The customer is not only buying the product but also the politeness and attention of the salesperson. Through the customization process the product becomes a story, a driver for conversation, and this can be partly what the customer is buying. Mass customization in the regard is a major enabler of the so called "experience economy".

18.1.2 Standardization and variant manufacturing as alternatives to customization

The paradigm of mass customization radically challenges the business process of a former mass producer. The sales process becomes more complex since more variants are offered to the customers. Marketing must understand customers needs more thoroughly and target its efforts to smaller segments than traditionally. When implementing mass customization the marketing and sales process may be supported by various tools like CRM systems and sales configurators. The product development process becomes also more complex as the variety requested by the customers somehow must be designed into components having manufacturing similarities before. In some cases the development of a product platform with modularized building blocks and well-defined interfaces is the solution, in other cases a parameterization allowing cut to fit. Finally manufacturing processes must be flexible towards the variations in the components. Venturing up on the road towards mass customization one must understand that the evolution form mass production to mass customization is a large and radical change in the way of doing business.

Companies that do not want (dare) to perform this change can use standardiza-tion as a tool to target market segments as an alternative to customization [5]. Standardization is pure mass production of a number of variants. The customer may experience the product fulfilling his particular needs since as like the customized product since there are choices. Standardization can for instance be seen in supermarkets, where toothpaste from the same manufacture can be bought in several variants. Similar, breakfast cereal comes in a large number of variants targeted at well-defined groups such as weight conscious woman, children or conservative men. Variation as a tool is most often used for low cost products. Benefits and drawbacks of variation include:

- The benefits of using variation compared to mass customization are low costs, no waiting time and rather simple production.

- The drawbacks are a small solution space compared to a customization system and the lack of flexibility to changing markets.

A McKinsey report [6] points to another alternative for the auto industry called "locate to order". Within a "locate to order" system a dealer can locate a product with the desired variant through the use of a nation wide database at all dealerships of the brand. Thereby the manufacturer can increase customer satisfaction similar to a customized solution without changing the production system.

18.1.3 Variety versus customization

Variety is the tool of mass customization. By using variation it is possible to create the link between the customer and the product. In advertising it is often seen that the major benefit of mass customization is that a product comes in a large number of variants [7]. But if variety should be of any value it must be applied with great care. Variety has to be considered from the perspective of value creation for each customer. A customer is only interested in the one variant that fulfills his need while all other variants are irrelevant. In some situations, the other variants are merely disturbing the customer from finding his particular variant – remember how it was looking through a pile of trousers to find exactly your right size. Therefore, variety is only a mean and not the end for achieving increased customer satisfaction when a company serves market with customers having different needs.

Thus, variety must be faced from the customers' perspective as only thereby the customer will perceive the adaptation as being of value. All too often variety is not relevant to the customer, even though it is expected that the customer will spend time making the choices regarding a commodity. Variation must have an effect that the customer can observe. An example of failed variation is Nissan which, according to Pine [4], offered 87 different steering wheels. This is supported by observations made at Toyota [8]. In the late 1980's Toyota set out to build a wide range of made-to-order cars. Reality revealed that 20% of Toyota's product varieties accounted for 80 % of the sales. Each variation comes at a cost and therefore too much variance will weaken the entire product program. In order to maintain the competitive advantages it is necessary be careful in the selection of attributes where customization is offered.

18.2   Factors influencing the expansion of mass customization

As argued earlier the business logic of mass customization is based on the added value of a product when customized. Therefore a thorough understanding of how customers perceive value is central for the successful application of mass customization. Otherwise the customer buys a mass-produced product. An example of customer value is the Vitamin Company Vitabiz [7]. The company was founded by Brad Oberwager. His sister had to undergo cancer treatment that required more than 15 different vitamin products. Each product had to be swallowed which was uncomfortable for the patient. Therefore, in this case customer perceived value was a reduction in the number of tablets that the patient would have to swallow. The result was a customization of portions, so each tablet would contain a mixture that would fulfill the patient needs. Thereby the patient would have to make fewer swallows. In the following, we want to examine some factors influencing the expansion of mass customization from a customer perspective.

Striking the balance of mass customization

Mass customization is a trade-off between cost and the scale of variation. In order to provide a sellable solution, the designer most master the balance between overall cost and the customer perceived value. This can only be done through a deep understanding of the customer and his perception of the product. Hart [9] describes the balance of this trade off as *customer customization sensitivity*. The uniqueness of the customers need is balanced towards the sacrifice (hassles, inconveniences, discomfort, long waits, product or service deficiencies, high cost, difficulty of ordering, lack of fulfillment options etc.). This balance is then seen with the perspective of the price sensitivity of the market.

The key to efficient mass customization is not only finding the right attributes to vary, but also locating the right point of customization. From the point of customization the customer can influence the configuration of the product. Finding the right point of customization is a dilemma. On one hand the customer perceived value is related to the level of influence on the product, but on the other hand costs must be kept down and therefore the point of customization must be placed as late as possible through postponement. Customized products can be divided, based on the point of customization, into the following segments by Mintzberg et al. [10]. Examples of customization classifications, as illustrated in Figure 1, may include:

- *Pure standardization:* Toothpaste, soft drinks.

- *Segmented standardization:* Stereo components like CD player, amplifier, TV, FM-receiver that is assembled for a HI-FI system

- *Customized standardization:* Car production where color, interior etc. are assembled according to a customer order.

- *Tailored customization:* Products where the customer can have materials or extra equipment of choice, but is constrained by the basic design of the product e.g. a house or a boat.

- *Pure customization:* Is often larger "one of a kind" projects where the customer is tied by few constraints; ship building etc.

Lampel and Mintzberg [11] argue that the mass customization trend is not towards pure customization, but more of a concentration around customized standardization, and as a result the variation will be reduced for some products.

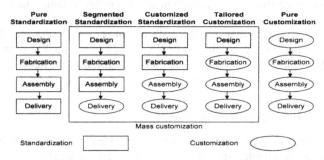

Figure 1: Division of products into groups based on the point of customer involvement (figure based on [10])

The customized product – basis for customer interest in customization

Consumers are not equal in size, nor taste or financial ability. Therefore there is a market for products adapted to reflect the needs and wises of the individual consumers. There can be numerous reasons for choosing a tailored product over a mass produced, but the main reason is that a product "cut to fit will" will increase performance and as a result the customer will be willing to pay an extra premium. An example of this could be clothes that will have a better appearance, if adapted to the individual measurements of the customer or a newsletter that contains filtered news based on personal preferences, and thereby creates value through timesaving.

When selling customized products, it must first be recognized, that it is not the product that is sold, but the solution to a need. The product can only be perceived as the carrier of a function and a tool for customer satisfaction. So the value of the product is closely related to the need that the product is targeting. In order to get the best value the need must first be understood. In some markets even a decoupling of the ownership of the product can be observed e.g. car rentals for commercial use. The rental company Hertz provides the cars to address the total transportation need of a company, but the cars and trucks are owned and serviced by the rental company. This will also be seen in consumer markets, and the first indicators of this are the increased leasing of durable goods. Taking the demateri-alization of the product even further might make possible to rent a car with out air-condition even though an air-conditioner is installed, so that the customer only pays for the use, but not the ownership.

Products benefiting from customization

In order for a product to be a candidate for mass customization, there must be a variation in the customer preferences, and the product must have a significant importance for the customer. Mass customization has had a devastating effect on some markets, e.g. housing, cars, and furniture. When Henry Ford launched his Ford T in the beginning of the 20th century there was no customization, and the alternative to the Ford T was a horse carriage. As alternatives started to show, variation started to grow. First it was the color then the shape, and today all car manufactured in the industrialized world offers at least variation in colors, chassis, engines, interiors, etc. this is because the customers wanted to influence a significant product, like a car. Another success story of mass customization is the clothing industry. According to Computerworld [12], 10 % of all output in the apparel industry were customized and customization was accounting for 30 % of profits. The basis for customization of clothing is the different sizes and shapes of the customers. It is clear that there is a willingness to pay extra to get an exact fit. Customization also opens up for an aesthetic adaptation to favorite colors, styles etc.

Drawbacks of customization

Investigations by Huffman [7] have shown that selling a customized product can be difficult because the many options and the number of choices can lead to frustration for the customer. In a developed mass customization market, like the automotive there are great regional differences, 90 % of all cars sold on the US market are without customization [13], compared to the 60 % of the German market [14]. This observation must lead to the conclusion, that there are some imperfection to the way that mass customization is applied. There can be numerous reasons why customers avoid customization, a few are listed here.

- *Lead-time:* Often customization will result in a lead-time since the product will have to be build to order. Therefore the customer must be willing to wait.

- *Cost:* There is often a premium connected to the customization process.

- *Time consumption:* Often the added value from customization do not match the time invested in the configuration process.

- *Uncertainty:* Customers often face uncertainty regarding the right configuration of the product, and the effect particular decisions will have on the final per-formance of the product.

The background for the large difference on the car market might be due to cultural differences. The American consumer sees to a large extend the car as a commodity. As a consequence the American consumer is priced focused and impatient. In contrast, German consumers consider the car of importance and are therefore willing to accept the drawbacks to a customized solution.

Costs of customization

As stated earlier, mass customization is characterized by the combination of mass production and customization, offering unique products at a low cost. Looking at the market it is clear that in most cases there are some costs generated by the customization process. Pine et al. [8] claims that mass customization comes at no extra cost, compared to mass production. This can be argued, but looking at customized products today it is difficult to find products where there is no extra cost. In general three major cost drivers can be identified:

- the implementation of flexible production equipment,
- availability of individual logistics,
- configuring, designing and selling the customized product.

Flexible manufacturing equipment is typically more expensive than mass production equipment, and the cost of this equipment most be distributed over the products, which thereby becomes more expensive. A flexible system has difficulties to be a match for the efficiency of a dedicated production line.

For products customized at a production facility, individual logistics will generate cost. Customer and product are linked before manufacturing. This is for instance the case for automobiles, where the customer creates a pull situation through production by ordering a unique combination of color, interior, engine etc. This results in a reduced capacity utilization or creates waiting time. Some automotive manufactures compensate for this by producing cars to be sold of the lot, whereas more exclusive manufactures will have to accept a lower capacity utilization, or a waiting time, since their customers expects a unique product.

The sales process of a customized product is typically more complicated than a standard product. Since the customer must be supported in the decision making process either by a configurator or a salesperson. Therefore the process of selling a customized product generates more costs than selling a product off the shelf, which does not demand any interaction with the customer. During the last ten years the cost of customization has fallen. It can be observed that some markets are dominated by customized products. As a result of the cost reduction which has been enabled by the application of industrial management methods ensuring cost efficiency.

Products not benefiting from customization

Failure on the customization market is often linked to a lack of customer interest in adaptation. For many products variation is of little or no value to the customer, since it's not noticed or maybe the trade offers are not attractive. This goes for most consumer goods. As humans, many of our need are uniform, and so are the products that addresses these needs. Because of this there is no value in adapting commodity product like a vacuum cleaner, hand soap, books or appliances. In most cases mass-produced variants will cover these needs. If we look away form this group of functional products, another interesting group of products can be added, products with artistic attributes. This is typically products sold on appearance, provocation, or products that are sold as an experience or an interpretation of a product. Examples of these products can be, design products, works of art, music, literature, travel, gourmet food etc. These products do not only have a function but also a statement, and this statement must not be compromised if the product shall keep its value for the customer. As wealth increases in society a proportionally larger part of consumption will be seen in this area.

When looking at the customized product value generated from variation has to be considered firstly. Radder et al.[15] have set up some basic questions regarding customer orientation helping to define the value of mass customization:

- Do the customers really have unique needs?

- Do the customers really care about more customization of their products/services?

- Do they really want more choices or will they be overwhelmed by the large variety?

- Are customers prepared to accept certain sacrifices in order to buy form an specific organization?

- Will they be prepared to pay more and/or wait longer for a customized product or service?

18.3 The customization process

Having reviewed the market for customized products, it can be observed that the barriers connected to mass customization are the customization process and the cost of customization. This observation is supported by Eastwood [16] who has pointed out two of the major obstacles implementing mass customization, cost and time. The goal of the customization process must therefore be the reduction of the customer sacrifice and cost through the customization process. The manufacturers must help the customer to configure the right product, since the manufacturer holds the most knowledge of the products and its utilization and thereby can eliminate uncertainty. When observing applied customization, it is a paradox that the customer is often let alone to design a product, which the customer does not have any experience with. The customer knows and understands his needs, but for the customer it is difficult to transform this need into a configuration of a product. If a manufacture wants to exploit the potential benefit of customization, it is as important to help the customer to get the right product, as to design the product it self. A poor but well adapted product can be as valuable to the customer, as a good product that is poorly adapted.

Mass customization increases the complexity in the selling process by offering choices and options. When selling a customized product there is a demand on knowledge of the product from the sales person. The customization process has show to be problematic. There can be many pitfalls in customizing a product, this process is often time consuming and generates errors. To support the sales process a configuration can be used to handle the rules and constraints of the product. The use of a configurator will push the knowledge of the manufacturer towards the customer thereby reducing structural uncertainty. By using configuration systems it is possible to guide the customer into a valid solution. A valid solution is not necessary the right solution for a customer and this is where most configurators fail. In order to achieve the best result the configuration must take its starting point in the need of the customer, instead of the structural configuration of the product. Choices and options are not necessary good; according to Zipkin [17] elicitation is one of the major challenges to mass customization. Using a traditional configurator the customer often feels insecure, hence it can be difficult to link the product options to the need. This can be a major barrier to the expansion of customization. Figure 2 shows the process that must be taken into account when designing the configuration process.

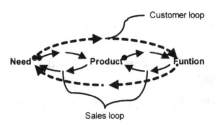

Figure 2:   Configuration process based on understanding of customers need

Configurators almost always addresses the product instead of the need, so in order to compensate for this a salesperson is often used as the expensive link between need, function and product, thereby reducing the competitiveness of the customized product. Not only are sales personnel a costly solution but also the lack of skill can inhibit the customization process. If a salesperson shall sell a customized product further education is often needed and for the retailers this investment is a barrier because of the high overturn of personnel in retail.

18.3.1 The customer as designer

Previously it is argued that the customer is not interested in the structure of a product but only in its function and thus only wants to express the need as an input to the configuration process. Nonetheless, one may reasonably claim that some customers have an interest in being involved in the structural configuration of the product i.e. being a designer. For one reason, the customer may not be able to express his need precisely and consider the configuration process as a trial-and-error process requesting a direct involvement. For instance, it may be difficult to express how a meal ought to taste.

Once however a certain meal is tasted, it is easy to say if this meal has the correct taste and then adjust the initial expression of the need according to this knowledge. It can be argued that the customer does not want to be involved in the actually configuration process, but rather wants to express some provisional needs and quickly wants to see if partly these needs can be meet and partly if the needs were correctly expressed. In other terms, ideally the customer wants to freely and quickly match the intended and actual behavior of the product, but how this match is realized by a structural configuration is not of relevance. Naturally, a reason for involving the customer could be the close Ping-Pong discussions between the one 'expressing the need'/'impressing the fulfillment' and the one trying to bridge these. This however is not to be confused with the customer being interested in the configuration process. For some products, it is feasible to utilize a trial-and-error process, in particular when the aspects of consideration easily can be prototyped, e.g. the aspect of shape handled by digital prototyping in CAD (computer aided design). For other products, where this is not feasible, an alternative may be to express a need using an existing product as a point of reference. For instance, a customer in search for loudspeakers may provisional express the requested sound as more light than loudspeaker A and less hollow than loudspeaker B. Though, a tricky part is then interpret the customers formulation of the recognized need. Finally, the two approaches may be combined e.g. in the configuration of a car, where the matching of the interior colors may be digitally prototyped. And the power of the engine may be described in relation to other cars – it may even be related to the present car of the customer.

Other reasons for a customer being interested in the structural configuration of a product may relate to a need for self-realization. A need for acting as a designer maybe the primary need, whereas the outcome of the act i.e. the product, only is a secondary need. Several reasons may exist for such an acting need. One reason may the act itself, since design acting may be joyful, challenging, educational etc. In an experience economy [18] customers may even be interested in paying extra for the possibility of acting, thereby turning the uninteresting experience of purchasing into a memorable event. Acting however always takes place on a stage, a stage set by others, so in reality the customer is not a designer, but merely an actor on a stage set by the product designer. More precisely, a precondition for the structural configuration of a product is the design of the modules and their interfaces. This is truly a design activity, since it causes the existence of new modules as well as a set of rules for how these modules may be configured. Therefore, a customer only acts as a designer within the domain of the pre-designed modules and the preconditions for configuring these modules. A customer may experience such acting as being a designer, but given a known and predefined solution space as well as a map on how to navigate in this space, one may question who is in control of the acting. This balance is also expressed by the CEO of Lego, who once reminded the company that they could design products addressing their customers either as architects or as assembly workers.

Another reason for acting, and thus being interested in the structural configuration of a product, may be a need to directly influence and control the surroundings i.e. to make a visible and concrete difference by being the primary cause for the existence of a product. In a world with mass-produced products it is reasonable to ask, if one really can be proud of owning a mass-produced product. Does a product become more interesting simply by a change of ownership? A change of ownership that from some points of view do not even take place e.g. when borrowing money for buying a car. In worst case, the only aspect worth being proud of when buying a mass-produced product is the ability to recognize and express a need and then the ability to earn the money for buying the product fulfilling this need. The ownership is not to be proud of, only what can be done with the ownership e.g. the pictures taken with a camera, but not the camera itself. The world of mass customization adds the dimension of allowing the customer to be proud of influencing the product and not only influencing what can be done with the product. This extra dimension may be of relevance, in particular for consumer goods, where the need for the function of a product can be fulfilled by most manufactures. Thus, such an influence may sufficiently be satisfied by expressing a need and then await the fulfillment i.e. the structural configuration of the product. In a world based on division of labor, most people are accustomed to not influence complete processes but only bits and pieces.

18.3.2 Design by customers

So far, only the situation of the customer being aware of a need has been discussed. Another situation may be that the customer is not aware of a need, i.e. a latent need, or that the customer does not even have a need. In this situation one may ask: Does a customer have a need that once recognized can be fulfilled by a product? Or does the existence of a product cause a customer to recognize a need?

In other terms, is the need the cause for or the effect of the product? The question is similar to the relation between a problem and a solution i.e. does a problem arise because of a solution or does a problem cause the solution? The answer to this question is of relevance, since it may be that the object of the design activity not only is the product but also the need of the customer, or more roughly speaking the customer himself. If this is the case, a precondition for mass customization is no longer present, i.e. individual needs, since the need of customers may be designed uniformly in larger segments. Furthermore, if the product prevails the need, it is difficult to customize the product before the need of the customer is known. The need of the customer however may be awakened by a standard product and then followed by his own customized product. In some situations a customer may not have a need, e.g. to be nobody who wants to be designed as somebody? Many products related to life style and fashion are based on such situations. Consider the customer who in relation to his professional work needs to be dressed in a business suit. The customer has no private interest in clothes, but recognize the unwritten rules on dress code in his business. Therefore, the customer accepts to be designed into whatever fashion that may be present, and the customer has no interest in customization except for the sizing of the suit, since the suit simply is a uniform to this customer. Although the customer has an interest in a product, he may still want his needs to be designed. Consider a visit to a restaurant, where you as customer not only want to buy some food, but also the chef's interpretation of how this food ought to taste. The particular configuration is part of the product that is bought, and the customer does not want to influence this configuration. Therefore, the value of a product may only have little correspondence to the value of the modules by which it is configured, since only the complete configuration of the modules has value to the customer. One may claim that no examples can be given on a customer who does not have recognized a need before fulfilling this with a product. This may be true, but it must be recognized that some needs are on a different level than first recognized, on a higher level where the product is at the lowest level in a means-end hierarchy.

18.4   Approaches to customer increased value

Buying a personal computer is often quite simple, what is difficult is getting the right personal computer. In order to illustrate two different approaches to this situation, two examples are described, Dell and Apple computers. If we look at the computers from Dell and Apple, we see that they are very similar in performance, but different in market approach. Apple believes that they know what the best computer within a certain price range is. On the contrary, Dell expects users to build their very own best computer that will be assembled by Dell according to the specifications of the customer.

Dell is possibly the one of the most recognized mass customization success. Dell offers the customer the opportunity to tailor a computer that will match the preference of the customer. A computer is by a nature a very modular product with standardized thereby providing the perfect product for customization. By using the modular approach, the assembly processes can be uniform thereby enabling the efficient production of a customized product. Thorough the application of direct sales channels and advanced supply chain management Dell have been able to eliminate the middleman. Not only have this reduced in a cost reduction but also an increased responsiveness to market turbulence. Dell holds very little inventory, consequently obsolescent is not a problem. This result has been made possible by a honed forecasting and a very short throughput time and a direct distribution. To compensate for the lack of retail outlets Dell has build an extensive service organization that can ensure long term customer satisfaction.

Buying a Dell computer at www.dell.com is technical, you will get the options of combining different processors, hard-disks, speakers, monitors, sound cards, ram etc. Through the designing process the customer will configure a truly customized computer between millions of variants. Through the past years Dell has profiled it self by offering mass customization, but the questions what value this customization have got for the customer. There is no doubt that the highly critical super-user will find value in a customized solution where as it is more doubtful if the average is better off than with an off the shelf product, and personal interaction in the configuration process.

At *Apple* the process of finding a computer is simple you choose a product family, and with in this family you will have 3 choices of pre-configured models, and the choice of color. Buying a computer at Apple is fast and simple, but the question is whether the customer gets the optimal computer, since there are few configurations to suit all kinds of different needs. The background for this approach is the assumption, that Apple knows the technology and therefore Apple knows how to get the optimal performance out of the product, thus there is no reason for configuration of the technology. In short technology is seen as enabling rather than the goal it self. Color and accessories are the only customer interfaces that are not related to the core technology of the product and therefore options are offered. It is obvious that the focus of the two approaches are different Dell is product orientated where as Apple is user orientated, with a strong focus on lifestyle aspects rather than technology.

For the expert user the Dell solution might be preferable, since a performance can be exactly matched the need. Whereas the average customers will have difficulties to know the consequences of combining a processor and a hard disk so for the average customer the variation only ad's uncertainty and the perceived complexity of the process might intimidate the customer. The less knowledgeable customer are able to express their needs through statements such as "I use the net, and work with pictures in document up to 100 pages" and this the dell configurator can not handle and this customer most go through the traditional and more expensive sales channel.

18.5 The future of customization

Customization is not a changeless paradigm, it will always reflect the development in society and technology. Therefore it is of great importance for its user constantly to develop the application of mass customization. In some markets mass customization have become the standard and the paradigm have started to mature. Customers are no longer impressed by thousands of variants, but are focused on the value created by the variance.

Given the central role of the customer, a thorough understanding of the situation arising from the meeting between the customer and the customized product is essential for a discussion on the future development of mass customization. According to future studies, we are moving towards a dream society where companies will have to differentiate themselves by creating a story [19]. Pine and Gilmore [20] describe the experience economy as the next progression of economic value after commodities, goods, and services. Others combine words like experiential marketing [21] and emotional values [22] pointing at a future with increased focus on values beyond the traditional product. These future values differ from the sole focus on the properties of the product, e.g. quality and performance. If these values, being the carriers of differentiation of future products, reside outside the traditional physical product, the preconditions for mass customization will radically be changed in the future. Future values differ from the sole focus on the properties of the product, e.g. quality and performance. It is also recognized that a discussion on the expansion of mass customization also must include aspects related to the future values of customers.

We see that software becomes a larger part of a product, and to day an average car holds more computer power than the first Apollo spacecraft. Software is used to optimize products and can replace much more expensive mechanical systems. Using software to achieve variance will become more widespread, the way this is often done is that the software is used to control the performance of the product, by doing this one product will be able to serve different customer segments. This can be seen, e.g., in cell phones where the printed-circuits of all models are the same but software controls whether a particular model is a standard or a state-of-the-art phone. The same can be found in Saab cars where one type of engine can have different characteristics through the use of software control and variations in turbo pressure, thereby replacing four engine variants. It is imaginable that the engine in the future can be upgraded to higher performance without any expensive mechanically alterations, only by changing software. As a result the basic car will become to a large extent a platform which may be rather updated than replaced. Using software to control the characteristics of a product moves the manufacturing process towards mass production. This form of customization, called adaptive customization [23], is an interesting interpretation of being customer centric: By moving back to mass production and less variety in physical manufacturing we can get even closer to true customization on the product and usability level.

Mass customization is a paradigm that has matured, we have seen many customized products that have created new levels of customer satisfaction. But mass customization is no panacea, the truth about mass customization has more aspects. Not all products will benefit from mass customization, and there are alternatives to mass customization. Customers are more interested in getting the right product, than a unique product and therefore pre-configured products are interesting for many customers.

If we look at the limitations to mass customization we often see that they are related to the customer, this is understandable since mass customization have been used as a tool for the manufacturer and not the customer. If mass customization should fulfill its true potential it must be turned in the direction of the customer. The key to this is for the manufacturer to understand the needs of the customer, and based on this provide a solution to a need. Thereby the choices, uncertainty and time vast can be eliminated, and the customer will get the right configuration of the product. Today we often see a tragic situation where the customer must make decisions that require a knowledge that is not possessed by the customer. This is damaging to both the customer and mass customization.

The business structure will change as a result of the increased competition. All parts for the fabrication most be efficient to ensure the overall competitiveness of the product. As a result of this manufactures will focus, at the same on the time customers demand for more and more complex products, which increases competencies demands on manufactures, and as a result of this companies will change, new skills must be build. Managing customization and supplier network can become the core competence of companies.

As mass customization is maturing it will move from being a novelty to being commodity. Experience has shown which markets are suitable for mass customization, and one this markets, new measures of differentiation must be found. Other concepts will take over the progressive role that mass customization have played. But this is no reason to forget the lessons learned since efficient customization might not be a mean of differentiation, but it will join the group of immortal qualifying attributes.

References

[1]   Toffler, A.: Future shock, New York 1970.

[2]   Davis, S. M.: Future Perfect, Reading 1987.

[3]   Hill, T.: Manufacturing strategy: Order-winners and Qualifiers, 2nd Ed., Burr Ridge 1993, pp. 43-76.

[4]   Pine II, B. J.: Mass Customization, Boston 1993.

[5]   Duray, R.; Ward, P. T.; Milligan, G. W.; Berry, W. L.: Approaches to mass customization: configurations and empirical validation, in: Journal operations management, 18 (2000), pp. 605-626.

[6]   Agrawal, M.; Kumaresh, T. V.; Mercer, G. A.: The false promise of mass customization, in: The McKinsey Quarterly, 3 (2001).

[7]   Huffman, C.; Kahn, B. E.: Variety for Sale: Mass Customization or Mass Confusion?, in: Journal of retailing, 74 (1998) 4, pp. 491-513.

[8]   Pine II, B. J.; Victor, B.; Boynton, A.: Making mass customization work, in: Harvard Business review, 1993, pp. 108-119.

[9]   Hart, C. W. L.: Mass customization: conceptual underpinnings, opportunities and limits, in: International Journal of Service Industry Management, 6 (1995) 2, pp. 36-45.

[10]   Mintzberg, H.; Ahlstrand, B.; Lampel, J.: Strategy Safari: A Guided Tour Through the Wilds of Strategic Management.

[11]   Lampel, J.; Mintzberg, H.: Customizing customization, in: Sloan management review, Fall 1996, pp. 21-30

[12]   Alexander, S.: Computerworld, September 1999, p. 54.

[13]   Fisher, M. L.: What is the right supply Chain for your product?, in: Harvard business review, 1997, pp. 105-116.

[14]   The Economist: Mass customization: A long march, July 2001.

[15]   Radder, L.; Louew, L.: Mass customization an mass production, in: The TQM magazine, 11 (1999) 1, pp. 35-40.

[16]   Eastwood, M. A.: Implementing mass customization, in: Computers in industry, 30 (1996), pp. 171-174.

[17]   Zipkin, P.: The limits of mass customization, in: Sloan management review, Spring 2001, pp. 81-87

[18]   Pine II, B. J.; Gilmore, J. H.: Welcome to the experience economy, in: Harvard business review, 76 (1998) 4, pp. 96-106

[19]   Jensen, R.: The dream society, 1999.

[20]   Pine II, B. J.; Gilmore. J. H.: The experience economy, 1999.

[21]   Schmitt, B. H.: Experimental Marketing, 1999.

[22]   Barlow, J.; Maul D.: Emotional value, 2000.

[23]   Gilmore J. H.; Pine II, B. J.: The four faces of mass customization, in: Harvard Business Review, 75 (1997) 1, pp. 91-102

Contact:

Carsten Svensson
Department of Manufacturing Engineering and Management,
Technical University of Denmark, Lyngby, Denmark
E-mail: csv@ipt.dtu.dk

Part V: Customer Centric Manufacturing

Process design, production planning and control for achieving near mass production efficiency

The term mass customization represents an oxymoron, and at no stage of the mass customization value chain is this more true than in manufacturing. The contradiction within the claim of producing high variety products with mass production efficiency poses a great challenge for manufacturing in any enterprise. Manufacturing for mass customization introduces multiple dimensions, including a drastic increase in variety, multiple product types manufactured simultaneously in small batches, product mixes that change dynamically to accommodate the random arrival of orders and the wide spread of due dates, and throughput that is minimally affected by transient disruptions in manufacturing processes such as breakdown of individual workstations. Solving the trade-off between customer centric manufacturing on the one hand (meaning high variety and fast responsiveness) and low costs, stable capacity utilization and high quality on the other necessitates incorporating systematic methodologies for manufacturing planning, process design and quality assurance in an integrated manner. Main enablers of customer centric manufacturing are modern flexible manufacturing technologies. They focus on batch production environments using multipurpose programmable work cells, automated transport, improved material handling, operation and resource scheduling, and computerized control to enhance throughput. The development, implementation, operative planning and control of these systems were the kernel of research on mass customization in the last three decades. But despite this history of research in the field, there are still many unanswered questions and new methodologies needed, as the chapters of Part IV will demonstrate.

Urbani, Tosatti, Bosani and Pierpaoli open the field in Chapter 19. The authors elaborate on the internal and external implications of mass customization and propose system capabilities which focus on flexibility and reconfigurability as a possible solution. They discuss the relationship between the principles of a mass customization system and the evolution of corresponding manufacturing systems. Several mass customization oriented paradigms are compared with a hypothetic, desirable evolution in market organization, providing an analytical approach. *Tsigkas, de Jongh, Papantoniou and Loumos* consolidate this discussion and present an innovative approach in Chapter 20 called "Distributed Flow Design and Development". This method should integrate product and process development. The objective is to boost the capability to sense quickly changing customer value requirements followed by the capability to rapidly transform these requirements and expectations in a variety of new product platforms and services. Their solution is closely related to the principles of lean manufacturing. Lean manufacturing and mass customization can supplement each other. The authors also introduce a new performance indicator to measure how fast an enterprise turns customer demands into value-adding mass customized products. A steady increase in this factor will lead an enterprise towards continuous, customer centric invention. While the first two chapters of Part V argue on the level of strategic production planning, *Lopitzsch and Wiendahl* address the important issue of planning and controlling a mass customization manufacturing system on the operative level. In Chapter 21 they present their approach of "Segmented Adaptive Production Control" which combines the advantages of a customer-oriented push system with the benefits of an efficiency orientated system of pull control. In doing so, their approach merges KANBAN and CONWIP control systems. The system makes it possible to control the manufacturing of parts manufactured in mass production as well as of customized parts being produced at the same work stations. Traditional control approaches relying on either push or pull principles are unable to face the trade-off between variety and efficiency introduced by mass customization.

The next three chapters address mass customization manufacturing from a broader perspective. In Chapter 22 Schenk and Seelmann-Eggebert comment on the complexity of implementing mass customization in an existing mass or serial production system. Pioneering examples of customer centric manufacturing often focus on newly founded enterprises. However, most firms will implement mass customization principles within an existing setting. Implementing mass customization is reflected in all parts of a company and consequently in the entire supply chain. As existing production and logistics systems have evolved individually, no standard solution can be offered for the implementation of mass customization into an existing production line. The authors name some of the resulting challenges. An important point is the training of the employees in manufacturing which must be able to respond promptly to changing demands, too. Modularization can be seen as a main enabler and principle of being customer centric efficiently. However, though many companies have gained experience with modularization, there is still significant confusion about managing the modularization effort. The cause-effect relationships related to modularization are complex and comprehensive. In Chapter 23 Hansen, Jensen and Mortensen discuss the impact of modularization in greater detail. Recognizing the need for further empirical research, the authors formulate a research framework with the purpose of uncovering the current state of modularization using the example of Danish industry. Finally, Mchunu, de Alwis and Efstathiou present a framework for selecting a best-fit mass customization strategy in manufacturing. The authors report about their findings in a large-scale empirical project and introduce a methodology for characterizing the mass customization capability of a manufacturing enterprise (Chapter 24). Their methodology emphasizes the collection and analysis of quantitative as well as qualitative data. A key component is the use of a field workbook to collect triangulated data. Together with three other tools presented in this chapter, it forms a framework that aids the selection of an optimal mass customization strategy.

19 Flexibility and Reconfigurability for Mass Customization

An analytical approach

Alessandro Urbani, Lorenzo Molinari-Tosatti, Roberto Bosani and Fabrizio Pierpaoli
ITIA-CNR Institute of Industrial Technologies and Automation, National Research Council, Milan, Italy

In the 21st century, companies are going to operate in a dynamic and challenging environment that requires new approaches to manufacturing. Mass customization is a general trend that is more and more widespread, being felt as the productive paradigm for the future. On the manufacturing point of view, much work must be done to develop adequate manufacturing systems meeting the new requirements, since traditional solutions (both transfer lines and flexible manufacturing systems) don't seem to be able to face the demands of mass customization. However, not only manufacturing area is involved in this evolution but also managerial and organizational aspects. For these reasons this chapter discusses the relation between the mass customization paradigm and the evolution of manufacturing systems. Several mass customization oriented paradigms are compared with a hypothetic, desirable evolution in market organization, providing an analytical approach to flexibility and reconfigurability as possible means of facing today's competitive demands.

19.1 Enabling mass customization by manufacturing

When we think about the evolution of manufacturing systems, we must also necessarily take into consideration the modifications which occurred in the socio-economical context. It isn't always easy to distinguish whether technology drives the changes in people life or, vice versa, whether new social needs are the key factors and stimulus for technological improvement or, at least, technological changes. In regard to manufactured goods the industrial revolution led roughly from a substantially craft production of customized goods to the mass production paradigm which dominated the main part of the 20th century. Mass production was enabled by technological factors like mechanization and line production as much as managerial factors like standardization and the fulfillment of economies of scale. The key principle of mass production is the ability to produce high volumes in order to answer to the requests of as many customers as possible at the lowest price.

Nowadays, customers attitudes have changed dramatically. During the last two decades customers' possibility of choice has grown in importance, leading to the idea of providing customers with customized products instead of the choice between product variants. Being able to combine customized products with high production volumes leads to the mass customization paradigm – satisfying (and not only supplying) as many customers as possible. The shift from mass production to mass customization seems to be caused by social reasons. This means that both technology and management must find the right answers to this new paradigm.

To obtain the competitive advantages promised by mass customization [2] two basic decisions are required. On the one hand – at company level – both the productive processes and the organizational structure have to be updated to provide companies with the desired versatility and responsiveness. On the other hand – at the extended enterprise level – the cooperation with other socio-economical subjects (workers, suppliers and even clients) has to be empowered. In the last decade efforts were made to introduce mass customization not only as a general necessity but also as a trend that is leading the evolution in many industries. However, much work has still to be done to approach such a wide and generic subject as mass customization analytically, above all to formalize its technological requirements [3]. In this context the chapter copes with mass customization and its enablers, in particular in regard to its manufacturing implications and needs.

19.2 Manufacturing demands of mass customization

Mass customization can be obtained in different ways [2]. It fits different business areas, from service to goods providers. Both services and physical goods can be customized. Inside both areas, the whole value chain (from R&D to distribution) must be involved to move to the mass customization paradigm. In this chapter, we are mainly interested in the manufacturing phase, in particular in the features of the manufacturing system that may fit the needed requirements best. The new necessity for manufacturers is to provide customers with products that fit their individual desires while maintaining high production volumes (Figure 1).

The concept of agility deals with the ability to follow customers desires and seems to be involved in the mass customization paradigm. In general terms, when we refer to agile production [4] we speak about the ability to adapt to market requirements which characterizes a whole productive organization. Much has been written already about agility and how to implement this concept in manufacturing firms. The salient features of agility [5] may be summarized in four main high level strategic points to pursue: (a) give value to customer's satisfaction; (b) give value to knowledge; (c) be ready to change; and (d) create co-operations (virtual enterprises).

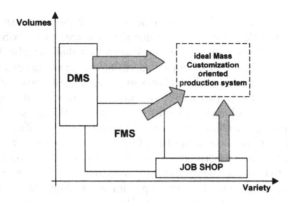

Figure 1: Mass customization in the volume/variety plane

These main directives are in contrast with the corresponding attitudes of mass production which are based on features like the exploitation of wide economies of scale, impressive but sporadic innovation campaign, definite separation between the design and the executive phase, great emphasis on efficiency and cost reduction. A general comparison between mass production principles and the principles of agile production inside the mass customization paradigm is summarized in Table 1.

Table 1: General comparison between mass and agile production

Mass Production	Agile Production
cost reduction to reach as many customers as possible	value to the single customer satisfaction
impressive, sporadic innovation campaign	value to knowledge
standardization, long development cycles for new products	ready to change
economies of scale	cooperation
stable demand	fragmented demand
emphasis on the single product	emphasis on the process

Focusing on manufacturing systems the concept of agility is deeply related to the ability to be ready to change. This means that the system must be able to produce wide and evolving mixes in an economically sustainable way. This must be true not only in a fixed, planned productive period, but also in different, unplanned periods that could be necessary to answer to uncertain demand evolution. The system must be able to continuously update the manufactured mix to follow customers desires.

FMC, FMS and FTL were built to partially cope with these necessities, providing a certain level of flexibility to the productive structure. This flexibility can be used to manage different parts inside one known mix that can also be updated according to system's properties, both as regards the volumes (through scalability) and as regards parts geometry, which allows to introduce new parts and to provide the customer with many desired model variants. Up to now flexible applications allowed medium-high volumes as shown in Figure 1. Nowadays the dynamic nature of many markets implies to modify the system during its life cycle. In such contexts flexibility in itself could not be sufficient and even become a drawback. In fact, complexity and expensiveness of flexible systems may represent a further obstacle to effectiveness in highly dynamic contexts, as it is very expensive to modify this kind of system. To come up with an analogy, flexibility can be associated with the concept of robustness. However as we know from control theory robust control is not always as efficient as an adaptive control scheme. For example, if you have to introduce a new part which the system is not able to handle in its current configuration, it is very expensive to adapt the flexible system to the new demand requirements.

The need of having the exact productive capacity exactly when needed led to the new idea of Reconfigurable Manufacturing Systems (RMS) [6]. A RMS is a system that can be easily modified (reconfigured) to follow demand requirements rather than use as many functionalities as possible to cover as many requirements as possible (which is a flexibility oriented approach). The system can easily evolve through different, following one another configurations able to face exactly its own productive goal: the user is not paying extra costs for extra functionalities he is not using. It means that the user may have at his own disposal dedicated, highly productive configurations that can be easily modified to follow customers requests. In this way it may be possible to face customers desires also when maintaining high productive volumes.

Accordingly, the more a system can produce wide mixes at different volumes through economically sustainable modifications of dedicated configuration, the more this system is reconfigurable. Modularity is one of the enabler of such a kind of manufacturing system: the ideal possibility to assemble the desired modules to provide the system with the desired functionalities allows to combine efficiency and productivity, while the ability to modify the available functionality provides the possibility to face wide mixes. This kind of manufacturing system may then be the most suitable to fit mass customization requirements under the manufacturing point of view.

19.3 New frontiers for manufacturing

As stated above, to obtain the competitive advantages promised by mass customization, both the organizational structure and the productive process must be updated. The emerging tendency by capital goods producer to become service providers rather than product providers helps to make the whole scenario more homogeneous inside the mass customization context and its related requirements. Also for manufacturing companies, in fact, some of the themes related to providing customized services become very current problems. Manufacturing companies need to become "customer oriented" rather than "product oriented", consequently changing their own organizational structure. The reasons for the supply of a global service rather than a traditional supply result from the fact that, for example, the market of machine tools or other production devices is more and more turning into a commodity market. As a result, manufacturing companies are following the idea of customer proximity to gain an advantage by providing something more than simple (standard) hardware, distinguishing their own supply from the others' by granting and possibly customizing a maintenance service to the customer. Even if this approach seems to be an enabler for the suppliers to survive in the current highly competitive and dynamic markets, it doesn't necessarily enable mass customization in itself.

Other emerging evolutions in the relationship between capital goods producers and their customers derive from the latter's request for a number of parts over a time horizon rather than the simple supply of manufacturing capacity. Such tendency can be considered part of the general trend toward outsourcing. Such a scenario, as much as the previous one, highlights the increase in the requested involvement of the suppliers in the productive process. There is, in particular, an identifiable trend to ask for services, which can be found in different fields, even out of manufacturing. The outsourcing of IT in such environments as banks or other businesses is a suitable example of supply chain management oriented to the improvement of the agility of a company [6]. The traditional implementation of one integrated, very complicated application to manage each product (loans for example) is going to be substituted by web based applications that are purchased from a suitable supplier, who is responsible for the maintenance and the necessary updates of the specific application. The final effect is to have adjustable products that can be, as a consequence, more easily customized.

Similar examples already exist in manufacturing. The first manufacturing industry where the evolution described above has taken place first is probably the electronic industry. In this industry, the contract electronic manufacturers (or electronic manufacturing service providers) are becoming responsible for the production-related activities relative to specific products, thus providing their customers with a kind of service that is similar to those previously described for IT. The possibility to use reconfigurable systems may be exploited in order to update the processes and, consequently, to provide adjustable services, resulting in increased ability for the companies to customize their products.

There are different possible configurations for the evolution of the supply chain inside the service oriented supply scenario [12]. Those previously described, in fact, are examples of desirable and actual evolutions but other solutions may be adopted inside the general idea of changing the way in which the responsibilities inside the supply chain are allocated. Inside the mass customization paradigm such evolution can be particularly desirable. In particular, the possibility to rent some manufacturing capacity inside the service oriented solution would represent a suitable solution. The optimal choice about the allocation of modules (according to the reconfigurability-oriented manufacturing scenario) has to be made by the service provider. Such choice has to be repeated whenever the necessity to update a process occurs. This process can result in improved customization ability for the customer of the service supply.

The scenario is desirable above all for SMEs, as they don't have great financial resources at their disposal. Together with the development of systems able to evolve during their life cycle and to adapt to market requirements (described above), renting the desired configuration of such a manufacturing system may thus be a "financial" instrument to mass customization. Hence, on the system provider side the manufacturer of the machinery could retain ownership and lease "production hours" or "products per month", taking responsibility for operation, programming, service, maintenance, etc. The customer (a material goods vendor) would pay this service.

As an alternative, a "system integrator" might act as a "technology broker", working as an interface and arbitrator between a company that needs a given productive capacity and a group of functionalities and one producer of modular macro-components of the production unit. This actor, probably supported by a finance and leasing company, would be responsible for the selection of the modules, their customization for the required process and integration. This new actor could rent a customized production capacity (related to a peculiar process) to the end user together with operation and maintenance services and, after the term in use, could disassemble it into basic "building blocks" and reuse them to assemble a new production capacity fitting a new end user requirements.

Maintenance and reliability (that are expensive and time consuming) become critical aspects in these evolutionary layouts. Maintenance, in fact, is part of the service provided, so that the provider is interested in minimizing it. Equipment must then be designed for maintenance, and reliability. Maintainability prediction techniques are desirable enablers for an effective service providing. Modularity and reconfigurability in manufacturing systems and system components must be considered as key enabler for such a new market layout. Besides the dynamic behavior they enable, the reusability allowed by modular components increases the potential effectiveness of such a supply. What is more, the possibility to have reduced ramp up times (key element for Reconfigurable Manufacturing Systems) to start new productions or update existing processes is a key enabler for such productive scenarios, which, as previously explained, may be a potential enabler for mass customization.

It must be noticed that in such a scheme the system provider becomes a process provider. This is coherent with the mass customization idea to place the emphasis on the process and its life cycle rather than on the product: many products are realized inside one process, and each process lasts longer then the product realized inside it. The kind of update in the organizational structure of the enterprise (suitable for the described market layout change) can be considered as a part of the extended enterprise approach [7] that seems to be a promising paradigm allowing enterprises, and in particular SMEs, to cope with the dynamic nature of the current global market and to compete with bigger organizations. In the following section we will present an analytical approach to support the decision which kind of manufacturing system should be used in dynamic contexts. Our approach is referred in particular to the machining processes.

19.4 An analytical approach

19.4.1 General analysis of flexibility and reconfigurability

When we need to produce a given part, the best solution from the point of view of cost, time and quality must be certainly looked for among dedicated solutions. A dedicated solution can be optimal according to the desired goals. When introducing flexibility to cope with wide mixes and/or volume fluctuations, optimality about each of these goals is lost. By using a reconfigurability-oriented approach it is possible to partially achieve the advantages of both dedicated and flexible solutions, i.e. both optimality and the ability to cope with varying situations. Obviously this is not always the best choice, but there may be conditions of mix and volumes in which it can be an effective solution [8]. Such an approach, anyway, cannot be considered completely new, as kinds of reconfigurability have often been considered in many kinds of systems. Scalability, in fact, can be considered as a reconfigurability-oriented property. This means that reconfigurability must be considered as one of the abilities which a system can be equipped with. In order to take system reconfigurability into consideration, system life cycle must be divided into successive production periods T.

Each one of these periods corresponds to a known *productive objective* (mix and volumes) which is realized in one *configuration*. As reconfigurability may try to combine the effectiveness of dedicated solutions with the ability to manage variable demands, two systems concerning aspects must be taken into consideration to get reconfigurability measures:

- The *efforts* to modify the system's and the system's components structure,

- the *efficiency* in functionality implementation.

The easier to modify the system and its components, the more reconfigurable the system itself. Being able to cope with demand changes through flexibility is an optimum under this point of view. In such a situation, in fact, no modifications are needed (we are not taking software modifications into consideration). But in order to be able to cope with expected and unexpected demand changes a flexible system must use very wide functionalities, which could also result to be underused. Furthermore, high flexibility is a very expensive feature.

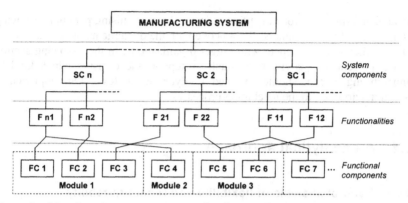

Figure 2: Manufacturing system representation for reconfigurability analysis

To analyze the ability to cope with wide demand fluctuations effectively one has to consider also the efficiency to implement requested functionalities. After identifying the system components to be analyzed reconfigurability analysis can be applied to each of these components to finally analyze the whole system (in an automated machining system, for example, the system components can be divided in five main components: machines, fixtures and pallets, tools and material handling system, warehouses). Each component can be associated with a general functionality that it has to realize. Each functionality (characterized by its own technological attributes) can be decomposed further into sub-functionalities.

As previously stated, a key-enabler for reconfigurability is modularity of the components. The modules (defined by the granularity implemented) and the relationship among modules which are used to work a mix during a period T define the configuration corresponding to period T. Each module is made up of one or more (functional) elements. By defining the correspondence between functionalities and the modules which implement them, it is possible to find general, high level reconfigurability measures.

19.4.2 Metrics for the modification effort

The efforts to modify the system's and the system's components structure can generally be measured by the ratio:

$$\text{Effort} = \frac{\text{Number of modules to be modified}}{\text{Number of total modules}}$$

An *intrinsic* measure for the effort must be based on how the modules are linked together and how they realize functionalities. An intrinsic measure, in fact, must only depend on system's attributes, and must not depend on one peculiar change which is requested. To get an intrinsic measure for the effort, then, we can evaluate what can be ruled out in the whole possible requested system changes. Modifying a system means to modify the functionalities implemented by its components. The functionalities implemented by one system's component can be changed under two points of view: change in the *number* of functionalities (for example, adding one functionality), and change in the *attributes* of functionalities. In this way we can represent the whole possible requested changes.

The effort to modify the number of functionalities (for each system's component) can be evaluated by the general index

$$RI_n = \frac{\sum_{i=1}^{I} IC}{I} \quad \text{where}$$

I is the number of interfaces, and *IC* is associated with the completeness of each interface, i.e. the number of layer which characterizes each interface with respect to a complete one.

The effort to modify the attributes (for each component) can be evaluated by

$$IR_a = 1 - \frac{\sum_{i=1}^{f} \left(\frac{\sum_{j=1}^{a} NMC_j}{a \cdot CT} \right)_i}{f} \quad \text{where}$$

a is the number of attributes which characterizes each functionality, *f* is the number of functionalities implemented by the component, *NMC* is the minimum number of the component's elements which can be ruled out to modify the jth attribute of the ith functionality, *CT* is the total number of component's elements.

Then, the following IR_i is an intrinsic index for the measure of the effort (for the ith component) depending only on component's nature:

$$IR_i = n_i \cdot IR_{n,i} + a_i \cdot IR_{a,i} \qquad (a_i + n_i = 1)$$

and the intrinsic reconfigurability index for the system is

$$IR = \sum_{i=1}^{C} r_i \cdot IR_i \qquad \left(\sum_{i=1}^{C} r_i = 1 \right)$$

with C being the number of components which the system is made up of, and r_i the weight that is related to the ith component and depends on the desired reconfigurability strategy [4].

IR can be used to *evaluate* one configuration's intrinsic reconfigurability, and to *compare* two configurations as regards their easiness to be modified.

19.4.3 Efficiency metrics

An intrinsic measure for efficiency cannot effectively be found. Efficiency in the implementation of functionalities must depend on the mix to be worked. It can be calculated as

$$Efficiency = \frac{NF}{AF}$$

where NF represents the number of needed functionalities and AF the number of available functionalities.

Regarding machines, functionalities can be decomposed in *movements* (which correspond to real part-tool contacts) and *reorientations* (part-tool relative movements without contacts). Needed functionalities are those exactly needed to process a part (for example drilling corresponds to one needed functionality). The number of needed functionalities can be calculated as the sum of the number of functionalities needed to process one part's faces, thinking that reorientations can prevent duplication of movements. A simple algorithm has been introduced to find out the number of needed functionalities based on these considerations. The number of available functionalities can be calculated as

$$AF = \sum_{i=1}^{S} A_i + R$$

with A_i being the number of axes used to process the ith (i=1,2,...,S) step of the part's cycle (each step can be identified by one fixture of the part or one machine which the part is processed on). R is the number of re-fixturings in the part's cycle.

If a configuration is used to process P parts, efficiency about the configuration's machines is

$$E_M = \sum_{i=1}^{P} \alpha_i \cdot \frac{NF_i}{AF_i}$$

The weights α_i can be calculated as $\alpha_i = \dfrac{V_i}{V_{tot}}$ to evaluate the ith part's relative importance in the mix. Similar analysis can be conducted for tools, pallets, the handling system and warehouses. In this cases the analysis starts from configuration and not from the parts to be processed.

19.4.4 Operative reconfigurability metrics

When we decide to work a given family of parts, we can decide to use one flexible configuration (FMS for example) which must be considered as the minimum effort solution (*IR = 1*) because it doesn't need to be modified, but may imply low efficiency E in respect to the family to process. A reconfigurable system would use different, following to each other, high IR, high E configurations. The index

$IRO = IR \cdot E$

can then be considered as an operative reconfigurability index (the higher IRO the easier to be modified and more efficient the system), which can be used to *compare* alternative configurations for one period T and to compare the effectiveness of different productive strategies for the system's life cycle. As a consequence, IRO may represent a support to the choice of the kind of manufacturing system to be used to face manufacturing requirements. In particular, it takes into consideration the nature of existing and or possible systems and the functionalities associated to a productive goal, providing suggestions about the level of effectiveness achieved in the use of a given manufacturing strategy by weighting system ability to be modified and the efficiency maintained during system life cycle. It helps to evaluate the system effectiveness in dynamic context, typical for mass customization where manufacturers are interested in following market dynamics efficiently.

19.5 Conclusions

In many markets the ability to follow customers desires is becoming of major importance. For companies dealing with a high number of customers achieving this goal is hard work. Mass customization requires updating the commercial environment both on the internal and on the external point of view. Both the organizational structure and the productive process have to be updated, using a process oriented rather than product oriented structure. This evolution includes also the manufacturing system. Thus, new ideas are needed in this field. These are internal changes. But even the market layout may be involved in such an context as companies have to update their relation with the other companies, too, modifying for example their relationship with the suppliers. Some evidence exists that a service based supply is a desirable configuration for the supply of manufacturing capacity. Combining such evolution with the implementation of reconfigurability in the manufacturing system can increase the ability of the companies to provided their customers with customized products, above all for SMEs.

In this chapter, we have analyzed both the internal and the external implication of mass customization, focusing on manufacturing as the field of research. Flexibility and reconfigurability were analyzed as possible solution for the new paradigm: the ability to shift from one productive goal to another while maintaining high efficiency is used as a decisional support in the analysis of manufacturing systems' effectiveness in dynamic contexts. The possibility to rent a productive capacity and the consequent market layout modifications were described as possible enablers for mass customization.

References

[1]   Camarinha-Matos, L. M.; Pantoja Lima, C.: A framework for cooperation in Virtual Enterprises, in: report from the European funded project PRODNET II, the ESPRIT project 22647, on www.uninova.pt/~prodnet.

[2]   Conklin, J. M.; Perue, B. A.: Extending Capabilities Through Contract Manufacturing, in: Proceedings of the Electro International 1994, Boston 1994.

[3]   Duguay, C. R.; Landry, S.; Pasin, F.: From Mass Production to Flexible/Agile Production, in: International Journal of Operations & Production Management, 17 (1997) 12, pp. 1183-1195.

[4]   Goldman, S. L.; Nagel, R. N.: Management, technology and agility: the emergence of a new era in manufacturing, in: International Journal of Technology Management, 8 (1993) 1/2.

[5]   Gunasekaran, A.: Agile manufacturing: enablers and an implementation framework, in: International Journal of Production Research, 36 (1998), pp. 1223-1247.

[6]   Hagel, J.; Brown, J. S.: Your next IT strategy, in: Harvard Business Review, 76 (2001) 5.

[7]   Koren, Y.; Heisel, U.; Jovane, F.; Moriwaki, T.; Pritschow, G.; Ulsoy, G.; Van Brussel, H.: Reconfigurable Manufacturing Systems, in: Annals of the CIRP, 48 (1999) 2.

[8]   Pine II, B. J.: Mass Customization, Boston 1993.

[9]   Son, S. Y.; Olsen, T. L.; Yip-Hoi, D.: Economic benefits of Reconfigurable Manufacturing Systems, in: Proceedings of the 2000 Japan USA Flexible Automation Conference, Ann Arbor 2000.

[10]   Tseng, M. M.; Jiao, J.: Design for Mass Customization, in: Annals of the CIRP, 45 (1996) 1.

[11]   Urbani, A.; Molinari Tosatti, L.; Pedrazzoli, P.; Fassi, I.; Boër, C R.: Flexibility and Reconfigurability: an Analytical Approach and Some Examples, in: Proceedings of the First CIRP International Conference on Agile, Reconfigurable Manufacturing, University of Michigan, Ann Arbor 2001.

[12]   Urbani, A.; Molinari-Tosatti, L.; Pasek, Z.: Manufacturing Practices in Dynamic Markets: Reconfigurability to Enable a Service Based Supply, in: Proceedings of the IMECE: 2002 ASME International Mechanical Engineering Congress and Exposition, New Orleans 2002.

Contact:

Alessandro Urbani
ITIA-CNR Institute of Industrial Technologies and Automation,
National Research Council, Milano, Italy
E-mail: a.urbani@itia.cnr.it

20 Distributed Demand Flow Customization

Alexander Tsigkas[1], Erik de Jongh[2], Agis Papantoniou[3]
and Vassilis Loumos[3]
[1] FlexCom. AT&P Ltd., Athens, Greece
[2] Tecnomatix Technologies Ltd., Herzeliya, Israel
[3] National Technical University of Athens, Athens, Greece

This chapter will present new collaborative tools and methodologies enabling the rapid development of new product platforms and their corresponding manufacturing processes. The intended development cycle times are months for automobiles and computers, weeks for consumer electronics products, and days for many others. The new battlefield of corporate competition is the capability and speed to create and transform "new" knowledge into personalized products and services. Lean and flow enterprises have the best chances for winning this battle. The proposed tools provide the missing link for the development of products and processes for mass customization. To address this challenge an integrating approach in product and process development is presented. The objective is to boost the ability in sensing quickly changing customer value requirements and expectations followed by the capability to transform these requirements and expectations rapidly in a variety of new product platforms and services. Doing so we will introduce a new key factor to measure the speed of an enterprise to turn customer demand into value-adding mass customized products. A continuous increase in this factor will lead an enterprise towards continuous invention which is translated into the speed to systematically and continuously mass customize products and introducing new ones. We call this enterprise an entropy enterprise.

20.1 Introduction

Of the many profound changes to which businesses must respond to succeed in today's turbulent climate, nothing is more difficult, more perilous, and more vital than being customer-centric. But the initiatives of most companies – measuring customer satisfaction, implementing quality function deployment, putting together cross-functional teams that span marketing, manufacturing and development; and so forth – are in fact *market*-focused, not *customer*-focused. In product development, even when individual customers are sought, it is done to create better averages rather than customized products and services.

The imperative today is to understand and fulfill each individual customer increasing diverse, desires and needs – while meeting the co-equal imperative for achieving low costs. The critical role for executives, managers and engineers with the product development community is harnessing new technologies, manufacturing capabilities and management techniques to efficiently serve each customer uniquely, [1].

The fact that customers, whether consumers or businesses are becoming more demanding about getting exactly what they need, when they need it, while increasing competitive intensity – arising in particular from the globalization and convergence of industries – dictates that, costs keep decreasing as well. Where companies used to pursue either a low cost or a high differentiation strategy, today companies are increasingly finding that they must adopt strategies embracing both efficiency and customization. Instead of mass-producing standardized products and services or incurring high costs to produce high variety, companies must combine the best of both strategies to mass customize their products. In order to achieve this target product development and product introduction needs to reach very high levels of flexibility. As main obstacle of increasing flexibility and further improved productivity, quality and reduction of product and services cost, emerges the fact that product engineering and manufacturing engineering work in sequential fashion towards delivering the product to the markets. It is not only a 'cultural' problem, it is mainly the lack of tools and education that prevents the adoption of a collaborative operating approach and cross disciplinary as well as intra-company and inter-company reusability of the developed knowledge. The consequences of this fact that are well known are: prolonged time-to-market or time-to-volume, since manufacturing processes, production procedures, tools, raw materials and components are defined and developed very late in the product delivery cycle. This time is non-value-adding time or waste in the lean production terminology that is extremely expensive.

On the other side, flexibility requirements are boosted by the customer perception of value adding products and services, where the customer, whether consumers or businesses, will be part of the design process thus creating a need for collapsing industry typical total value cycle time. Total value cycle time is defined by the time it takes to deliver the product to the customer after a value requirement has been assessed. There is a lot of work being done currently with respect to mass customization, but most of them, we are aware of, concentrate in the product development and do not take the integrative approach of product and process development. Even simultaneous or concurrent engineering approaches do make the distinction between product design, development and manufacturing processes. We take the innovative approach to introduce the concept of the design of a product at the process level. This is the highest degree of integration of the engineering disciplines and tasks. We believe that design at the process level will be fundamental for the delivery speed of mass customized products and will help Global Enterprises in their steady efforts towards continuous innovation.

The assumption done at this point is that an enterprise that looks for adopting the proposed method is already a lean or demand flow manufacturer or at the minimum they consider seriously to transform their business and production operations from being functional into being process oriented. Furthermore, since mass customization means to build on demand and not to forecast, this method provides the ability to global manufacturers to design anywhere and to build anywhere the customer needs the added value provided. Therefore, it is thinkable that this approach will enable the rise of 'distributed demand flow customized manufacturing' with fewer giant plants and more small and flexible ones assembling from prefabricated modules closer to the end-user to become a reality.

The organization of the remaining chapter is as follows: We begin with the challenges of today, manufacturers are faced with in their battle to reduce the total value cycle time for mass customization. We continue with our proposal about how to measure mass customization performance and then we describe a design and development method based upon the approach for a process level product design. Finally we suggest a road map for the design of mass customized products for demand flow or lean enterprises.

20.2 Problem definition

Today frequently employed product development and industrialization practices are the major obstacles and barriers that systematically slow down total value cycle time for mass customization. In particular, the use of the waterfall model where product development and manufacturing process development, is based upon the sequential processing of development phases is suitable for the era of mass production but is totally ineffective for the era of mass customization and build on demand for the new economy society. In fact, product development and industrialization processes take the collaborative effort of many departments in an organization in order to win the time-to-market competition. Furthermore, the products themselves usually consist of parts that must be designed, tested, integrated, and tested again to meet customer requirements before these are released to production.

The interaction between these two systems, the parts forming the product and the people and teams involved in the design process, forms the basis of product design knowledge acquisition and further reuse not only for mass customization purposes. Reusing this past product design knowledge helps to speed up the product development process, thus leading to high quality products, reduced development times and predictable project deadlines. As Bell et al. state in order to design a product development process that delivers quality products better, faster and cheaper, three major issues must be addressed, these are [4]:

- The ability to capture, understand and manage the interactions occurring in the system of the product and the system of the people involved in the design process.

- The ability to capture and analyze all kinds of system interactions as early as possible during the product development process when the most important decisions about the product design process are made and the cost of changes is at its minimum.

- The ability to ensure that the product design and relevant process are closely aligned with the product requirements for value generation for the customer.

Since mass customization means to build on demand and not to forecast, global manufacturers need to have the ability to design anywhere and to build anywhere using the collective knowledge existing in the enterprise. It is thinkable that the rise of 'distributed demand flow customized manufacturing' with more and more small and flexible plants assembling from prefabricated modules closer to the end-user will be a reality in the near future.

Current methods in the design and development of products and manufacturing processes not making use of the collective knowledge suffering frequently from the so-called 'corrective action syndrome'. Action should be proactive and not reactive in order to eliminate waste that is non-value adding work within the total value cycle time. The most expensive way to do customization is to try and perform custom products somewhat quickly, although the products are not designed for effective customization. This is shown on the 'reactive' side of Figure 1 below [1]. Thus the challenge is to achieve *timely* and *efficient* customization of products or mass customization as shown at the 'proactive' side of the same Figure.

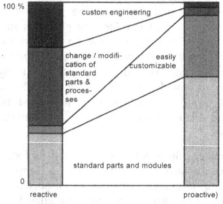

Figure 1: Customization contribution for reactive and proactive modes

Desired therefore, is a mechanism for the successful management of the knowledge generated from the above mentioned issues, overcoming the main hindrances all knowledge management initiatives face, some of them being:

- The design and control of product development processes is often an ad hoc effort based on the individual managers experience, with insufficient attention paid to communication structures. This leads to the facts that (a) knowledge developed in one company cannot effectively travel to another – let us not forget that product design usually involves more that one company; and (b) joint knowledge creation, basically the collaborative sharing of knowledge to create new process innovations and implementations, is not achieved.

- Minimum attention is paid to the documentation of system level knowledge coming out of information processing that usually resides within certain company employees.

- Ethics of knowledge sharing, meaning that companies are ethically compelled to share process expertise.

By providing, planning and engineering tools to design and document the manufacturing process and web-based collaboration tools to allow the various participants in the extended enterprise to take part in planning the manufacturing process, the proposed methodology helps overcome the above mentioned barriers and obstacles. These new collaborative technologies can dramatically reduce the total value cycle time of new product introductions while increasing effectiveness of collaboration between the service provider and the customer. In addition this can be achieved in a distributed manner via high bandwidth satellite and fiber optic telecommunication capacity by putting three-dimensional images in for example New York, London, Hong Kong and Tokyo simultaneously to suit Global Enterprises.

Mass customization, in our view, should liberate itself from the obvious opposite of mass production and become mature and independent. Mass production is based upon 'old' knowledge gained in the past that repeats itself through the use of it in 'old' products and services. But competitive advantage is not any more gained through the repeatable use or exploitation of 'old' knowledge [2, 3, 7]. On the contrary we believe, that competitive advantage is gained through the creation and immediate exploitation of 'new' knowledge. New products and services are developed and sold that could be one-of-a-kind based upon customer individual and personalized demand. Once this 'new' knowledge has been created it becomes already 'old'. That initiate a new cycle of creation, instant exploitation and customization of new knowledge (Figure 2).

Figure 2: Perpetual new knowledge transformation cycle

The question is therefore how we can develop a capability of *instantaneously* transferring and converting the newly created knowledge into value adding personalized products and services for individual people and not simply for individual markets any more. Mass customization should be evolving as the natural answer to this question. If we then were to give the definition of mass customization this should be along the following axis: *It is a set of methodologies and strategies that target at the massive exploitation and customization of newly created knowledge in order to develop product and services for individual customers.*

Mass customization should target simply at the mass transformation of 'new' knowledge into individual products or services, thus customizing first knowledge rather then the product itself. There is a fundamental difference between *knowledge* mass customization and *product* mass customization. With the latter we normally mean to produce products or services with a 'planned' variation of theme. With the first we mean to produce products and services with virtually no limitation on the variations. This requirement imposes of course new challenges in the manufacturing process design. Pine, Victor and Boynton [9] suggest that the passage from invention, the creation of new knowledge, to mass customization should pass through the phase of mass production (Figure 3).

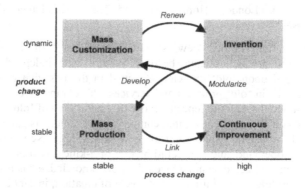

Figure 3: Competitive Reality: the development from invention to mass customization [9]

We strongly oppose this suggestion, since mass production is steadily based on the exploitation of 'old' knowledge and this by definition does not provide a competitive advantage [8]. On the contrary we believe that the enterprise that can steadily create and customize new knowledge into personalized products and services, will survive in the future. We call this enterprise *'Entropy Enterprise'*. This enterprise sees its competitive advantage in the quick adaptation not only in new business environments directed from external market conditions but, mainly in changes that alone is introducing into the market place in order to create new differentiation factors and capabilities for continuous profit creation.

Our contribution here is aiming at initializing a discussion and also offer an approach as to how manufacturing processes can be designed in order to be able to transform themselves with competence into whatever the new situation requires, imposed by the new knowledge exploitation. Collaborative practices and technologies for new knowledge creation and application will have two effects on product and manufacturing process development. They will enable many activities to be done that could never have been done before, and will allow activities that could be done to be completed much faster, more efficiently and more effectively. Included in this latter group will be the ability, through virtual reality and other simulation technologies, to rapidly examine and test new product concepts without having to build the physical components. Combined with lean product and lean process design, the potential time saving in new product development and transfer to production are enormous.

However the requirement of offering value-adding variety to the customers imposes noise to the business environment and naturally incurs costs if this noise is not controllable and drives business operations out the steady state as this is used to operate. These costs reflect the cost of variety. The requirement therefore is a business environment to stabilize to a new steady state with the minimum on overshooting. The speed with which a business environment can deploy value-adding variety and stabilize to new steady states will characterize the continuous invention or the entropy enterprise. Figure 4 illustrates the variety deployment cycle, striving for shorter cycle times towards continuous invention.

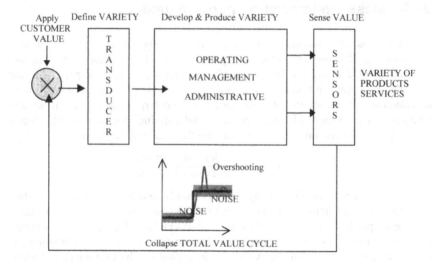

Figure 4: Variety deployment cycle

In Figure 5, an industrialization process is illustrated. The example illustrates the transition from a certain level of variety state A to another level of variety designated with state B. The cost of variety is oscillating around the steady state of level B. The requirement is to effect the transition from A to B with the minimum costs and these costs should be known already during the design and development phase. This will allow manufacturers to look for ways to reduce costs long before the product comes into the industrialization phase.

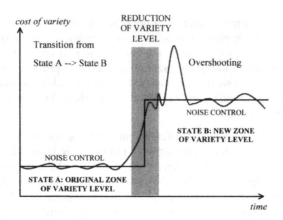

Figure 5: Cost of variety

20.3 Mass customization performance

A new key factor is introduced in order to measure the speed of an enterprise to turn customer demand into value-adding mass customized products. We measure this speed by defining the factor of value-adding turns. Thus, a persistent and continuous increase in this key factor will lead the enterprise towards continuous invention which is translated into the speed of the enterprise to systematically and continuously mass customize their products and introducing new ones. This factor is defined as the division of days per year and the total value cycle time:

$$\text{value turns} = \frac{\text{working days per year}}{\text{total value cycle time}}$$

This factors may be defined by product family and based upon the product platform, such as automobiles and computers to be measured in months, consumer electronics products in weeks, and many products in days. The target is to increase the value turns factor and in order to do this we need to focus on following the basic elements of lean that are: value (as identified by the customer), the value stream (value set in order), flow (turn on the value stream, synchronized events and drive with a tact time), pull (have the material available and triggered by Kanban signals), perfection (revisit your value stream). The focus of lean is simply removing non-value activities. The objective is to make use of demand flow manufacturing principles and to apply them to design approaches including collaborative engineering integrated product development and process design, empowerment, design for manufacturing and assembly, ownership knowledge based design, risk reduction, asset utilization, and continuous improvement.

20.4 Proposed methodology

The methodology presented here includes the two main activities in the development of new products, new components and manufacturing processes: (1) the design and development process, and (2) the design documentation and data management.

(1) The design and development process

The proposed method shrinks significantly the total value cycle time, thus reducing considerably the total capital immobilized in the development of new products and the deployment of manufacturing processes. This method departs from the classical waterfall model for product development, based upon the sequential processing of development phases and uses the 'flow rings' model shown in Figure 6, allowing parallel processing of the necessary activities.

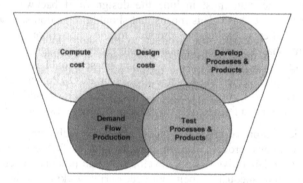

Figure 6: Flow Rings Model

In the following, we will give a short description of this paradigm and its benefits compared to the Waterfall model:

- The model does not show a series of phases. On the contrary it models a series of cycles which, in and of themselves, also constitute a single large cycle. Product development is thus a continuous process rather than a linear path between two points.

- The model has no starting point. A development project may start at any point along the rings. For example a project can begin from the testing phase of a current product and the corresponding processes (i.e. reliability, or manufacturability test). Components, especially off the self readily available in the market can and should be tested as soon as they become available in the market. Once certified they can be incorporated into the design and development phase. In fact a task should be incorporated in order to explore newly marketed third party components and technologies (tools and machines) on an on-going basis for future products and certify their quality and performance to the current target-costing model.

- Likewise a project can start in the prototyping of a product and move back to the specifications and design phase. This is especially useful in migrating or improving an existing product (i.e. end-of-life) can be quickly prototyped to construct a benchmark from which changes can be specified in order to fit a new costing model.

- A unique feature of the 'Flow Rings' paradigm is backtracking, or the ability to move back and forth along two adjacent rings without disturbing the overall flow of the project. This is no more apparent than in the Design and Development phases when the integrated process and product design team continuously reshapes the prototype to fit the chosen value-adding variety target cost model.

However, despite the versatility of this model, there are certain paths one cannot take. For instance one cannot move back from production directly into development. The path must include the design or if backward movement is desired into the test phase. This means that should production still see a need for improvement after releasing the product to the shop floor, new design and development phases must be initiated. Conversely if design errors were to be found during production start-up, the testing phase should be fired-up again to initiate an ECO (Engineering Change Order) or PCO (Process Change Order). Likewise the path from development to production should always include 'testing', as there is no feasible way to move between the two phases directly.

The cyclical nature of this R&D model combined with component-based-tools and the continuous process pattern permits and in fact encourages (parallel development) collaborative design, development and production set-up can all be taking place simultaneously, albeit, not necessarily working on the same product or subsystem. In the air conditioning industry for example, the following scenario is possible: while developing a new model, the design of a new coil subsystem may be initiated while the design and development of a tubing assembly is well under way, and while testing of the capillar is going on.

Figure 7 illustrates the time relationship of the activities in the two models, the traditional 'Waterfall' model and the 'Flow Rings' model. It is evident that new product development time is shorter in the case of the application of the 'Flow Rings' model in comparison to the traditional 'Waterfall' model.

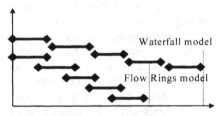

Figure 7:   Time dependency of the development cycle

(2) Design documentation and data management

Development of new products to be produced in a flow environment requires a new approach and product design documentation to be adopted in the development organizations. This approach integrates the traditional way of design and development in decomposing a product in functional groups and delivering subassembly drawings and bills of materials with the use during the development phases of the manufacturing processes existing in production. We name this approach *distributed flow design and development.* Distributed flow design and development uses the flow rings model and it concerns design and development at the process level.

Distributed flow design and development categorizes products into process families. Any product that falls into a process family will be in a multilevel process and in a functional world in a multilevel BOM structure. Looking at process families associated to products one can look at different kinds of fabricated components, gears and housings and so forth. These can be casted, be formed, they can be welded kinds of assemblies that relate to different process families that these kinds of products would fall into. When getting right down into this process family related to a particular product, the process family really is a series of processes in many cases multilevel processes. This is valid even related to assembly.

Products that are assembled and tested belong to process family definition, that relate to those kinds of products going though different kinds of assembly, machine assembly, hand assembly, test and so on. Therefore every product is a series of processes, often multilevel processes. These processes can have an identification that relates directly to the product. Each process consists of eight factors that can be grouped into three clusters, as shown in Table 1.

Every process to be developed is tying into the graphic work instruction identification relationship. In the distributed flow design and development process, the approach is therefore not to use a multilevel product structure in order to define a product in a functional way any more. A necessary condition in applying this approach however is to flatten the BOM structure and convert it in a simple structure of part number and a pile of parts.

In the following, we give a brief description of the distributed flow design and development process.

- Look at the related process identification to the products based on the product family. Each process-identification has a Graphical Work Instruction-identification underneath it.

- Look at the Graphical Work Instruction identifications. These Graphical Work Instruction identifications represent TAKT grouping of work and quality and have components underneath them.

- Go from top to bottom level, from the product to the components. This is the single level BOM pile of parts. These are the components at which engineering changes (ECO) and material systems are interested in (Figure 8).

In the middle of the multilevel process structure, there exists only the process identifications, the Graphical Work Instruction identifications, and this is where the process change order (PCO) has to apply.

Table 1:   Factors of process identification

Group	Factor of process identification
TAKT process grouping of standard work definition	*TAKT* is the time volume relationship calculated as the rhythm, beat or cadence for each process of a flow line and used to establish resource definition and line balancing.
	Resource is a person, machine, and/ or office equipment that can carry out work tasks required producing a product or performing an administrative activity.
	Line Balancing: The process of establishing individual operations approximately equal to a TAKT time amount of work. The elements of work to be performed at each operation is determined by the standard work definition divided by the TAKT time
	Process is defined with a physical location where a logical grouping of resources performs sequential work tasks. A process in manufacturing is a combination of resources (people and machines) that convert material toward the completion of a product. All resources in a process must have the same TAKT time.
	Standard Work Definition is a definition of the required work and identified quality criteria to build a product in the specific production process.
TQM validation and verification	*TQM validation checks:* At any point on the standard work definition where an element of work contains process variability, the one correct way to perform that element of work must be identified and inspected for compliance by the operator.
	TQM verification checks: Upon performing the TQM validation check, a secondary inspection of the element where variability occurs must be performed by the operator to assure the work has been completed.
Graphic work instructions	*Graphic Work Instructions* are a set of graphical representations depicting the work in motion including the TQM checkpoints and self-checks performed at an operation. The identified work elements are derived from the standard work definition and equal a TAKT time amount of work.

Thus, the design of a new product would be initiated by looking and deciding in which process families the product would fall into. Define the products based on process families and then move on to understand the work and quality to each of these processes to make up the product. That is actually how engineers would document the product. Today they use subassembly drawing; in the future they would use process level drawings, which of course is going to tie into work, Graphical Work Instructions drawings. It does mean that while design can be based upon designing with Graphical Work Instructions, engineers will still have to use the CAD system, in order to drive engineering analysis, finite elements analysis, understand what the product design is going to be and the product specification. However engineers do have to come back to that process definition to be able to document the product and actually to produce the product. The big difference from functional design is the way the product is documented and managed.

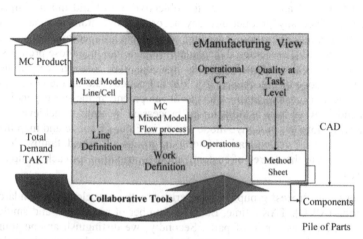

Figure 8: Mass customization at the process level

Further benefits in speeding up development cycle of a product, is to use *flexible specification for lean supplier partnerships*. The flexible specification is the specification of a part to a supplier, showing the non critical attributes of the part in addition to the critical attributes. Today most engineering drawings who go to the supplier will show tolerances shown down to the bottom of the drawing saying + or - 5/1000 unless otherwise stated. That means that all tolerances are set at the drawing. In a flexible specification engineers would not do that. Instead they would show areas that do not fall under a tolerance specification and they would let the supplier process drive where the tolerances or the specifications would go for that specific area of the part. One example might be surface finish, surface roughness. In many cases engineers will let the supplier process drive those dimensions to tolerances and not go ahead and restrict them by putting their full power in an engineering drawing.

20.5 Design of mass customized products for demand flow enterprises

Based upon the methodology described above the objective is to design mass customized products that add value in the eyes of the customer. To be able to timely and efficiently mass customize products, build them on demand, or excel in niche markets, products must be built flexibly. In this way we will reduce the noise in the system and build value- adding variety on demand. Manufacturing flexibility is one of the key process infrastructures needed for mass customization. This can achieved via lean production and/or demand flow manufacturing. Our approach in product design.

In demand flow or lean environments the objective is to build mixed family processes that produce any product any day according to customer demand. A mixed family is a family definition consisting of products grouped together based on similar manufacturing processes and standard times to produce those products. There exist simple tools to allow achieving this objective, however, describing them is out of the scope of this chapter [5]. We will concentrate on the design of products for effective customization and the development of the corresponding flexible demand flow or lean manufacturing processes. In order to achieve this objective we will adopt the proactive mode shown in Figure 1 above and map the four different zones of customization into four groups of mixed family flow processes according to the degree of customization contribution for each group of these processes.

Firstly, every flow process group is characterized by the four elements Standard Work Definition (SWD), TAKT time, Balance (number of resources), and single level Bill of Materials as a pile of parts. Secondly, we distinguish among four groups of mixed family flow processes: standard processes (no change), adaptable processes (configurable), customizable processes (change existing), and engineer-to-order processes (add new/replace old). Doing so, the matrix in Table 2 comes up.

Table 2: Mass customization process map

Process Group	Standard	Adaptable	Custom	ETO
TAKT	no change	no change	change +/- 20%	change >+/-20%
SWD	no change	small variations +/- 10%	big variations +/- 20%	change >+/-20%
Balance	no change	no change	no change	may change
BOM	no change	small changes +/- 10%	big variations +/- 20%	new parts
Contribution to product	60 %	25 %	10%	5%

The above map will help the product design and the manufacturing engineer to set up the design parameters for every product to be introduced into the Value-adding chain. The objective is, on one hand to define and build processes that are standard and there is no variation or change with respect to the basic characteristics and on the other hand to offer value-adding variety through various processes disposing different degrees of customizability. The customer acquires a personalized product that can be one-of-a-kind by means of a timely and efficient customization concept and methodology. In the eyes of the customer this product looks 100% personalized and in fact is. However, the way we have build and designed the product for the lean enterprise is a mix of standard and less standard processes that will allow the production environment to adapt quickly and efficiently to different customer demanded variety.

With the help of the simulation tools from Tecnomatix described underneath product design and manufacturing engineers will be able to accurately validate and adjust the design parameters in order to achieve the desired results achieving the cost and time targets set by the enterprise. In using and reusing the gained knowledge from similar products a continuous decrease of the Total Value Cycle Time will be achieved, since the objective is to increase the ability of the manufacturer to deliver value to the markets they operate and compete. Table 3 below gives some indications of the total value cycle time for different product platforms.

Table 3: Indicators of mass customization performance

Total Value Cycle Time	
Cars	80 (the 3 day car)
Air Condition Units	120
Consumer Electronics	240
Fashion Clothing	560 or more

20.6 Supporting IT tools

A set of tools developed by the authors of this chapter should help to implement and deploy the proposed methodology. Using this tools will equip manufacturing and engineering companies with a fully integrated infrastructure for the collaborative lean design of mass customized products and manufacturing processes that will seamless plug into an existing or future lean or flow environment. The result will be to virtually collapse time-to-market and time-to-volume current time benchmarks per industry segment or otherwise said, the total value cycle time. The tool set (marked as *"Tecnomatix fMPM"*) provides global manufacturers with an environment where their own planning departments and plants and their manufacturing contractors as well as customers can collaboratively create and operate optimal manufacturing processes. The solution provides:

- Planning and engineering tools to design and document the manufacturing process and allow product design on the process level.

- Web based collaboration tools in order to allow the various participants in the extended enterprise to take part in planning the manufacturing process.

- An fMPM portal allowing relevant parties to access and share all of the manufacturing process information over the web.

The major benefit lean and flow manufacturers and lean design engineers will gain from this environment is to be able to test lean and flow production execution of a product already during the design phase. This brings an enormous advantage on every side of the enterprise since, product design can be tested in very early stages for manufacturability against existing flow/lean processes, or new flow/lean processes that need to be designed, developed and deployed. Workstation lean ergonomic design and Kanban design takes place collaboratively with the development of the product. This is what we call 'Distributed Demand Flow Customization' infrastructure. New interactive communication technologies, like the internet, are fully exploited by this information system infrastructure. Networked across multiple users of product engineering and manufacturing engineering and production departments or even factories, suppliers and customers can form a virtual global supply chain organization and breaks the usual barriers that traditionally exist across the Industry. The lean extended global enterprise can then become reality. The whole set consists of the following tools:

Planning and engineering: The planning and engineering tools support the assembly planner's workflow. These integrated tools are specifically designed for this task and are thus replacing the use of generic, non-dedicated and non-integrated tools like Microsoft Excel, Project, Visio etc used by many planners today.

Process definition: Using the assembly sequence planner (fMPM-Assembler) the planner can visually, using 3D CAD data, find and validate a collision free sequence of assembly for the product. If such data this sequence serves as the basic operations scheme and can also be created without the use of 3D data. Once this scheme (pert), or operations tree, is created, the planner, using fMPM-Planner will allocate a resource and a part, thus creating the basic definition of the manufacturing process. We call this definition the basic electronic bill of process (eBOP).

Electronic Bill of Process: The eBOP is a description of the manufacturing process, representing the way this product is going to be manufactured, assembled and tested for quality. The eBOP is the basic token of information exchangeable between central planning departments, plants and contractors.

eBOP-Browser: Allows users to search, navigate and view eBOPs in real time and at different levels of detail – from a bird's-eye view for managers, to detailed 3D information for designers. Incorporating real-time, dynamic streaming, eBOP-Browser users can now interact with and exchange 3D manufacturing information over the Web, even over low-bandwidth networks.

Process design and analysis: Further, using the various eM-Planner applications the planner will allocate times to each operation, layout the assembly lines (in 2D), check their performance (throughput, resource utilization, bottlenecks etc), balance the lines and analyze the production costs.

Detailed workplace design: If needed, the planner will use fMPM-Workplace to detail the assembly lines, in 3D, including equipment selection and positing, human simulation, assembly time optimization, ergonomic analysis and final line performance analysis.

eM-Work Instructions: Users can quickly and easily generate and publish work instructions for shop-floor management, maintenance and training. Using drag-and-drop standard or customized templates, these instructions can include live links to the eBOP, as well as vivid graphics and snapshots, notes, reference files, annotations, attachments, hazards, safety procedures and rules. Accessed by any standard browser, eM-Work instructions ensures complete and up-to-date availability of the pertinent documentation and instructions required to enhance productivity, efficiency and quality on the shop floor.

Collaboration: The entire data created is captured in the fMPM Server (fMPM), a dedicated web-based server allowing the collaboration over the process design for multiple users at multiple sites. Using the fMPM, distributed planning teams can work together to design the process. This replaces a fare share of the use of faxes, phone calls, emails and face to face meetings while designing the manufacturing process. Not only time and travel are saved. fMPM provides a disciplined, common, way for a global organization to design its flow manufacturing processes. Once designed and captured on the fMPMS, the process can be easily archived, accessed and retrieved for re-use in following projects.

Transferring of production process made easy: By creating one structured data base and using a set of software tools describing the process in a highly visible way, the transfer process becomes clearer and faster. Using the eMS it is also easy for the receiving party to make modifications to the process, receive approval from the original planning department and create the new documentation, thus shortening time to volume considerably.

Data Sharing: While the fMPMS serves as the main data distribution and sharing platform for data creators, e-Manufacturing uses the web to provide the fMPM Portal as the main data sharing and distribution tool for those who need only to browse the information. This information includes, but is not limited to, electronic work instructions, eBOPs and various reports on the design of the manufacturing process. In the future, the fMPM Portal will also be the gateway to viewing the status and progress of the actual production process.

20.7 Conclusions & further research

Since mass customization means build on demand and not to forecast, this method provides the ability to manufacturers to design and build anywhere what a particular customer demands. Therefore, this approach may enable the rise of 'distributed demand flow manufacturing' with fewer giant plants and more small and flexible ones assembling from prefabricated modules closer to the end-user.

Furthermore, the advantage of using the same simulation models for real time production planning and execution as well as for operator on demand training purposes is a side benefit that should not be ignored. Finally, the proposed methodology can easily be a serious starting point for a standard for evaluating company's product development capabilities. The approach offers also the tools available for assessing best practices during the product development process, thus providing feedback required for possible reengineering of existing processes and a mechanism for companies to rigorously assess product development improvements and compare their performance with peers in industry.

However, further research is suggested in the following areas:

- The development of expert systems for the suggestion of optimal processes and reusable modules for mixed model mass customization process maps.

- The development of optimal mass customization process maps for different product platforms.

- The development of new costing models in order to reflect more accurately of the cost of value-adding variety in mass customization [6].

References

[1]   Anderson, D. M.: Agile Product Development for Mass Customization, 1998.

[2]   Dixon, N.: Common Knowledge: How companies Thrive by sharing what they know, 2000.

[3]   Krogh, G. V.; Ichijo, K.; Nonaka, I.: Enabling Knowledge Creation, 2000.

[4]   Bell, D.; Giordano, R.; Putz, P.: Inter-firm Sharing of Process Knowledge: Exploring Knowledge Markets, in: Xerox Palo Alto Research Centre and Centre of Innovation in Product Development-Massachusetts Institute of Technology, 2000.

[5]   Taiichi, O.: Toyota Production System: Beyond Large Scale Production, in: Productivity Press, 1998.

[6]   Hicks, D. T.: Activity Based Costing, Ed. 2, 1999.

[7]   Davenport, T.; Prusak, L.: Working Knowledge, Boston 1997.

[8]   Souder, W.: Managing New Product Innovations, Lexington 1987.

[9]   Pine II, B. J.; Victor, B.; Boynton, A. C.: Aligning IT with New Competitive Strategies, in: Luftman, J. N. (Ed.): Competing in the Information Age: Strategic Alignment in Practice, New York 1996, pp. 73-96.

Acknowledgments: We would like to thank Arie Rochman at Tecnomatix and Professor Kayafas and Associate Professor Loumos from the National Technical University of Athens for their suggestions regarding the integration of knowledge management into the subject of mass customization.

Contacts:

Dr. Alexander Tsigkas
FlexCom. AT&P Ltd., Athens, Greece
E-mail:flexcom@hellasnet.gr

Erik de Jongh
Tecnomatix Technologies Ltd., Herzeliya, Israel
E-mail: erikj@tecnomatix.com

21 Segmented Adaptive Production Control

Enabling mass customization manufacturing

Jens R. Lopitzsch and Hans-Peter Wiendahl
Institute of Production Systems and Logistics,
University of Hanover, Germany

In order to generate product variants not only by customized assembly of standard parts but by individual manufacturing processes, an innovative production control is necessary. In this chapter, the methodology of the Segmented Adaptive Production Control is presented to perform this task. The approach combines the two basic control principles push and pull. Using the mass-production character of the pull principle and merging it with the customer-oriented push principle, the basic approach of mass customization is applied to production control. The sophisticated system allows controlling the manufacturing of both, parts manufactured in mass production as well as customized parts, at the same work stations.

21.1 Customer orientation – a rising challenge without control?

In recent years, customer orientation became more and more important for the manufacturing industry. Especially, enterprises with mass production are obliged to put strong emphasis on customizing their products by variable production concepts. Obviously, the fundamental aim of efficient production of high volumes has still to be considered. Otherwise, efficient cost reduction could hardly be combined with the manufacturing of high-quality products. Nevertheless, the crucial challenge for modern companies is to satisfy individual customer orders and varying customer needs. This trend might have been satisfied by designing modular product concepts in the last years. In the future, it will be an additional challenge in matching customer demands by individualizing strongly the production processes. Pursuing the two objectives – customer orientation and efficient mass production – two main approaches have to be considered: First, flexibility can be increased by reengineering and designing flexible work stations and machines, jigs and fixtures. Second, mass customization principles can be supported by pure organizational changes within the manufacturing process: modifying the logistics by inventing innovative principles of production control.

Implementing principles of mass customization into production control, classic systems of production control reach their limits. Offering customized manufactured products in short delivery times leads to an increasing number of variant-specific buffers and to raising inventory costs on the one hand. On the other hand, the higher product variance decreases the single order volume and, by this, the manufacturing lot sizes. The time slice of set-up processes increases. This chapter proposes an innovative way of agile and adaptive production control, to avoid raising costs and allow a high logistic performance in terms of low inventory levels, short delivery times and high delivery reliability. The system presented is developed at the Institute of Production Systems and Logistics, Hanover, Germany. Our research was funded by the Deutsche Forschungsgemeinschaft (DFG), the central public funding organization for academic research in Germany.

21.2 Production control for mass customization

21.2.1 Fundamental organizational structures

Basically, mass customization requires high flexibility of the production and the production organization. Companies can ensure market success only with flexible reactions to customer demands, to fluctuation of quantities and to new competitive products. Analyzing production organization and designing suitable production control systems under these pre-conditions, two main organizational options must be distinguished (see Figure 1). Each alternative determines directly the basic draft of the production control system.

On the one hand, one can divide the production into segments for standardized high-volume production and segments for customer-individual low-volume production (*segmentation*). On the other hand, the process chain can be split into anonymous pre-production and customer-specific end-production (*separation*). Additionally, a mixed form of these possibilities can be identified: Dividing only the pre-production into a customer-specific segment and an anonymous segment. To investigate whether the organizational structure is able to support mass customization, the two main options are briefly be analyzed.

(1) The principle of segmentation: The main idea of segmentation is to define parallel working production areas. A criterion for this segmentation is for example the order volume. The objective of mass customization can be supported by practicing mass production and customer-specific production in parallel – one segment for mass production and a second segment for customer-specific production. For this, it is necessary to duplicate every process step and to provide all machines in both segments. As well, the product range has to be designed for its allocation to different segments – a high-volume standard program and a low-volume product program for customization. With this realization of mass customization the satisfaction of customer demand on individualized products is limited without additional investment in flexible manufacturing technology. Trying to avoid both, the necessity of double investments and the limitation of customer-specific quantities, there is a need to enable mass customization via production control.

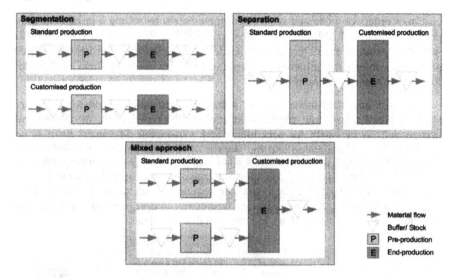

Figure 1: Basic options for the organizational structure of production

(2) The principle of separation: Additional to producing parallel in two segments, production can be split along the process chain. Starting first with customer-anonymous production, the second part of the production is customer-specific. Separating the production into pre-production and end-production, all products have to be designed according to principles of modularity. On that condition the final product can be customized easily within the last process steps of the end-production. A classic example for this is the end-production consisting of assembly processes only. Standard parts are pre-manufactured and stored. The customization is realized by customer-specific assembly.

One of the most important decisions to make is the definition of the customer decoupling point. This point identifies the change from customer-anonymous production (pre-production) to customer-individual production (end-production). Without having a clearly defined decoupling point, the strategy of separation can not be applied. But, the realization of the anonymous pre-production requires abstaining from all customer orders that include only small variations from the standard manufacturing process. Once again the need becomes clear to enable individualization via the production control system.

21.2.2 Basic production control principles

Bearing in mind the characteristics of the main organizational options, the production control must be sensibly adjusted. In general, two main control principles have to be taken into account – the push-principle and the pull-principle (Figure 2). Looking at the production from the order point of view, either the specific customer demand triggers the manufacturing and assembly (push) of customer orders or a demand is satisfied immediately by selling a product already produced (pull). This approach can be transferred to both, entire companies as well as separate areas of production. Orders can be customer orders or internal production orders (e.g. a part-manufacturing order, an assembly order or an order for a single manufacturing process).

The most important advantages of a decentralized pull controlled system are short delivery times, low requirement towards the complexity of the control system and no necessity for scheduling and sequencing. However, there is no possibility for customization and the storage of finished goods causes additional expenses for inventory and warehousing. In contrast, a push-system has no immanent need for storing goods and the production can totally be customized. But, every push-system requires a sophisticated module for order release and inventory control to assure short and constant delivery times.

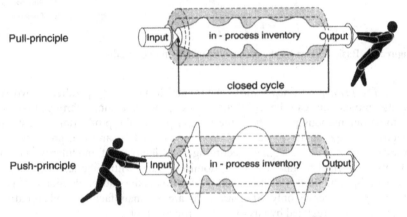

Figure 2: The main principles of production control push and pull [1]

Searching for control systems that realize the conceptual ideas of push and pull, the two most widely known approaches are the MRP system and the KANBAN system. Because of extensive discussion about them in the past [2, 3, 4] we will not go into further detail here. Knowing about the disadvantages of both systems, an innovative control system has been presented by Hop and Spearman [5]. Their proposal of a CONWIP control consists of the main idea to keep the work in process on a constant level (CONWIP – Constant Work In Process). It is possible to produce customer-oriented with an easy information system like KANBAN. Because of predefined control loops for the order-specific manufacturing with CONWIP, the quantity of controllable variants increases in comparison to KANBAN. Furthermore, the KANBAN specific buffers for all variants become redundant. Beyond that, short and reliable delivery times are guaranteed by limiting the quantity of released orders with a card system. The disadvantage in comparison to KANBAN is the necessity to put effort on sequencing and on inventory control inside of a CONWIP loop (Figure 3).

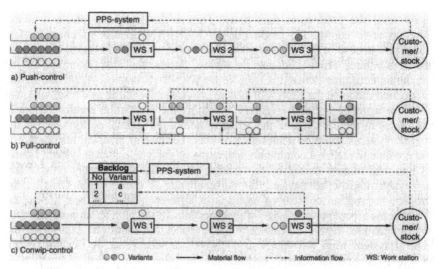

Figure 3: A comparison of the control principles of push, pull and CONWIP

Applying push, pull and CONWIP control within the organizational options presented, the variety of possible combinations is limited . *Customized production* can be realized only by push or CONWIP control. Because of pre-production and storing of finished parts, it is not possible to use KANBAN control. The advantage from short delivery times of KANBAN loops within the customized production can not be taken. Also, the requirement of mass customization to take use of directly available standard parts cannot be fulfilled. Instead, *the standard production* can be controlled via each of the three principles with their respective advantages and disadvantages. Summing up, there is no control principle that allows breaking the organizational structures and enabling mass customization. Neither a push system or a pull system nor a CONWIP system can solve the identified problems.

21.2.3 Deficits of today's production control systems

The production control systems presented before are well known and proven approaches to support short and stable delivery times (KANBAN) as well as customized production (CONWIP). Although each system has its clear benefits, there is no chance to take advantage from both systems at the same time if not in a parallel way for example like proposed above. The resulting deficits can be summed up as follows.

(1) Fluctuations of demand: None of the systems can respond on fluctuations of the customer demand for example within the life-cycle of products. Usually, over a product life-cycle the sales follow a bell-shaped curve. After launching a product, quantities sold are low. They increase up to the maximum and decrease

again until the production runs down. During the launch-up phase it is not appropriate to control the production by KANBAN because of normally low quantities. Instead, a CONWIP system would offer good possibilities. In contrast, the high-volume production directly before, during and after saturation of the maximum consumer demand is suitable for KANBAN control. Again, during the production rundown the CONWIP system provides advantages as a result of lower costs for warehousing and inventory. Adapting the production control system in the course of the life-cycle to the actual situation, the optimal performance of the logistic systems could be achieved. However, because of practical restraints it is not possible to realize a continuous adaptation of the whole control system. Losses of time and money caused by this change process would probably prevent any achievable logistic advantage. But the challenge is even bigger. Usually, the production program consists of products in different phases of their life-cycles. That means a continuous change of their portions of the production program. Due to this situation, there is no chance to adapt production control by exchanging the control system from KANBAN to CONWIP or vice versa. A flexible control enabling the optimal logistic performance in spite of mid-term fluctuation of demands is missing.

(2) Single work stations: A second example, that is not yet controllable taking advantages from both, KANBAN and CONWIP in different control loops, are single work stations with high capital expenditure. They become weak points for the logistic system. The necessary separation of the production into one KANBAN-controlled section for shortest delivery times and another CONWIP-controlled section for customer-individual production cannot be realized, if there is just one single, highly expensive work station. Regarding manufacturing cost, it is unprofitable to duplicate these stations and to accept lower utilization rates. Nevertheless, both segments have to use this work station. A combination of the two different production control systems is necessary at this point.

21.3   A new approach: Segmented Adaptive Production Control

21.3.1 Main idea

The main idea of our new approach is the combination of the KANBAN and the CONWIP control system. While high-volume production is controlled with classic KANBAN loops, the bigger part of variants is controlled with the CONWIP system (Figure 4). By this, different control principles are applied within the same production section. Using cards as a basic control device, there are no fundamental difficulties to combine KANBAN and CONWIP. The KANBAN card is always related to a specific variant, the CONWIP card is a variant-unspecific card defining a certain work load of the production. Both cards are circulating within the production accompanying and defining production orders. Using the

KANBAN loop, parts are produced, they are put in stock and they can be withdrawn according to a customer order with a minimal delivery time. The CONWIP loop allows producing customer-specific parts. Limiting and controlling the quantity of cards of the control system, the maximum load of production can be defined and reliable delivery times ensured. It is possible to use this combination of KANBAN and CONWIP control either to control an entire production process from the first process step until the shipment to the customer or to control just a production section – for example a section of pre-production before an entire customer specific end-production.

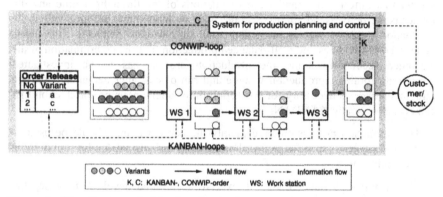

Figure 4: The Segmented Adaptive Production Control based on KANBAN and CONWIP

Still, several preconditions have to be considered for the application of this adaptive control system. Originally, this application has been designed for the job shop production of series of a large number of variants with a directed material flow. Starting from this initial production concept, it becomes clear, that first of all the Segmented Adaptive Production Control does not work with interlinked synchronized machines. Furthermore, set-up times should be as short as possible to allow small lot sizes and the immediate switch between KANBAN and CONWIP production orders. As for all KANBAN-controlled systems it is important to have a steady demand of KANBAN-controlled parts in the short-term. A production controlling instrument can respond to long-term and mid-term fluctuation of demand with a switch of the chosen control system.

21.3.2 Steps for implementation

Putting the proposed control system of Segmented Adaptive Production Control into practice, certain steps have to be completed. First of all, it is necessary to investigate the organizational structure for the production of all main components of a product. Possible structures have been mentioned above. It has to be determined, whether the production should be segmented in two or more parallel working sections, whether it could be separated into customer-anonymous and customer-specific production (customer decoupling point) or whether a mixed approach promises the best results. The pros and cons of segmentation have been presented in detail by Wildemann [6].

Determination of the customer decoupling point

The advantages of defining a customer decoupling point are obvious. First, the delivery time depends only on the throughput time after the decoupling point. The shorter this time is, the smaller are fluctuations of the throughput time and the higher is the delivery reliability. Second, with an anonymous pre-production, the production program can be decoupled from the customer demand. This means better adapted lot sizes and well-balanced production. For the determination of the customer decoupling point, the production process has to be investigated concerning routings, production volume and run of customer demand. There are three main aspects necessary for the definition of anonymous pre-production that can be controlled by KANBAN:

- The production process creating mainly the parts' variants should be used as the first customer-specific process because less variants have to be stored in the decoupling stock.

- The quantity demanded should exceed a minimal level. Otherwise, the inventory turnover rate of the decoupling stock is low.

- The customer demand at this point should be steady to avoid a high inventory level buffering the deviation of demand and production.

After analyzing the production process and defining customer decoupling points for all components of a product, the results can be aggregated to an entire control structure. This structure shows for each component, if there are sections of at least partly KANBAN-controlled pre-production with the decoupling point to switch from anonymous to customer-specific production. Within the customer-specific production and parallel to partly KANBAN-controlled sections, the production can be controlled by CONWIP. For mass customization purposes it is helpful to control as many variants of a component and as many components of a product by KANBAN to benefit from short delivery times (see Figure 5).

Allocation of product variants

After determining the customer decoupling points, each possible variant of a product component has to be allocated definitely either to the KANBAN-controlled part of the control system or to the CONWIP-controlled part. For this purpose, always the same crucial questions have to be answered.

- Does the throughput time related to customer-orders permit an on-time delivery according to customer requirements?

- Does the production work precisely to permit high delivery reliability?

- Is it more expensive to carry higher capital costs caused by a higher inventory level in stock (finished parts, storage space, storage equipment) or to carry the additional costs caused by higher order administration expenses, by less customer orders due to longer delivery times and by eventually a higher portion of set-up time due to smaller lot sizes of customer-order specific production.

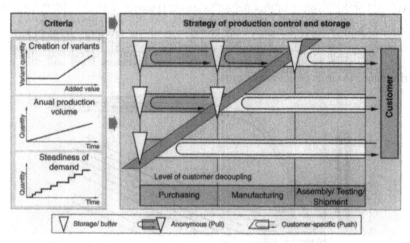

Figure 5: Criteria for the determination of control and storage strategy (adapted from [1])

To find an answer to these questions and to decide whether a variant is controlled by KANBAN or CONWIP, different criteria have to be taken into account. Some fundamental important facts to know are:

- the production volume of each product variant and each component (refers to inventory turnover rate and minimal lot size)

- material cost, manufacturing cost and cost of storage space, size of the variant (refer to economical stock sizes and minimal lot sizes)

- throughput times of pre-production and end-production (refer to delivery time requirements and minimum inventory levels)

Determination of buffer levels

One of the most important parameters of the Segmented Adaptive Production Control is the determination of the buffer levels. Basically, the levels for both control elements, KANBAN and CONWIP, can be defined by the quantity of cards. That means the quantity of KANBAN cards and CONWIP cards of each variant and each controlling loop. Determining the inventory level, throughput times as well as reachable delivery times are defined.

First, the lot size of each variant has to be calculated. For producing as flexibly as possible, the customer demand should be transferred to customer-specific orders without increasing the lot sizes. KANBAN lot sizes can be defined by classical lot-sizing algorithms [7]. However, it is important to restrict the maximal lot size. Work stations that are manufacturing KANBAN as well as CONWIP parts must not be blocked in case of too large KANBAN lots. Otherwise, the flexibility of production decreases. For the actual dimensioning of the inventory level it is recommended to use the funnel model and the theory of logistic operating curves that have been developed at the Institute of Production Systems, Hanover, Germany [8, 9, 10]. The logistic operating curve visualizes the correlation of the inventory level of a work station with performance, and throughput time (see Figure 6). It is possible to define a target operation point with short lead times, low inventory level and still, high performance in terms of utilization rates.

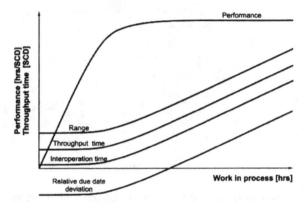

Figure 6: Logistic Operating Curves

After determining the inventory levels at each work station, the total inventory level can be calculated aggregating all single work stations. This inventory level has to be transferred to the quantity of necessary KANBAN and CONWIP cards. Summing up, there are four steps within this process:

1. Determination of the target inventory level of each work station

2. Aggregation of all target inventory levels of single work stations to the entire pre-production

3. Allocation of the total inventory level to the CONWIP and KANBAN loops

4. Allocation of the KANBAN cards to the single KANBAN loops and variants and eventual increase of the cards quantity.

In the beginning, it can be necessary to increase the quantity of KANBAN cards slightly. After a certain start-up period without major disturbances the quantity should be reduced step by step. The fine tuning process can be initiated within the normal operation. Based on the results of the determination of inventory levels, the expected throughput-times for CONWIP parts have to be calculated. These times can be used as a set-up value for scheduling processes.

21.3.3 Continuous controlling

Once in operation, the high performance level with regards to delivery reliability, delivery time, inventory level and utilization rate has to be assured durably. It is most important to oversee and to control the allocation of variants to the KANBAN and to the CONWIP loops. The mid- and long-term fluctuation of demand, for example according to the life-cycle, has to be taken into account. Variants have to be switched from KANBAN to CONWIP or vice versa. For this, a system for the logistical production controlling is needed. A sensible controlling process can be divided into six steps [11]:

1. Objective: The different objectives have to be defined. Examples are throughput times, inventory levels, utilization rates, delivery times, overall delivery reliability.

2. Objective quantification: After defining the objectives, they have to be quantified (for example an on-time delivery of 95 percent). These definitions serve as references values. Generally, they are determined for every single work station. Values that have been determined within the implementation phase have to be considered.

3. Actual situation measurement: The operation has to be evaluated based on actual process data.

4. Comparison: All quantified objectives have to be compared with the actual situation. Inadmissible deviations have to be marked.

5. Analysis of deviation: It is necessary to find out the cause for every deviation to plan corrections of the process.

6. Measures: The logistical production controlling should provide support for taking useful measures of correction. An eventual measure can be the switch of a variant from KANBAN control to CONWIP control or vice versa. It is important to priorities all measures concerning use and effort.

Implementing the instrument of logistical production controlling, the capability to ensure a continuous high performance can be guaranteed.

21.4 Capabilities of the new system

The proposed production control system allows taking advantages from both, the push as well as the pull principle. On the one hand, the push principle is introduced via the CONWIP control. Products can be produced according to the customer demand and a short delivery time as well as a high delivery reliability are assured with the limitation of work load by CONWIP cards. On the other hand, the pull principle is realized with KANBAN control. The decoupling strategy of the customer leads to short delivery times, to low efforts on scheduling and central production control and to a market-decoupled production with optimal lot sizes and an equalization of demand peaks.

The combination of KANBAN- and CONWIP-control loops leads to the Segmented Adaptive Production Control that provides advantages with regard to planning and control of the production and to factory planning as well as regarding the product life-cycle. This system enables highly flexible reactions concerning operational restrictions. It covers long-term and mid-term fluctuations of the customer demand and it offers a unique control solution for work stations that have to be used for both, standard mass production as well as customized production.

Production planning and control: The important potential is the simultaneous control of two different part ranges. Parts that are manufactured in mass production and parts that are customized according to customer demand are controlled with an adaptive system by KANBAN and CONWIP. It is possible to increase flexibility of production not only by separation with an anonymous pre-production and a customer-specific end-production but also by deliberate individualization of the pre-production.

Factory planning: Looking back to the part flexibility by the segmentation approach presented above, it is now possible to manufacture both, KANBAN parts and CONWIP parts at one single work station. Investment-intensive work stations can be loaded with orders from each segment. From the control system's point of view there is no difference between segmentation with one shared work station and a fully combined production of KANBAN and CONWIP parts without segmentation. The flexibility to define segments increases.

Product life-cycle: The basic problem – fluctuations of customer demand within the product life-cycle – is solved with the KANBAN-CONWIP combination. After launching a product with a high production volume, product parts will be controlled in KANBAN loops. If the customer demand decreases, the product parts are switched to the CONWIP control. Depending on the market situation the control system can be adapted to the requirements of each product variant. Especially the trend towards shorter life cycles emphasizes the importance of this possibility.

In the very near future, a slightly modified type of the Segmented Adaptive Production Control is foreseen to control the power-unit production of a company featuring standard and customized products in the south-west of Europe.

References

[1]   Eidenmüller, B.: Die Produktion als Wettbewerbsfaktor, Köln 1995.

[2]   Petermann, D.: Modellbasierte Produktions-regelung, in: VDI Fortschritt-Berichte, 20 (1995) 193.

[3]   Rohde, V.: MRPII und Kanban als Bestandteil eines kombinierten PPS-Systemes, Fuchsstadt 1991.

[4]   Wildemann, H.: Flexible Werkstattsteuerung durch Integration japanischer KANBAN-Prinzipien, München 1983.

[5]   Spearman, M. L.; Hopp, J. W.; Woodruff, D. L.: A Hierarchical Control Architecture for Constant Work-in-Process (CONWIP) Production Systems, in: Journal of Manufacturing and Operations Management, 2 (1989), pp. 147-171.

[6]   Wildemann, H.: Die modulare Fabrik – Kundennahe Produktion durch Fertigungs-segmentierung, München 1988.

[7]   Burmeister, M.: Auslegung der Verbrauchssteuerung bei vernetzter Produktion, in: VDI Forschritt-Berichte, 8 (1997) 658.

[8]   Wiendahl, H.-P.: Fertigungsregelung – Logistische Beherrschung von Fertigungs-prozessen auf Basis des Trichtermodells, München und Wien 1997.

[9]   Nyhuis, P.: Logistische Kennlinien, Habilitation, Universität Hannover 1999.

[10]   Yu, K.-W.: Terminkennlinie – eine Beschreibungsmethodik für die Terminabwei-chung im Produktionsbereich, in: VDI Fortschritt-Berichte, 2 (2001) 576.

[11]   Hautz, E.: Logistik-Controlling – Voraussetzung für eine konsequente Nutzung logistischer Potentiale, VDI Bericht 994, Düsseldorf 1992.

Contact:

Dipl.-Ing. Jens R. Lopitzsch
Institute of Production Systems and Logistics, University of Hanover, Germany
E-mail: lopitzsch@ifa.uni-hannover.de

22 Challenges of Mass Customization Manufacturing

Michael Schenk and Ralph Seelmann-Eggebert
Fraunhofer Institute Factory Operation and Automation (IFF),
Magdeburg, Germany

While the basic idea of mass customization with all its different facets excites marketing directors and CEOs, mass customization is still not yet a world wide standard. The reason for this can be seen in the complexity of implementing mass customization in actual existing mass or serial production. Pioneering examples of mass customization often focus on newly founded enterprises or on exclusively set up production lines. However, existing products, production and logistics systems have evolved individually. Thus, no standard solution can be offered for implementing mass customization into an existing production line. Questions such as which product, which feature and how many features should or could be individualized, remain. Implementing mass customization is reflected in all parts of a company and consequently in the entire supply chain. Therefore the manufacturing site needs to be redesigned in order to face the new challenges. Transport times have to be reduced within and between production lines. Producing lot sizes of one implies a need for high flexibility of the machinery, therefore an increased investment which must be planned thoroughly. Workers must be able to respond promptly to specific demands, too, hence resulting in a stronger need for special education programs and tools.

22.1 Mass customization as a reaction on today's market needs

Globally interconnected markets offer enterprises only marginal leverage for improving their operating results, since they are responsible for the increment of global competition. This is due to the growth of travel's frequency and easiness, the increasing speed of communication and the fact that products are sold worldwide [1]. Classic strategies of differentiation are no longer sufficient in many industrial sectors. In order to face a more demanding market as well as consumers the enterprises must manage to build individual and enduring relationships with their customers. As a result the highest directive for modern enterprises must be to ensure consistent customer orientation, always and everywhere. There are no markets anymore, only individual customers [2].

Modern multimedia applications such as the internet are spreading throughout the world and gaining popularity and can be used to enhance the power of mass customization. Individual users and business partners can satisfy their individual needs in the internet. A mere click signs a contract. When compared to the classic approach of mass production, this new development gives rise to additional challenges: the interconnectivity between worldwide electronic data interchange and the actual production and transportation of parts and goods. Having to ship and produce individual items adds to the classic challenges, especially those concerning time prolongation and cost increments.

The strategy of mass customization provides a way out of the dilemma between customer demand for the highest performance on one hand and customer demand for low prices on the other hand [3]. This strategy which was originated in and taken from marketing requires providing every customer with exactly the desired products and services. Moreover, this must be done without greatly exceeding the price of a comparable standard product [4]. Thus the advantages of mass and serial production can be combined with those of single item production, resulting in mass or serial customization. Consumers expect outstanding products and services with all the desired features at a very reasonable price and all of this as quickly as possible. Thus in both the consumer and the capital goods sectors, the enterprise's goal must be to recognize the customer's demand faster than the customer himself does and consequently to provide the right products when the customer desires them and for a reasonable price [5]. In this situation, mass customization offers enterprises the chance to act instead of react.

Some products have been built according to the customers specifications even before the concept of mass customization was conceived. Among these we can refer to the naval or civil constructions. Even if the production process only takes place after the client poses his order, it does not mean that mass customization is their strategy. In fact, the main aspect of this approach is the change from a traditionally mass produced well to a personalized one. Such a change implies that the whole way of perceiving the production process needs to be adapted, that means, modernized. This piece of writing intends to present a view of how this alteration can be done, not forgetting that there is no unique or key solution. Introducing such a strategy is not an isolated task, but an enduring process that affects all the parts and people involved.

However, many enterprises are still little aware of the extent to which the strategy of mass customization can be applied in practice. The few examples of successful mass customizers focus on new business ideas often connected with internet start-ups. In terms of production, the most celebrated pioneers have built up new lines where exclusively individualized products are made. Most of them made large investments and have not yet earned their return on investment (ROI) or have gone out of business. Today, however, most executive managers and decision makers recognize the necessity of being able to offer their customers first-class service. Hence, for some, mass customization appears as a welcome deliverance. This implies satisfying every customer demand and therefore frequently contributes to rapidly increasing complexity in all internal and external logistical business processes. In this context, compared with the classical meaning, the definition of logistics is enhanced. Apart from the mere transport of products and goods, this includes the steering and control of production lines, the holistic information flow within and outside production and the planning of the entire supply chain. The hasty individualization of products and process structures can however lead to a development past the market and thus past the customers. Before beginning far-reaching measures for restructuring and carrying out extensive investments in new machinery and IT solutions, enterprises must become aware of which form of individualization is right for them and their specific customers. External partners such as the Fraunhofer IFF perform world wide valuable educational work and offer approaches and methods which competently accompany the enterprise on its way to mass customization.

22.2   Theory and practice of mass customization

A company needs to face two basic difficulties before implementing mass customization. First, if it is really necessary to implement such a strategy. That is, if it is actually profitable on a sales perspective to provide each customer with an individualized product. And second, if there is a possibility with the current production capacity to deal successfully with those orders; and, if not, which changes could or should be made.

The impact on the sales needs to be accurately determined in order to quantify the implications on the outcome of having such a strategy. After this step, and if it proves to be viable, then the company needs to assess the influences on the manufacturing site of the company. In fact, even if it proves to be more profitable on a sales perspective to introduce mass customization, it does not mean that it should be at all initiated. Two more studies need to be carried out: the product and the process feasibility. The product feasibility concerns the opportunity to personalize the product. Usually the classic approach to implement mass customization is the modularization of the product. But some products cannot be modularized due to their specific characteristics and particular design.

Modularization offers the advantage of creating a wide range of possible combinations and therefore results in a broad palette of choices for the customer [6]. Depending on the product type, the selections can reach up to several million different options for the initial standardized product. Clever modularization also offers the possibility to separate standard production from customer specific production. For the production of the individual module or in cases when it is not possible to implement modularization or an order cannot be fulfilled by using the standard modules, the flexibility of the production needs to be extremely high. In that case, the company must be able to quickly respond whether the request is feasible or not. Of course, declining customers' orders is not the policy of strongly competitive companies. For that reason, a different approach needs to be considered. To deal with this type of request the production line must be completely flexible in order to respond quickly and effectively. Its design should be easily changeable and the workers should have all the necessary skills to cope with the situation. Preferably, the line workers themselves should be able to make all the necessary arrangements, which would obviously reduce the lead time considerably. Individual production islands can be introduced. Parallel to standard production, they contain highly trained and flexible personnel and/or machines.

The process feasibility relates to an intensive study of the manufacturing process. In most cases of this strategy's implementation, the process needs to be fully restructured and re-designed, i.e., re-engineered. It is important to refer that a flexible manufacturing process is the key factor for a successful mass customization implementation. The configuration process needs to be approached differently depending on whether the particular company is an established one with a mass production strategy or a start-up. In the former, production and distribution need to be redesigned and adapted to the new demands, while, in the latter, the configuration needs to be fully created. There is, of course, no strategy that fits every company, product or production and logistics process. Every case requires individual study and planning in order to find the most cost-effective method that still allows to provide uniquely individualized products and services [7].

Nowadays, the internet is changing the way business is done and has proven to be very valuable in implementing mass customization. This is achieved by allowing the customers to place their orders in a normal company web site. It is less time consuming, and therefore cost saving, to have the customer's specifications directly on the homepage, which are afterwards dealt within the company itself.

The inclusion of the entire supply chain is made possible by applying internet and intranet configurations (business to business, customer to business). The end customer's data can thus be distilled and all supply chain partners receive the needed information. In fact, it is impossible to conceive a customization strategy without sharing all the information among the various functional departments of the organization and the several supply chain parts [8]. A frequent drawback is the difference between the data type introduced by the customer and the one existing in the company, which can also be different from the data type of all supply chain partners. This diversity of information causes delays in the logistics cycle as well as mistakes due to false interpretations that result from a constant conversion of data.

Another important aspect is the amount of data the customer introduces when placing an order. This should be as extensive as possible in order to hasten the entire process. Assuming that the customer places the orders in the internet, the Web site should include as many features as possible, but still be easily comprehended by the user. Currently, there are several companies such as Lego and Reflect.com that have internet based customization and that are using specific software for individual customer configuration. One of the most popular examples is the opportunity to configure some features of a car such as colors, airbags and sunroofs in the internet. Companies such as BMW and Ford allow customers to introduce their specifications on the web, and these are then sent to a local car dealership which contacts the customer. Other specifications can be made, but not through the internet, only by contacting the actual car dealership [9]. Some kitchen manufacturers make it possible to manufacture the furnishing for an entire kitchen in approximately two hours according to the customer's specific demands. This can be done by using modern technology that automatically transforms customer data into production data.

22.3 Implementation challenges of mass customization

The change to mass customization can often only be effected by using existing production capacities. In fact, by using highly flexible machines and tools it is easier to deal more successfully with the production time when producing lot size 1. But it requires a large investment that SMEs are often not willing to make due to the uncertainty of the outcome and to the cost efficiency. Still, it is possible to respect in most of the cases the demand of the SMEs of not increasing the current production capacities when moving into mass customization, as will be shown. It is of uttermost importance to refer that the process of moving into mass customization does not imply abdicating completely a mass production strategy, since an interaction amid both leads to knowledge creation and organizational learning [10].

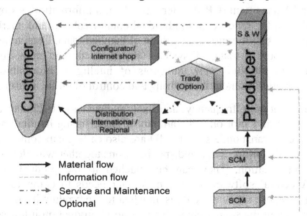

Figure 1: Logistics cycle for mass customization

Following the strategy of mass customization, different construction plans have to be generated for each order/customer and individual production information has to be distributed. The clever assortment of modern production and information technologies combined with an intelligent logistics system can thus make mass customization a strategy accessible for every company. The planning and design of the holistic value chain however most likely has to be adapted, redesigned or customized in its orientation following the logistics cycle of mass customization as shown in Figure 1.

Employees have to be trained to understand and accept this new production method. Consistent modularization of products and manufacturing processes is strongly recommended, since this standardizes the different parts of a product. The traditional rule 'the more standardized, the lower the costs' has not ceased to exist after all. Thus classic single product or variant manufacturers in the consumer and capital goods business are offered opportunities to sustainable reduce manufacturing costs and complexity costs.

22.3.1 Increase in information intensity

Generally, two initial situations can be detected on the way to mass customization–single item production (mainly capital goods) and mass production (consumer and capital goods). In both businesses, an increased volume of data is emerging, produced by a sharply increasing frequency of recurrence of the 'New Configuration'. Classically, one order contains several products. In mass customization one order is one product for one specific customer. Thus the information's complexity is not only generated by the increase of orders controlled but also by an increase of information. This information, such as the client's name and address, needs to be handled along with each order. An even larger potential of mass customization can be utilized by introducing Customer Relationship Management (CRM). Here the client's information is stored, altered, updated and used for specific and individual client treatment and marketing. This of course enhances furthermore the volume of handled data. Thus in order to cope with a large amount of data, IT systems chosen to support mass customization must be able to introduce new concepts of intelligent information logistics. Information logistics refers to the steering and control of all the information flows.

On this account radio frequency identification devices (RF-ID) can as well help to significantly reduce not only the amount of paper being handled but also data being processed. Transponders, as RF-ID are also called, can contain information about customized working steps and specifications together with delivery date and distribution information. They can be attached to the products or assembly carriages. A central software system no longer needs to compute or distribute information. Specific consumer data is attached to the product itself already while in production along the supply chain. The required production information can be read manually or automatically. Transponders make it possible for automated production to be steered and controlled not by central software but by the product, i.e. decentralized. When equipped with read/write abilities, transponders can furthermore serve as quality control media. Life cycle data can be stored on the RF-ID. This includes, for example, the address of the owner, maintenance dates as well as the personnel that serviced the product and service information that describes maintenance or repair steps. The amount of centralized information needed is therefore drastically reduced and an extra value is added to the product.

The information included in the product guarantees not only that it reaches the right customer but also provides data concerning all the people involved in its production process. Thus responsibility for a defect can easily be detected so that, in the event the product turns out to be defective, it is easier to find where the responsibility lies. Pilot projects in the automobile and the aircraft industry show additional values arising through the usage of RF-ID. Not only while being manufactured the chip can stay on the product life cycle but also afterwards in order to monitor the entire product life cycle. Maintenance and service checks can be stored as well as customer behavior.

22.3.2 Material logistic challenges

In addition to the information logistics also materials and parts have to be at their place on time. New logistical requirements arise for the classic standard and mass production. These are the result of the new link of the new configuration and production start, which can only take place after the reception of the customer order. There can no longer be buffer storage to keep inventory in reserve, one of the most quoted advantages of mass customization. What is more, an optimization of the logistics is required by the reduction down to production lot size 1. In classic lot size cost calculations with the aim of defining an ideal lot size, *storage size* and *tool changing time* build the most important factors. Having the 'ideal' lot size already defined, mass customization adds the critical factor *lead time* to the calculation. Surpassing a certain time of wait several customers might withdraw their orders and/or surely not buy again.

The impacts on the supply chain can be seen not only in the use of intelligent machines and universal control of the production flows, but also in spatial aspects. Today, customers can buy classic products and take them home. Therefore, in the case of mass and serial customization the waiting period for the end customer must be held to a minimum in order to be able to exploit the desired sales and market opportunities. In an ideal way producers and suppliers are geographically 'close' to each other. Thus long during shipment processes along the supply chain and distribution times can be held short.

22.4 A manufacturing model for mass customization

An approach to investigate the challenges occurring when changing from mass production to mass customization is based on the model shown in Figure 2. Sequential and parallel logistical processes can be examined separately or conjointly. For the sake of simplification, the model is restricted to only three cost centers in each direction. The overall value will differ when more or fewer cost centers are investigated but the general tendencies and relations will remain the same. Exponential tendencies can therefore be detected easily. Simple mass customized structures are examined with this model. The 'production model' is used to compare the initial situation of mass or serial production with that of simple or complex mass customization.

Thus, simulations are performed at that point in the same production line where an increasing number of individual parts are being produced next to the normal mass production. Overall cycle time, machine load and inventory buffer are the parameters investigated. In order to simplify the simulation, the machine's obsolescence is not considered as a variable characteristic but it is set to zero. These simulations have as a background the manufacturing of individualized bicycles. The different cost centers refer to the production of the different parts of the product. As it was previously pointed out, the model is limited to three production phases. Therefore it is assumed that the remaining parts that constitute the final good are not in-house produced but outsourced, hence not customized. The parts that can be customized are the frame, the handle-bar and the saddle.

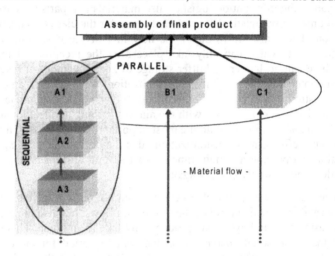

Figure 2:   Sequential and parallel cost centers for mass customization

22.4.1 Single individual parameter

Simple individualizing procedures are characterized by a single individualizing step. In this case, the individualization occurs only once in the entire process. Applied to the model, this could take place in one of the cost centers marked with A1, A2, A3, B1 and C1. The ideal solution would be to delay that step in order for it to take place in one of the following cost centers: A1, B1 and C1. Locating the only individualization step just before the final assembly primarily avoids complexity costs arising from planning, labeling, individual transport, etc. Due to production characteristics, sometimes it is necessary that the individualization step occurs in an earlier phase. For instance, in this model that would be represented by having the individualizing step in the cost center A3. In such a case of an early individualization step in the process chain, the aforementioned factors (planning, labeling, individual transport) might cause severe time prolongation and thus result in extreme cost increases and customers disappointed because of the possible late arrival of the product. Nevertheless, the plus in sales figures due to a 'new idea' and the exclusive market position may well equalize if not exceed the negative results. This of course depends strongly on the individual situation of the enterprise, the product and the market focus.

The postponement of the customization phase implies sometimes that the final manufacturing stage occurs in the warehouse facilities. Therefore the latter must be able to provide a flexible service. The outsourcing of warehouse services or the use of third party logistics providers present themselves usually as the best and sometimes the only solution for such a demand and also to avoid stocks' obsolescence [11]. Third party logistics providers are targeting postponement applications as an extension of their service portfolio [12].

22.4.2 Multiple individual parameters

When multiple individualizing steps are projected, more complex structures and logistical challenges arise depending on market needs and product possibilities. In this simulation it would mean having random individualizing steps in, e.g., four cost centers: A2, A3, B1 and C1. The overall production time varies greatly and does not permit forecasts of delivery times for the customer. The average production time tends to grow. Some of the inconvenient conditions can be overcome using classic methods such as outsourcing or re-engineering, which has been proving to be a way to help the manufacturers meeting specific customers demands [13]. Individual algorithms for order management and creating transportation lots that differ from production lot size 1 can help to relieve production. In special cases, the production lot size does not necessarily have to be size 1. Some production processes allow simultaneous treatment – cleaning, bathing, painting and such - of different products.

Experiments with the model show that a sequential individualization of two or more steps influences the production more than multiple parallel steps do. This influence is reflected in the increase in the process time and frequent buffers, leading production in exactly the opposite direction it was meant to head in using mass customization.

22.4.3 Individual parameters in supply chains

The initial model can be very easily expanded in order to reflect the situation in a supply chain. The former cost centers of the model are then seen as individual enterprises that are integrated into the holistic logistics cycle as shown in Figure 1. The parameters investigated change into the material flow between enterprises, information and data exchange, the degree of cooperation between enterprises, and combined planning procedures. For individual enterprises, production cooperatives and supply chains, intelligent product planning can lead in advance to moving the point of individualization closer to the final assembly and can thus help avoid difficult and expensive logistics solutions for mass customization. Modularization of the products and processes in this context makes production in standard variants and lots possible for individual enterprises of a delivery chain. Only the product modules with individualized parameters are then affected by the increased requirements, which altogether burden the supply chain less. In the case of modular product layout, it is essential to define the interfaces of the modules exactly. Modifications, brought about by customer specifications in the module, may not violate the standards of the interfaces agreed upon.

After applying the model in Figure 2 on all possible combinations and variations, one specific trend was detected influencing the implementation of mass customization in existing production lines. Figure 3 does not scientifically exact represent all values but shows the tendency overall of increase in all dimensions. The dimensions measured [D] consisted of the overall cycle time, machine load and inventory buffer. Using structures of production lines and sequential individualizing processes, the value of the dimensions measured [D] shown in Figure 3 increased significantly when only up to 25% of all parts produced in one production line were individual.

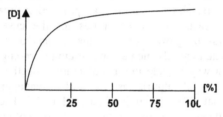

Figure 3: Increase in product dimensions

Augmenting the number of individual products being produced in a hitherto exclusively standard production line to a net result of 50% individual products and 50% standard products results in only a slight increase in relation to the initial increase. The values measured are of course valid for individual production and the standard production conducted parallel. That kind of behavior leads to a firm conclusion for the implementation of mass customization: running try-outs in an existing mass or serial production proves to be a complex task to accomplish. When faced by such drastic increases as in the model, an initial pilot phase seems only reasonable when a continuation of the project is seriously considered, the market possibilities have been thoroughly evaluated and the feasibility is beyond doubt.

22.5   Flexibility in the manufacturing site

As it was previously mentioned, in order to successfully introduce a production system oriented to the customer it is absolutely necessary to have a high flexibility in the manufacturing site. This demand ought to comprise the equipment, the production layout and the workers. When the equipment flexibility is the focus, a previous study must be undertaken, not only to evaluate its influence in the production process but also to avoid unnecessary investments. Such a study is individual for each company, since it depends mainly of the product itself and the company's output.

The production layout must provide the workers with a working-friendly environment and reduce the time in-process. Such demands are also generated by the classic mass production strategy. A carefully planned production layout prevents future changes which incur in onerous delays. Having a high life quality in the working site proves to be valuable since it influences directly the profits by increasing the productivity and the frequency of original suggestions alongside with a quality concern throughout the company.

Workers flexibility results primarily from an engagement in achieving the defined goals. Considerable knowledge of the innovative methods is essential. For that reason education is a crucial element. Producing according to customer specifications implies creating unique products and consequently finding one-of-a-kind solutions. It would be quite time-consuming and expensive to thoroughly discuss which solution is the best every time. If these solutions come directly from the line worker, they can be implemented immediately. Workers usually expect higher payment in return for acquiring other skills and for accepting different working habits. Sometimes workers do not respond in a very positive way to the new competence requirements and to the higher level of responsibility. Therefore job enlargement schemes may not be easily implemented or not even at all feasible. This initial inertia can be reduced by using interesting educational tools, for instance, virtual three-dimensional and interactive simulation programs (a more thorough explanation of possible education and training tools is presented further ahead). This kind of equipment makes it possible for workers to be in close contact with the type of environment they will work in, enabling them to participate in a wide range of hypothetical scenarios. Apart from being extremely interactive, it is also cost saving, since this kind of software is usually less expensive than real machinery. What is more, the return on investment when a machine has been used for educational purposes is lower than that when the equipment has been used in the production process itself.

Visualization techniques can also be used for research. Simulating a production line is less time consuming and more cost effective than building one. Features can be easily changed and/or added without altering the current production line, thus resulting in not having to stop or delay the production process. Not only do the line workers have to be fully aware of the new strategy but every one inside the company has to be fully committed in order to implement such an innovative technique. Special business games can help to point out the demands of a mass customization strategy to the entire staff of a company or even a supply chain. These educational tools are employed to create a general awareness of the upcoming challenges. By solving cases the staff involved understands and better supports future changes that might occur in their own process structure.

The customer no longer plays a passive role in the production process. The customer's participation is absolutely required and most welcome. More then ever the customers are making sure that their needs are met in the service delivery process [14]. In order to make individual specifications, the customer has to be aware of all the possible choices available or else a feeling of confusion can arise related with the complexity existent in a wide assortment of options [15]. Therefore great emphasis has to be placed on the information provided to the customers. Sometimes some specifications can be impossible to fulfill, this means that the customer must have the information about what not only can be but also what cannot be chosen. Therefore, education of sales personnel has to be emphasized, since they must be able to provide the client with all the information necessary, not only about the product itself but also about the features of the production process. By knowing how this process works, the sales personnel can immediately verify if a certain request is viable or not.

22.6   Education and training tools

Implementing mass customization implies a strong commitment from all the employees as well as motivation and strong knowledge of the process. Business games and visualization techniques are a valuable help to teach and train the employees that ought to be considered as the most important assets. Several tools have been developed by the Fraunhofer Institute IFF to be applied for the education and training of the personnel. These tools are visualization techniques and business games for the implementation of mass customization. As an example, the business game ELISA$^{®}$ is presented.

This business game intends to present a simple case that can be used both as an education tool but also as a way for all the personnel to realize the implications of mass customization in the manufacturing site. In the game, the customer initially introduces an order, which always includes a standard part and an individualized one. Therefore it is clearly related to the mass customization strategy. In this particular game the order is a string of ten letters, of which five are standard, i.e. they are included in each order, and correspond to the word ELISA. The other five compose an independent word that varies according to each order.

- ROUND 1

    Customer: The customer must inspect the quality of the products, match products with the order number and take over the delivery schedule in the list of orders. It is necessary to assign numbers to every order, i.e., to every word and enter this in the list of orders.

    Assembly: After taking the customer's order it must be attached to a table rack, which is the physical support of the letters. Next the production process follows.

    Work Station 1: In this work station the letters A, B, C, D and E are worked off. This means that if the words in the order include these letters then they should be attached to the table rack. The ones to be included in the personalized word should appear in the front while the ones to form the standard word ELISA should be attached on the back side. A maximum of 14 letters can be processed per cycle. When this capacity is reached ("capacity of the machines"), the work must be stopped until the beginning of the next cycle. As soon as the capacity limit is reached, no new orders may be processed. The current order remains on the work surface. Afterwards, the table racks worked on should be directed to the next work station.

Work Station 2: For this work station, letters F, G, H, I and J are worked off. The procedure is similar to the one described for the previous work station.

Work Station 3: Same course of action, only the letters change (K, L, M, N and O).

Work Station 4: Here the letters are P, Q, R, S and T.

Work Station 5: The letters are U, V, W, X, Y and Z. Since it is the last manufacturing phase, it is important to inspect the quality of one of every product for the completeness and correctness of manufacturing. Return defective parts without delay to the source of the defect and from there they must yet once again run through the entire supply chain.

- # ROUND 2

In this round the procedure is in all similar to the one described in the Round 1, except the letters being worked off in each work station. After a statistical study of the frequency of use of each letter, a new distribution of the letters among the work stations was found, which proves to be more efficient since it levels the work load in each work station.

- # ROUND 3

Customer: In all the rounds the customer can either select a word from the word list or create another five-letter word. The selected words must be crossed off from the word list in order to avoid repetition.

Work Stations 1 to 4: Following the order, as many existing letters as possible have to be worked off. The capacity of the letters, which can be processed per cycle, is limited. It is essential to agree upon the level of the capacity limitation at the beginning of the round. As soon as the capacity limit is reached, no new orders may be processed. The current order remains on the work surface.

Assembly: The quality of one of every product (rubber letter pads) must be inspected for the completeness and correctness of manufacturing. Return defective parts without delay to the source of the defect and from there they must yet once again run through the entire supply chain. For each order the player takes one table rack from the corresponding stock and attaches the rubber letter pad with the customized word to it. Next, the player takes rubber letter pads with standard word ELISA from the KANBAN holder and attaches them to the back side of the table rack. The empty KANBAN holder should be directed to the work station 5 and the finished table racks to the customer.

Work Station 5 – Production on Call: When an empty KANBAN holder is available, further work steps must be carried out. With the help of the available letters, the word ELISA has to be attached to every rubber letter pad. Independently, the quality must be checked.

Order Center: The final phase of all the rounds is located in the Order Center. An order is to be deposited in a work station, which can process parts of it. Attention has to be dedicated into keeping the load of all work stations as uniform as possible. While taking the orders from the incoming store arriving from the work stations, an inspection for quality must be carried out. During the checking of the orders, if further necessary work steps are detected as being necessary, they should be passed to the corresponding work station. Once again, attention has to be devoted into keeping the load of all work stations as uniform as possible. The finished orders are to be passed on to the customer. The cycle is completed only when all orders have been processed.

This business game introduces a clear and simple practical approach to mass customization that enables the players to have an insight into the strategy's core. Several rounds with different manufacturing structures are presented, each with diverse outcomes and therefore advantages and disadvantages. Thus helping not only the employees to understand the basics and implications of mass customization but also for the company to realize the possible changes that might be implemented in the manufacturing site.

22.7 Conclusion

Innovative internet applications in e-commerce facilitate the efficient automation of individualization in mass markets. Thus media changes can be avoided and a swift fulfillment of the customer order is guaranteed. In contrast to the classic internet shops the user is then offered a genuine added value through individual product consulting and really appropriate products. The inclusion of the entire supply chain is made possible through the application of internet and intranet configurations (business to business, customer to business). The data of the final customer can thus be distilled and all partners of the supply chain receive the needed information. And still, on account of diverse reservations and a lack of local examples, Mass customization is still not yet spread very widely neither in Europe nor worldwide. This is happening against the background that mass customization is regarded globally as one of the best strategies to maintain and consolidate the market position of the enterprises in the future.

References

[1]   Verwoerd, W.: Value-Added Logistics: The Answer to Mass Customization, in: Hospital Material Management Quarterly, 21 (1999) 2, pp. 31-36.

[2]   Zeleny, M.: Customer-specific value chain: beyond mass customization?, in: Human Systems Management, 15 (1996) 2, pp. 93-97.

[3]   Pine II, B. J.: Mass Customization, Boston 1993.

[4]   Piller, F. T.: Kundenindividuelle Massenproduktion, München/Wien 1998.

[5]   Stuart, A.: All for One, in: CIO, 8 (1994) 2, pp. 50-58.

[6]   Duray, R.; Ward, P. T.; Milligan, G. W.; Berry, W. L.: Approaches to mass customization: configurations and empirical validation, in: Journal of Operations Management, 18 (2000), pp. 605-626.

[7]   Radder, L.; Louw, L.: Research and concepts: Mass customization and mass production, in: The TQM Magazine, 11 (1999) 1, pp. 35-40.

[8]   Jiang, P.: Segment-based mass customization: an exploration of a new conceptual marketing framework, in: Internet research: electronic networking applications and policy, 10 (2000) 3, pp. 215-226.

[9]   Hibbard, J.: Assembly Online, in: Information Week, 729 (1999), pp. 85-86.

[10]   Kotha, S.: Mass-customization: a strategy for knowledge creation and organizational learning, in: International Journal of Technology Management, Special Publication on Unlearning and Learning, 11 (1996) 7/8.

[11]   Gooley, T. B.: Mass customization: How logistics makes it happen, in: Logistics Management and Distribution Report, 37 (1998), pp. 49-54.

[12]   van Hoek, R. I.: The Role of Third-Party Logistics Providers in Mass Customization, in: The International Journal of Logistics Management, 11 (2000) 1.

[13]   Gilmore, J. H.: Reengineering for Mass Customization, in: Cost Management, Fall 1993, pp. 22-26.

[14]   Samenfink, W. H.: Are You Ready for the New Service User?, in: Journal of Hospitality & Leisure Marketing, 6 (1999) 2, pp. 67-73.

[15]   Huffman, C.; Kahn, B.: Variety for Sale: Mass Customization or Mass Confusion?, in: Journal of Retailing, 74 (1998) 4, pp. 491-514.

[16]   Schenk, M.; Seelmann-Eggebert, R.: Mass Customization für den Produktionsalltag, in: ZWF, 4 (2001).

[17]   Seelmann-Eggebert, R.: Mass Customization facing logistics challenges, in: International ICSC Congress, Australia 2000, pp. 174-177.

Contacts:

Prof. Dr. Michael Schenk
Fraunhofer Institute Factory Operation and Automation, Magdeburg, Germany
E-mail: schenk@iff.fhg.de

Dipl. Ing. Ralph-Seelmann Eggebert
Fraunhofer Institute Factory Operation and Automation, Magdeburg, Germany
E-mail: seelmann@iff.fhg.de

23 Modularization in Danish Industry

Poul Kyvsgaard Hansen[1], Thomas Jensen[2] and Niels Henrik Mortensen[3]
[1] Department of Production, Aalborg University, Aalborg, Denmark
[2] Danish Technological Institute, Center for Production, Taastrup, Denmark
[3] Department of Control and Engineering, Technical University of Denmark, Lyngby, Denmark

There are many pre-requisitions to make a mass customization strategy efficient. Speaking about the product dimension modularization seems to be an essential factor. However, though many companies have gained experience there is still a significant confusion about managing the modularization effort. In general, the phenomenon of modularization is not well known. The cause-effect relationships related to modularization are complex and comprehensive. Though a number of research works has contributed to the study of the phenomenon of modularization it is far from clarified. Recognizing the need for further empirical research, we formulate a research framework with the purpose of uncovering the current state in Danish industry and to identify tentative managerial implications.

23.1 Modularization and mass customization

Mass customization is defined in this chapter as the "development, production, marketing, and delivery of customized products and services on a mass basis" [1]. Ideally, customers can select, order, and receive a specially configured product. This means major shifts in operating methods throughout the organization – engineering, marketing, manufacturing – including the whole supply chain. For the new product development function, the successful implementation of mass customization has major implications such as: the compelling need for parts standardization and product modularization.

This chapter explores in particular the phenomenon of modularization. In many cases, modularization is presented in an inherently positive way. Based on the frequency of these positive cases, it might be induced that modularization is the universal cure of any competitive weaknesses experienced by manufacturing companies today. Definitely, modularization has big potentials regarding competitiveness but it is our experience that it needs to be configured in a way similar to other strong means. A decision on the degree and the shape of the specific modularization effort is a managerial task that can be compared with many other important managerial decisions. We suggest that modularization is considered as a contingent and configurable managerial means for manufacturing companies. If a means is to be configured one needs to know more about the adjustable parameters and the contingency factors. Secondly, one needs to know about or to set the specific goals. Therefore, one needs to know where the effects of the modularization effort will occur and how these effects can be measured. Often the effects sought by one part of the organization will conflict with the effects sought by other parts of the organization or external partners. Consequently, a balanced effort is needed.

In general, the phenomenon of modularization is not well known. The cause-effect relationships related to modularization are complex and comprehensive. Though a number of research works has contributed to the study of the phenomenon of modularization it is far from clarified. This chapter should contribute to the clarification of modularization by taking a managerial viewpoint and by including a number of empirical observations from an ongoing survey in Danish manufacturing companies.

23.2 Aspects of modularization

23.2.1 What is modularity?

In recent years modularity is discussed intensively [2]. This is related to the potential benefits from utilizing modularity in products and/or in production. Baldwin and Clark [3] define a module as an unit whose structural elements are powerfully connected among themselves and relatively weakly connected to elements in other units. Clearly there are degrees of connections, thus there are graduations of modularity. This understanding is clearly inspired by Simon [4] who defines a "nearly decomposable system" as a certain kind of hierarchical system in which the interaction among the subsystems are weak but not negligible.

The background can be summarized in two properties: (1) in a nearly decomposable system the short-run behavior of each of the component subsystems is approximately independent of the short-run behavior of its other components, and (2) in the long run the behavior of any of the components depends in only an aggregate way on the behavior of the other components [4]. Though Simon does not use the term modularity in his work, the concept is equivalent. The concept of modularization has strong strategic impact. The idea of product modularization is widely recognized as a major success factor in terms of meeting economic and commercial goals of a product program. Convincing examples can be found in the automobile industry [5] and in consumer electronics with Sony, Black & Decker, and Hewlett Packard [6, 7] as the most outstanding example. The effect of modularization can be interpreted as an encapsulation of complexity. When the task of developing and managing a system is exceeding the human capabilities one way of managing a complex system or problem is to break down the system into manageable parts. By encapsulating parts of a product by means of a module, the complexity can be reduced to handling and specifying the interfaces between modules.

Organizational barriers are important explanations for the difficulties of managing modularization efforts. It is generally recognized that the motivation for modularization has to be sought in other organizational units, for example:

- Product modularity reduces costs in the product life cycle due to the possibilities of economy of scale in production.

- Product modularity reduces delivery time due to the possibilities of late product "baptism".

- Product modularity enhances speed in the product development process due to the possibilities of distributing the activities and due to the inherent structure supporting the project management.

- Product modularity enhances speed in the introduction of new product variants due to the reuse of components and structures.

- Product modularity enhances the variety due to the flexibility in configuration of the final product.

- Product modularity enhances organizational flexibility due to the ease in communication of the product structure.

- Product modularity enhances organizational learning due to the inherent structure for storage of knowledge.

- Product modularity reduces risk in product realization process due to the exchangeability of modules.

Many companies experience significant problems in realizing these potentials and we interpret this as an incomplete and fragmentary understanding of the phenomenon of modularization. In the following, we will introduce a research framework aiming at creating a more comprehensive understanding of the phenomenon of modularization and thereby creating a basis for managing the efforts. Our framework considers the impacts of modularization from three perspectives:

- A *specific product-production perspective* focusing on products, activities, and knowledge.

- An *organizational functional perspective* focusing on the sales, product development, and production function.

- A *time perspective* focusing on strategic management, planning, and realization perspective.

The three perspectives with three subdivisions form a research framework for investigating the phenomenon of modularization. The research framework can be illustrated as a cube (see Figure 1) indicating the relationship between the perspectives. In the following, we will shortly discuss the three perspectives.

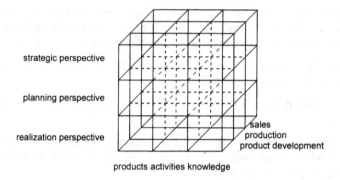

strategic perspective

planning perspective

realization perspective

sales
production
product development

products activities knowledge

Figure 1: Research framework for studying the phenomenon of modularization [8]

23.2.2 Specific product-production perspective on modularity

Modularization can be defined as an encapsulation of complexity. This complexity can refer to products, activities, and/or knowledge. By dividing a product into structural modules, i.e. building blocks, each module can be seen as a black box with well-defined interfaces. Though each building block may hold a number of complex functions, the final product can be configured relatively easily. One example are the DeskJet printers produced by Hewlett-Packard. For a printer, some of the functional elements are "store paper", "store output", "deliver ink", "supply DC power", "communicate with host computer", etc. These functions are realized in a few building blocks with well-defined interfaces [9]. Benefits resulting from this principle include:

- *Rapid product change.* A new product can be derived from existing products by changing modules, i.e. by changing only one or a few modules.

- *Higher product variety.* Modular products can be more easily varied without adding much complexity to the manufacturing system, i.e. the Swatch watches.

- *Component standardization.* The same component may be used in multiple products and thereby lead to lower unit costs and higher quality, i.e. Hewlett-Packard's thermal ink delivery unit.

- *Better manufacturability.* One important feature of design-for-manufacture is reduction of the number of parts. By dividing the product in building blocks, the integration of components can be considered within the block.

Another aspect of modularization refers to the development and/or production activities. The responsibility for the detail design of a module is usually assigned to a relatively small team within the company or to an outside supplier. If the realization of a function is assigned to two or more teams the coordination is likely to be substantially more complex and challenging compared to the situation where teams are designing modules with well-defined interfaces. Hence, the coordination of the development activities can be supported by the modular product structure or the pattern of a complex activity chain can be encapsulated by definition of a product module. Similar arguments count for the production and assembly of the products. This correspondence between activities and product structure is likewise found by Baldwin and Clark [3]. They claim that the design activities (task structure) and the design structure (the description of the artifact) have to be identical in order to have an efficient design process.

The fourth aspect of modularization is knowledge. Because development and production activities are often organized into development teams focusing on specific components and related technologies, the structure of the knowledge processes within a company is likely to reflect the structure of the product. This gives potential benefits in terms of focusing on critical know-how. Apart from this dependency among different types of structures, a more general issue in this context is the level of articulation, i.e. to what extent it is possible to make knowledge explicit and thereby possible to share in the organization. This is embodied in the discussion of knowledge as a continuum from tacit to explicit [10].

Thereby modularity and knowledge together constitute a knowledge map as basis for improved organizational learning. Furthermore, this understanding is likewise applicable to an individual level as to a group or system level. Finally, this discussion of tacit versus explicit knowledge is also relevant in relation with IT systems. A common issue often discussed in relation to all three aspects is the risk of blinding the company. Henderson and Clark [11] state that established companies often fail in case of a structural change. When the product structure is embedded in the organization the operation of communications channels and information filters both become implicit and therefore difficult to change, i.e. making the organization blind to changes. In other words, the structures have enabled a certain routine or pattern of activities and knowledge structures within the organization.

Table 1: Differences in orientation between organizational functions [12]

Function	Degree of departmental structure	Orientation toward		
		time	others	environment
R&D	Low	Long	Permissive	Science
Sales	Medium	Short	Permissive	Market
Production	High	Short	Directive	Plant

23.2.3 Organizational functional perspective on modularity

Organizational functions do not have identical images/requirements of an ideal modular product. However, the degree of awareness of these differences between the functions and the explicitness of the functional images/requirements is largely determining the degree of success of the modularization effort. The generic differences between the organizational functions are documented in a number of studies like the classical work by Lawrence and Lorsch [12] (see Table 1 above).

23.2.4 Time perspective on modularization

Several time horizons are important when planning and applying modularization. At the *strategic level*, the product platform or product architecture must explicitly specify the product modularity. Automobile makers define product architectures for car models, specifically defining the interfaces between the car body and its mechanical components, as well as interface between the major mechanical components themselves. Mixing and matching body styles and major mechanical components then leverage several product variants. General Motors, for example, has established a modular product architecture specifying about 70 different body modules and about 100 major mechanical components like engines, power trains, and suspension [13]. Similar examples are seen at Ford and VW.

Effectively implemented the product platform or architecture provides the ability to leverage product variations from a modular product architecture relatively cheaply and quickly improves the responsiveness in dynamic markets. A company may use the platform as an ability to leverage test products to improve its knowledge of customer preferences. An important success factor seems to be whether the company is able to decouple the product development from the technology development [14]. As an additional consequence, modular product design may lead to better understanding of the state of a company's current know-how and thereby provide a more effective strategic management of new technology development. However, this requires a change in the way companies in general divide their focus on the development processes and the technology management effort.

The *planning perspective* reflects the tactical realization of the intentions laid out by the product platform or architecture. At strategic level the modularization is mainly intentional. At the planning level, the intentions are realized considering the physical necessities (functionality, strength, robustness, etc.). Comparable conditions apply for both product and production development. Additionally, the considerations apply for both internal and external collaboration partners. The major tasks related to modularization are defining and developing the modules and interfaces. Product modules can be defined from a functional view or from a structural view. The interfaces define, for example, how one component may be physically connected to another (the attachment interface), how power is to be transferred between components (the transfer interface), how signals will be exchanged between components (control and communication interface), the spatial location and volume occupied by a component (spatial interface), and various ways in which the functioning of one component may generate heat, magnetic fields, or other environmental effects that must be accommodated by other components (environmental interfaces) [15].

At planning level, similar considerations apply according to modularity in production equipment in terms of defining and developing production modules and interfaces. For production development, as well as product development, relations to existing or potential suppliers must be considered in parallel. An intention of modularization may by realized in numerous ways. As indicated above the benefits of modularization are generally realized in other organizational functions than the product development function. Therefore, the effectiveness of the modularization effort is highly dependent on the explicitness and purposefulness of the induced dispositions on these functions.

At the *realization level*, the product is physically developed and produced. This may be indirectly to a customer order (to stock) or directly according to a specific customer order. In terms of modularization the strategic intention, the specific modules and interfaces are tested according to operational goals. These goals may be speed in deliverance, flexibility according to customers, speed in establishing new product lines, etc. The most frequent problems occur due to conflicts between or within the product and process architecture. Other frequent problems are market dynamics where changes in customer requirements make the product or delivery conditions inadequate. Companies in such dynamic competitive environments tend to see themselves in a continuous reconfiguration of development, production, distribution, and marketing.

The basic driver for coping with such dynamics is organizational flexibility. The modular principals applied to products may also apply for the organization. That is, focusing on the formation of organizational units, i.e. teams, and the interfaces between these units, i.e. the coordination structure. The realization level is the ultimate test of whether the strategic intentions can be realized. Consequently, the realization level is the crucial link for closing the loop: product architecture – development process – production. When done systematically this creates the conditions for initiating a continuous improvement process of the whole product realization system. Furthermore, this provides a framework for learning about technologies and markets as well as the internal processes of the company.

23.3 Empirical observations in Danish companies

The empirical study aims at mapping the effects obtained by modularization effort in Danish companies. The study has been conducted as loosely structured interviews based on the perspectives described in Figure 1. For each company one product series or product family has been selected in advance and interviews have been made with the sales function, the production function, and the product development function. After analysis of the initial interview data, we plan to make more in depth studies of the critical or generally interesting issues uncovered during the interviews. Our ambition is to cover at least 30 Danish companies. Yet, 11 companies have participated and the following observations are some of the qualitative results from the study so far. In the following, the initial observations are clustered according to general themes. We will present some of our findings as "snap-shots" in the following to give readers an indication how companies are dealing with the phenomenon of modularization.

(1) Considerations about volume and variety: In Company A, the modularization effort was triggered by a product launch from the major competitor. Apparently, the competitor had succeeded in developing a modularized product that seemed cost efficient. The economical success of the subsequent development project relied on a volume of 20,000 modules pr. year. Many in the organization questioned whether this volume could be realized but the autonomous project group kept the volume in their calculations. When the product and assembly concepts were developed the project was handled to a new group for implementation. During the implementation, there were significant problems in realizing both the product and assembly concept. This led to a rise in cost and when afterwards the estimated sales volume fell to approximately one third, the estimated cost of the modules increased by 100%. The project was considered as a failure and kept the organization from considering modularization for a number of years.

In Company B, it was decided to modularize a whole product family. The decision reflected the problems of updating a program of dedicated product variant. The design of the platform and subsequent development of the product variant required much more resources than estimated, and the organization questioned afterwards whether the project could have been less costly and maybe with equal commercial success if only part of the product family had been included. Though the project isolated is considered a limited success there is a trust in the organization that the lessons learned have a strong strategic effect. On the other hand, it is acknowledged that no one really knows the cost and effects of the project.

(2) Product configuration as a driver: In Company C, Company D, and Company E the initial initiative to a modularization effort has been a product configurator. In all cases, the sales function has taken an active role. In Company C, the reason for the initiative was an acquisition of a competitor. Following the acquisition there was a need to renovate the product program. With a simple spreadsheet program the company succeeded in setting up rules for configuration of the product and thereby renovating the product program. The initiative is considered as a basis for continuing the modularization effort including production and new product development.

In Company D, the modularization effort was initiated as part of a general renewing of a product family. The result is a product family of which each specific product can be configured with a paper-based system. A major part of the project has focused on creating a dynamic bill of material for each order. This means that there are no pre-defined standard products – each delivered product has its individual number generated automatically. Yet, there is no computer-based configurator and there is no integrated link to production preparation. Both efforts are considered profitable and there are specific plans of creating links to production preparation. The modularization effort has additionally lead to outsourcing of parts of the production.

In Company E, the modularization effort was initiated simultaneously by the product development and sales function. Due to a combined effort by the two functions, the company has succeeded in bringing down the time for order preparation from 600 hours per order to 50 hours per order. Yet, the production function has not been involved and it is assumed that the potentials in focusing on more or less automatic production planning have higher potential commercial benefits than the order preparation.

(3) Limits for variations: In Company F, the modularization effort led to a significant increase in the number of product variants. This increase in the number of variants created problems for the sales function. One problem was the different interpretation of functions in the product development function and in the sales function. Another problem was to update the competence in the sales function to support the customers in their choice and configuration of the project. A separate problem was that the functions of the modules and the physical modules (structure) in many cases were separated. This made it impossible for the production function to take advantages of the modular structure in the production. When the company realized these problems, it was decided to reconsider the whole product development project again. The following renovation project aimed to solve the problems according to production and sale with a minor effort. The lessons learned from this project were documented in a "Fundamental Law of Modules" to be used in future projects.

(4) No limits for variation: Company G has succeeded to renew one of their major commercial products by creating a modular structure of which the configuration of the functionality is software driven. When a product is sold the customer is given the full functionality of the product with the given modules for a test period. When the test period has expired the functionality that the customer does not want is blinded with a software key. The products can be serviced over the internet. The sales function considers the product as a major success. However, the production and the product development function consider the product as a minor success due to the cost and the problems of the project (see below under "Dual focus on modules and integration"). Company H delivers primarily for OEM customers. In a period from 1985 to 1988, they developed a new and fully modular family of products. The basic idea of this product is that it is sliced where each slice (structure) holds part of the functionality of the final product. By adding or modifying slices new variants can easily be created. Today, this product is still competitive – every variant that has been delivered can still be delivered and new variants emerge frequently. There is no immediate plan to renew the modular structure of the product family. The development plans concern the integration to production and production preparation.

(5) Ownership and participation: Most companies set up teams with a broad representation of the organizational functions. However, the autonomy and the time horizon vary between the companies. In Company B, the core team of the

product development project is consistent during the whole process until after the sales release of the product. There are considerations of increasing the responsibility of the team to including decisions about which variants to launch. This will radically change the way that the project team is measured – from project cost to realized business result. In Company A, the ownership of the product development project is handled to another team after the conceptual phase. There was no personal overlapping between the two teams. In the other participating companies the project ownership changed at least once during the product realization project.

(6) Preparation or renovation: Some companies put their primary effort in the preparation of a modularization effort and others in the renovation of the existing product program. Company E has decided to establish a separate department with focus on renovation of the existing product program and on establishing the integrated link to production. This new department will focus on the supply chain both internally and externally related to both customers and suppliers. Company I has included an architectural phase at the beginning of each product development project. In this phase they define the modules and the interfaces of the product. The suppliers are involved in this phase and this opens the possibility of outsourcing parts of the product development project.

(7) Differentiation between product and technology development: Most companies include some degree of new technology development in their product development projects and most companies experience problems when doing so. The most frequent problems are increase in project duration as well as the project and product cost. For some companies this is crucial (e.g., Company A) and some companies explains this as strategic expenses that will be capitalized in future projects (e.g., Company B).

(8) Technological platforms as frames for modularization: Company K has defined a number of platforms that include important knowledge about products, production and logistics. These platforms are intended to facilitate reuse between projects and secure sufficient expertise regarding the content. Within the more stable areas as production and logistics this has proved valuable. However, within the more dynamic fields of electronics and software the experiences are less convincing.

(9) Limitations in traditional cost structures: Many companies point to the problems with traditional cost evaluation methods. Modularization efforts can hardly be justified with these measurement techniques. Company H and Company I have considered to introduce total cost principles, but the considerations are still at a very initial stage.

(10) Dual focus on modules and integration: All companies point to a potential trap when working with modularization. The focus on modules reduces the focus on the problems of more integrative character. These are cooling, EMC, etc. Company G explains it this way: "We forgot all about what we were really good at". The consequences are additional work and most often additional components.

(11) Tentative implications regarding mass customization: Our initial empirical studies point to several contingency factors and parameters of importance when companies are to set up a modularization effort (as indicated in chapter 4). In the final part of this chapter, we will discuss a few of these factors and parameters related to more comprehensive case studies in some of the participating companies.

23.4 Foci on modularity in practice

After clustering some initial observations according to general themes in the previous section, this section gives a deeper insight how different companies deal with modularization. Each case has a specific focus on different success factors of modularization.

23.4.1 Company I – Focus on architectural phase

Company I develops and manufactures audio and video equipment. The company has for a number of years been focusing on defining core competencies and outsourcing parts of the traditional production. This has led to a significant increase in the efficiency of the whole supply chain. Following the success within production of components the company considered applying similar methods in new product development. This included outsourcing as well product development as production of components. The final assembly was still to be done internally. In the automobile industry, this approach has proven to be successful but the batch sizes in company I was much smaller. For some years, the company had outsourced part of the software development but the experiences were often negative. Some of the internal software people pointed to the role of ownership as the problematic part. It was felt difficult to create this feeling of ownership among the software suppliers.

Considering these experiences the company defined a new approach in the product development process. The initial concept development was done internally. As soon as top management approved the concept the involvement of pre-selected partners were initiated. This phase was named the "Architectural Phase". During the architectural phase the product specifications were only loosely and mainly qualitatively defined. The main focus was put on defining the modules and the interfaces in particular. The architectural phase was split into five smaller phases: 1) Involvement of partners, 2) Development of modules and interfaces I, 3) Evaluation, 4) Development of modules and interfaces II, and 5) Evaluation and accept. In each of the development phases there was required at least three alternatives. During the evaluation phases the different stakeholders (assembly, service, purchase, logistics, quality assurance, test, etc.) were confronted with the alternative solutions. The fact that the solutions were physical and that there were alternatives made it possible for the stakeholders to comment and judge the different solutions.

After the final accept of the architecture the traditional product development process began. During this process the different partners formulated and negotiated specifications. Since each partner were responsible for defining the specifications of their part of the final product there was created an ownership of these specifications – and thereby a more thorough responsibility of their part of the final product. The result of the process has been a product with 10 well-defined modules. Three of these have been fully developed by external partners. External partners produce seven of the modules. The product can be assembled manually without use of any specialized tools.

Company I is so convinced about the power and the way of handling the architectural phase that they have launched a training program aiming at training a new category of employees: Product Architects. A new project has recently been launched within the core products of Company I. This project are to follow the same principles as the one reported above. By doing this the company can increase the capacity of their product development function and thereby add more features for customization.

23.4.2 Company G - Customer specific configuration after production

Company G develops and manufactures analytical instruments. Lately a new modular product program has successfully been introduced to the market and this case focus on the background for initiating the development of the new products and the lessons-learned from the process. The previous product program faced some challenges regarding customer specific configuration during sales and operations. First, some of the modules in the product could not be reconfigure after production. Some of these modules were hardware based, e.g. the power supply. Other modules were software based, e.g. the language of the user interface. In a dynamic market were forecasts are made with uncertainties, this caused situations where products on stock in one country could not be relocated to other countries. Second, the products were not flexible for upgrading/downgrading of functionality while in operation, and the functionality that could be reconfigured, could not be changed by the customer or the dealer. Third, the product had to be tested as a complete assembly at the company and then disassembled and shipped in subassemblies. Most often, the dealer decided to do a second test.

Given these challenges, the company decided to develop a new product program based on the concept of having modules at stock and having the dealer to assemble and test the complete product. In the following, some of the experiences from adapting a modular approach are summarized. Naturally, the decision of postponing the assembly process of modules to the last step in the supply chain imposed new working tasks on the dealer, especially for small dealers with a little staff. Therefore, for those dealers who did not have an interest or capability of assembling the product, the company also had to offer an assembled product with an increased price. Furthermore, the technicians have to work on a modular level and not on a component level. The modular concept also allowed some modules to be bought from suppliers, e.g. the power supply. Other modules, e.g. a measurement module, could be taken directly from another product program of the company, although they were not developed with this application in mind. Therefore, the development process revealed that in future projects the reuse of modules between product families is also to be taken into consideration.

The company had to spend more time on the development project than planned. At the time when the product should have been launched to the market, competitors launched new products. Therefore, the company had to force a pre-introduction to their core customers. The first time investment in the development of a modular product program may be larger than normal, whereas the investment in the next development is smaller, since the development can be concentrated to the modules being renewed and not the entire product. However, as this case illustrates, in a market with competitors rapidly introducing new products, the first step may put the company in a fragile position were it cannot respond quickly to market changes.

The development project was structured in phases with a core team from production development, marketing, and quality. Compared to other projects, the concept phase was given more attention, since it also involved the identification of the modules and their interactions. Nonetheless, in detailing some incidental interactions caused the product not to work properly. A critical interactions was the heating from a step motor that interfered with the regulation of the temperature of a measurement area. Another critical interaction was EMC that required shielding of electric circuits. Altering and adding components to the modules solved these problems. All together, the solutions to these detailed problems added substantial cost to the product, which was considered critical. The lessons learned were that traditional engineering virtues on detail design is difficult to foresee in the conceptual design phase. This counts in particular the identification of interactions that are not directly related to the primary function of a module, e.g. heat transfer, vibration, electric and magnetic fields, etc.

Regarding cost, another experience from the modular approach was an increased redundancy in internal functions and thus increased direct cost of the modules. The redundancy was mainly caused by the modules that included controlling and supplying function e.g. the electric circuit for controlling a step motor, that had the capacity to control more motors, but only were exploited to control the motor inside its own motor module. The flexibility to easily upgrade/downgrade the functionality of the product resulted in a solution, where customers get the full functionality in a test period and later only the purchased functionality is available. If the customer later wish to upgrade/downgrade, software key can open or close up this functionality. This flexibility was feasible for the functionality that could be controlled by software, but it caused some hardware never to be in function, such as valves. Other functionality of the product could only be predicted, so a part of the cabinet was made empty for reserving a volume for future hardware upgrading.

23.4.3 Company H – Modularization in product and production

Company H develops and manufactures a product program of electronic regulators for OEM customers. The current product program is the fourth generation in a development over two decades that in some periods have been driven by the market and in other periods by the technology. From originally only one product family covering all applications, the current product program now consist of several product families targeting various applications. A challenge in this development has been to fulfill the customer's demand for compatibility with previous generations. Furthermore, the development has resulted in more than 200 parameters for configuring the regulation, while the first generation only had 15 parameter. Consequently, man-machine interaction has become a competitive parameter. Now, the sales representatives also need to be more knowledgeable on the product program, with an increased focus on understanding customer need, since the number of variants has increased.

The development process of the last generation has taken place in teams, were a set of core team members have been 100% dedicated to the project and co-located throughout the whole process. The team members came from various functional areas, e.g. production who amongst others defined rules for design for manufacture, e.g. few assembly directions, integration of more functions on the same component, and late creation of variety. The core of the product is an electric circuit composed by two distinct modules based on different technologies. Historically, in every step between the various generations of the product program, a shift in technology can be observed in one of these modules. Therefore, the interactions between these two modules have always been well defined. Given the rapid development in the two core technologies of the product, reuse of modules over time is not considered as relevant as reuse of modules between the families of the particular generation. Nonetheless, a shift is now taken towards 'planned reuse', since in earlier generations some of the product families were actually unplanned spin-offs.

One of the principle for the modularization effort has been to create the variety by software and not by hardware as in previous generations. This has expanded the earlier reuse of knowledge, to also reuse of modules. In some cases, the hardware modules also implement the variety, e.g. the power supply that exists in three variants. However, the spare part for the power supply has been developed as a specific module that can be configured to any of the three original variants of the power supply. This has a radical effect to the service technicians, since now they only have one spare part for the power supply. An interesting aspect of the effort to modularize is that not only the product has been considered but also the production. Similar to a product platform, a production platform exists with standardized assembly lines handling trays with well-defined test interfaces. Furthermore, the packing process of the product is modularized, were only a few variants of packing blocks can be combined and positioned in pairs to contain all the variants of the product. Therefore, each packing block is a module with several interfaces represented by different cutouts in the block, were only the relevant interface becomes active when the block is oriented relatively to the product.

The manuals for the product are an example on the preferable allocation of variety in software. Half an hour before the product is going to be packed at the end of the assembly line, the printing on the specific manual is started. The manual is printed language specific and it only describes the functionality relevant for the particular product. When the product is positioned in the packing blocks, the manual is positioned in specific cutouts in the packing blocks when it passes the printing area.

Some of the concerns regarding modularization have been how to successfully pass the transition phase to a modularized product program. On one hand a company may run the risk to spend a long time to develop a new product program while being of the market, or on the other hand lower the risk by faster and smaller incremental changes that of course lead to less innovative products. Therefore, risk management is considered important in modularization, but it is still unclear which factor to include, e.g. the factor that an error in a core module will cause the whole product program to fail. Another concern relates to the maintenance and development of the product platform, since the ownership in the organization is not yet clear. Third, tools for calculation of total cost are requested, since otherwise it may be difficult to justify the modularization effort from a traditional economical point of view. Finally, tools for synthesis of product platforms are requested, since various tools exists for analysis of an existing platform, but no specific tool have been found for the creation of a new platform.

23.4.4 Company A – Volume as critical success factor for modularization

The last company to be discussed develops and manufactures a mechatronic product. This case focuses on the development of a product program that has been on the market for the last 10 years. The development of the product program was initiated by a new technology that made possible radical changes to the design of the product. Some of the competitors had applied this technology to their product programs apparently with success. Therefore, the company decided to have a autonomous project group design a new product program based on the new technology while also consider the potential of modularization. The core module of the product is a power converter, which traditional had been manufactured in many variants. However, the new technology made possible only one variant of the module, which then could be couple in parallel with similar modules to increase the total effect of the product. This modular concept of a parallel coupling between the power converter modules seemed so promising, that the company decided to plan for also being a supplier to their competitors with this module. Therefore, cost calculations for purchase of components where based on a high volume and new production technologies, e.g. extrusion, were suddenly economical feasible at this high volume.

After having finished the design, the project group handed over the design for production planning. Unfortunately, several problems were encountered during production planning, e.g. a spring assembly solution that at first had to be replaced by a self-cutting screw, but since the tolerances of this solution was not satisfactory, a solution with treading had to be chosen at a much higher cost. Furthermore, the tolerances of the extrusion were not satisfactory, so an extra planning operation was required, again at a higher cost. Even worse, the estimated volume fell by a factor of 10, so most of the purchase orders had to be renegotiated at a higher price. Altogether, the cost of the product was increased with 100% and the company is now designing the next generation of the product program with cost reduction as primary purpose. Although the power converter module was designed to be coupled in parallel, a cooling problem arose when more modules were assembled in the same cabinet. The reason was that the necessary volume around the module for ventilation was not considered part of the module, so this volume was taken up by other modules in the final assembly. This case illustrates the challenges in defining the space ownership of a module as well as the interactions with other modules that is not based on simple spatial conditions, e.g. the air channel in the cabinet that interacts with the air flow output from the module. Nevertheless, even apparently simple interfaces may cause design problems. The base cabinet of the product could be configured with or without a top cabinet. This variety caused two variants of the base cabinet, one for mounting of a top cabinet and one without. Although the inner of the base cabinet was similar, the variant of the cabinet was "baptized" early in the production process, which was opposite to the preferred production concept of the company.

Regarding the power converter module, the company realized that the intervals between the total effects delivered by the various parallel couplings of the modules were too large. Therefore, the design department introduced a shunt solution for reducing the effect, but the sales department was never told that they actually was selling products that with a simple technical engagement could deliver a higher effect. The logistic concept of the modularization effort was based on having modules on stock at the dealers and it involved the customers for the final assembly of the modules. However, several problems were encountered by this concept. First, the tolerances of some of the interactions were not possible to achieve, causing customers to experience halfway in the process that the modules could not be assembled. At first the company solved the tolerance problem by assemble and adjust the modules at the factory for taking them apart again before shipment. However, recognizing the cost of the interactions, e.g. wires and connectors, the company decided to skip the modular structure. Second, sometimes modules were missing in the delivery to the customer. Third, the assembly of some of the modules could not be done by a single person. Therefore, customers started to include the cost of having employees not being able to work - since they were to assemble the product – into the total price of the product. Finally, the dealers could not afford to have the most expensive modules in stock.

The service concept of the previous generations was based on replacement of defect components. However, the modular structure of the new generation made possible a service concept based on having the customers/service technicians to replace defect modules. However, the service department was not ready for both a new technology and a new service concept on a modular level. Quickly the company lost their widespread competencies on servicing on the component level, which then instead was concentrated in a global service center. Furthermore, a regulation required a certain kind of test to be performed when at least one of two core modules were replaced, so in these cases the customers could not do the service themselves.

The company also developed a configuration system for supporting the sales process. However, the company realized that the customers did not have much knowledge on how to configure the product, and they were not interested in the configuration of modules, but only the product as an entirety. Therefore, the company is now shifting towards a more function oriented approach to sales configuration. By this approach, the prices of the modules are less visible to the customer during the configuration. This causes a beneficial after sales situation for the company, since modules can be charged at a higher price when they later are to be sold to the customer as spare parts. Regarding the use of the sales configurator, the sales persons have requested to be able to override the rules in the configurator. In some cases, they believe to have more knowledge on how to configure the product as well as which configuration is best suited for various markets. Furthermore, in general the configuration system suggested more modules in the configuration than necessary to fulfill the customers needs, which lowered the customers trust in the sales person and thus the credibility of the company. Besides a lack of maintenance, these conditions may explain why the first configurator of the company was given up after a while. Although this company experienced several problems in their modularization efforts and consider the project as a failure, they altogether believe these efforts have helped them to remain competitive, and they now recognize how their competitors encounter the same problems in their modularization efforts.

23.5 Conclusions

In this chapter we argue that the concept of modularization is an important factor when trying to realize many potential business objectives – including mass customization. Our empirical study has so far supported the initial idea of modularization being a highly configurable phenomenon. The companies included have treated the problem of modularization in very different way. Some of these ways have proven to contain elements of generality – a generality that might add to a theory about modularization.

During the empirical study, we have identified a number of contingency factors and parameters for configuration. However, these can still hardly be combined into theory. Though some success stories do exist, the greater picture of modularization in a manufacturing context still lacks a lot of elements and deeper insights. In order to contribute to this research we have proposed a research framework that would facilitate an empirical state-of-the-art study of Danish companies.

References

[1]   Pine II, J. B.: Mass Customization, Boston 1993.

[2]   Andreasen, M. M.; McAloone, T.; Mortensen, N. H.: Multi-Product Development – Platforms and Modularization, Technical University of Denmark, 2001.

[3]   Baldwin, C. Y.; Clark, K. B.: Design Rules – Volume 1: The power of modularity, Cambridge 2000.

[4]   Simon, H. A.: The sciences of the artificial, 3rd Ed., Cambridge 1996.

[5]   Baldwin, C. Y.; Clark, K. B.: Managing in an age of modularity, in: Harvard Business Review, 75 (1997) 5.

[6]   Sanderson, S.; Uzumeri, M.: Managing product families: The case of the Sony Walkman, in: Research Policy, 24 (1995), pp. 761-782.

[7]   Meyer, M. H.; Lehnard, A. P.: The power of product platforms, New York 1997.

[8]   Andreasen, M. M.; Hansen, P. K.; Jensen, T.; Mortensen, N. H.: Research framework for studying the phenomenon of modularization, Working paper, 2001.

[9]   Ulrich, K. T.; Eppinger, S. D.: Product Design and Development, 2000.

[10]   Nonaka, I.; Takeuchi, H.: The Knowledge-Creating Company, 1995.

[11]   Henderson, R. M.; Clark, K. B.: Architectural Innovation, in: Administrative Science Quarterly, 35 (1990), pp. 9-30.

[12]   Lawrence, P. R.; Lorsch, J. W.: Organization and Environment: Managing Differentiation and Integration, Boston 1967.

[13]   Kacher, G.: Spy Report: SAAB, Caddy, Sport-utes and an Engineering Revolution, in: Automobile, September 1994, p. 15.

[14]   Sanchez, R.: Strategic Product Creation: Managing New Interactions of Technology, Markets, and Organizations, in: European Management Journal, 14 (1996) 2, pp. 121-138.

[15]   Sanchez, R.: Towards a Science of Strategic Product Design: System Design, Component Modularity and Product Leveraging Strategies, in: Proceedings of the Second International Product Development Management Conference on New Approaches to Development and Engineering, European Institute for Advanced Studies in Management, Brussels 1994.

Contact:

Professor Poul Kyvsgaard Hansen
Department of Production, Aalborg University, Aalborg, Denmark
E-mail: kyvs@iprod.auc.dk

24 A Framework for Selecting a Best-Fit Mass Customization Strategy

The MC Data Acquisition Framework approach

Claudia Mchunu, Aruna de Alwis and Janet Efstathiou
Manufacturing Systems Group, University of Oxford, United Kingdom

This chapter presents a methodology for characterizing a manufacturing enterprise's mass customization (MC) capability. The methodology is essentially a case study that emphasizes the collection and analysis of quantitative as well as qualitative data. A key component of the methodology is the use of a field workbook to collect triangulated data. The application of this tool to the case of a durable consumer goods manufacturer is critically reviewed. This analysis gives rise to three additional tools that are separate, but complementary to the original field workbook. One of these tools, the MC Data Acquisition Framework is described in detail. It is a dedicated data collection tool, which improves upon the effectiveness of the original field workbook. The other proposed tools, the MC Focused Site Tour and the MC Competency Profile are briefly outlined. These tools form a framework that aids the selection of an optimal mass customization strategy.

24.1 Introduction

Mass customization is understood to encompass the processes by which companies apply advancing technology and management methods to cost-effectively provide product variety, flexibility and customer responsiveness [1, 2]. The concept has been proposed as a potential strategy for competitive advantage [3]; however, the wide scale adoption of mass customization is hindered by the lack of validated operational strategies to customize on a mass scale [4, 5]. Mass customization is a fairly new concept and it is evident from surveys of the literature that this concept has not been sufficiently studied from an operations perspective. There is insufficient research to assist an enterprise desiring to introduce this 'new paradigm' in the form an explicit strategy [3, 4]. Furthermore, practitioners are discouraged by the lack of industrially usable tools such as frameworks to research and understand the operational issues associated with mass customization [6].

This research aims to address this. The chapter critically reviews a methodology developed for conducting research on mass customization in manufacturing enterprises. The research reviews a tool developed and applied at one manufacturing site. Based upon the findings of the analysis, the authors offer the 'MC data acquisition framework' as a facilitative tool for improved efficiency of mass customization data collection in manufacturing enterprise. The chapter is organized into six sections. After introducing our research motivation in the next section, we will present our research methodology and the proposed framework. The following sections will discuss the application of this framework to a case study. Conclusions are contained in the last section.

24.2   Critical success factors of mass customization

Based upon the preliminary review of the mass customization literature there are only the beginnings of research activity into the operational aspects of implementing an explicit strategy of mass customization. The *Mass Customization for Manufacturing Enterprises Project* aims to address this need. The project is a major three-year project funded by the Engineering and Physical Sciences Research Council (EPSRC). It is a collaborative effort between the universities of Nottingham and Oxford to develop models that assist the analysis of operational issues raised by mass customization. Its specific objectives are

- to identify and classify mass customization strategies, their applicability and success factors,

- to develop approaches to analyze the business case for adoption of mass customization,

- to develop and deliver industrially based tools, in the form of conceptual models and prototype decision support software to aid the selection of mass customization strategies and to aid the design of corresponding systems,

- to generate a validated knowledge base to underpin mass customization principles.

A preliminary step in achieving these objectives is to identify or develop methods of studying the critical issues within manufacturing operations. Critical issues have been conceptualized [6, 7, 8], but methods of studying and confirming these have not been offered. Armed with a systematic research methodology, researchers and practitioners alike would be able to study how mass customization is achieved strategically and operationally in different sectors and contexts. In the following we will review the issues thought to be critical to success with an explicit strategy of mass customization and critiques one method of studying these critical issues. We will then propose an enhanced tool for conducting research within manufacturing enterprises. The framework is one that guides the process of researching and profiling the operational characteristics thought to support success with strategies of mass customization. The enclosed research draws on and builds upon previously developed work. Mchunu et al. present an influence diagram [6]. The influence diagram was developed via a qualitative systems dynamics modeling approach. The process drew upon information gathered from literature reviews, surveys of industrial sectors, manufacturing site visits and the research-ers' own experience of the topic. The diagram is a conceptual model of the underlying structure of the system enabling a manufacturing enterprise to mass customize. The analysis of the influence diagram identified five factor variables as having priority for mass customization success.

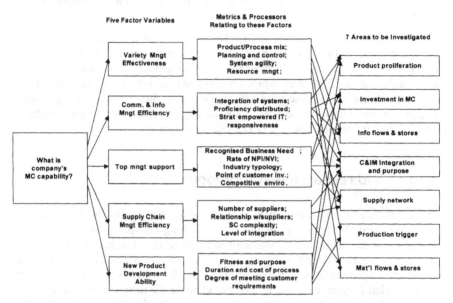

Figure 1: Investigating the factor variables most likely to determine a manufacturer's mass customization capability

These factor variables and the approach to investigating them in the field are illustrated in Figure 1. From the above analysis, five factor variables are proposed as having particular impact on an enterprise's ability to mass customize:

(1) Effective product variety management: The ability for an organization to manage variety and meet fluctuating customer demand while holding relatively minimal RG, WIP and FG inventories. Furthermore, variety is defined as the variation of an operation's output over time in terms of the range of offerings, the volume, mix, and timing of output over time.

(2) Communication and information management (C&IM) efficiency: The ability to manage knowledge and information for competitive advantage; the level of integration of the functions along the value chain as evidenced having an explicit strategy, having hardware and processes that support communication and having this need recognized at the strategic level.

(3) Support from top level management: The degree to which top-level man-agement has shown support for MC and incorporated it into the strategic objectives of the company.

(4) Supply chain management efficiency: The degree to which a company can dynamically pool resources to meet the changing interests of customers efficiently and effectively; with decreasing vertical integration, this is increasingly accomplished through relationships with collaborating organizations. For example, the level of complexity of the supply network, the lead-time on purchases and the quoted vs. actual lead-time are example measures used in the assessment of this factor variable.

(5) New product development ability: The ability to satisfy customer interests by efficiently (time and resources/cost) developing a concept into a realizable solution that is both economic and effective. New product development ability takes into account the time and resources required to proceed from initial concept of product to actual launch and is also concerned with the fitness for purpose of the new products. The proportion of new products introduced of the total number of products within a given period of one aspect of new product development ability.

24.3   Field research methodology

The key feature of the research methodology was the focus on fieldwork. Fieldwork yields the opportunity to observe real manufacturing enterprises in their complex contexts. The developing theory was applied in the reality of an industrial setting [9, 10]. The approach was both inductive and exploratory. Saunders [11] explains that an inductive approach is best applied in situations where the researchers seek to 'get a feel of what is going on, so as to understand better the nature of the problem…In this scenario, the data helps to formulate or further develop a preliminary theory.' The researchers had a preliminary theory, which emerged from the analysis of the influence diagram. However, the diagram was only conceptual. The aim therefore was to use the theory as a guide to focus the fieldwork on the aspects or characteristics about manufacturing enterprises likely to have the most significant impact on mass customization. These are the processes, if studied in detail, would greatly inform the development of mass customization theory.

In addition to being inductive, the case study approach was also exploratory. The researchers sought to let the preliminary theory focus the research effort, but at the same time to remain open to the emergence of entirely new relationships and theory. At no point did we set out to test the correctness of our preliminary theory. Rather we sought to use this theory to focus the research process. A key stage in this focusing exercise was the development of the research questions.

24.3.1   Research question

The overall research effort sought to unveil certain characteristics of the organization and to develop a profile for further analysis and cross comparison with profiles developed of other companies. We sought to understand, for example:

- What is the site's line layout and how does it facilitate agility and the management of variety? How does the production scheduling process enable the company to be responsive to unique customer orders?

- How many different product types, Unique Stock Keeping units (SKU's) are there? Is this number increasing, decreasing or staying the same? Are these SKU's grouped into product families?

- How do the information systems facilitate the management of product variety and customer responsiveness?

- How is what customers want for current or future orders captured? How are individual/unique customer orders tracked through the production process?

- How do you facilitate short lead-times from its suppliers? How are lead-times from its suppliers being reduced? How integrated is the supply network?

- What is the rate of new product introductions? Is this rate increasing or decreasing? How does the new product introduction process facilitate short time-to-market?

The researchers were also concerned with testing the usefulness of the field workbook as a tool facilitating the collection and analysis of key data in the manufacturing context. For the purposes of this chapter, the research question was: *How well does our methodology assist in gathering the data necessary for establishing an MC capability profile of a manufacturing company?* This question was articulated in order to make explicit the fact that this study was piloting a process to be repeated in subsequent case studies.

24.3.2 Case study approach

Gaining insight into the underlying reasons for success and/or failure with MC was the larger objective. Given the absence of an established knowledge base of scientific principals underpinning mass customization, an in depth study of a small sample of cases was recognized as the most appropriate approach. The aim of studying several cases was to portray an accurate profile of manufacturing enterprises in their rich contexts. Our case study followed a multi-method approach [6]. Existing theses, reports and literature written about the company were analyzed as secondary sources. And primary sources in the form of hard data as well as soft data were collected through the processes of interviewing, document analysis, and observation. To facilitate the research process, a purpose-built field workbook described in the next section was developed. The research sought concurrently to acquire data using this field workbook, as well as to test its usefulness as a data collection and analysis tool.

The researchers were guided by the theory proposed in the previously developed influence diagram, but recognized that the diagram was a conceptual one. In order to guide the actual fieldwork, a purpose-built field workbook containing describable metrics was developed. The field workbook incorporated the seven metrics presented in Figure 1 above and it detailed how to investigate these metrics. It aimed to facilitate the collection of relevant data and information about the enterprise.

The field workbook was divided into a number of sections. The first two sections contained general information about the company, such as contact details, number of employees and industrial sector to which it belongs. Section two also contained the process of choosing a product or product family to focus on. As it would be impractical to get sufficient depth of detail on every product, the researchers choose, in conjunction with the main contact from the company, a subset of products that followed a path assumed to be representative of what is typical in the company. Section three was dedicated to researcher notes. It contained the objectives of the research, an overview of the data collection methodology and a spreadsheet for logging data as it was collected. Section four was the main section of the workbook. It was comprised of data collection tools aimed at gathering the targeted data in a form that could be readily analyzed as it was collected. These data collection tools were organized into seven sub-studies each with its own aims and specified method:

1. Product type, variety, and proliferation with the product line of interest.

2. Company's investment/commitment to mass customization.

3. Flow of material within the company with flows and stores indicated.

4. Flow of information within the company with production trigger(s) indicated.

5. Purpose, degree of integration and efficiency of the communications and information management system(s).

6. Network diagram of the supply systems and its degree of integration.

7. Point and quality of customer involvement/ production trigger.

Table 1 presents two examples how data for these sub-studies were collected to give the reader a flavor of the kinds of data collection tools and questions comprising the field workbook. For each sub-study there are aims, method and additional tools for the application of the method. The two examples in Table 1 give an indication how the workbook was organized to facilitate the collection and preliminary analysis of data. In the next section, we report on the application of this methodology to a case study.

24.4   Application to case study

24.4.1 Case study background

The company chosen to trial the MC Data Acquisition Framework operates within the durable consumer goods sector. The durable consumer goods sector includes goods such as domestic and office furniture, appliances and equipment; sporting goods, toys and other merchandise such as photographic equipment, PC's, watches, and eyewear, that individual consumers may be expected to purchase and use over a period of years. Manufacturers within this sector usually supply a mass market with some fashion-conscious niche markets. As a result these companies are characterized by traditional mass production manufacturing systems and relatively dependent upon set-up. The sector tends to be inventory intensive and there is an increasing reliance upon linking of information technology to processing technology.

Table 1:   Two examples of data collection for sub-studies

Example Data Collection Tool	
Sub study Product Proliferation: Product Type, Variety, Proliferation within Product Line of Interest	*Sub study C&IM:* Communication and Information Management Process
Aims: ▪ to describe the product type ▪ to describe the degree of variety ▪ to index the degree of proliferation	Aims: ▪ to index the degree of integration of its communication and information management systems ▪ to describe the purpose of its communication and information management systems
Method: ▪ collect current product lists and lists from one to two years ago for the chosen product or product family. ▪ use the enclosed questionnaire to interview main contact person and marketing manager	Method: ▪ refer to the collected samples of: ▪ 30 customer orders ▪ 30 supplier invoices ▪ interview any two of: IT manager, production manager, IT officer. ▪ follow-up on leads from above interviews. ▪ create a process flow diagram. ▪ locate the enterprise on the C&IM flow chart
Application : ▪ Name/Title/Function of interviewee ▪ What is the total number of products offered on this product line? ▪ List the main products offered on this product line ▪ Can you discern a skeleton product(s) (standard product, which is/can be customized)? If so, how many product variants are there for each skeleton? ▪ What type(s) of products are on this line and how complex are they?	Application: ▪ Name/Title/Function of interviewee ▪ If there is a C&IM strategy, collect a copy of it. ▪ What is the yearly spending on C&IM (Accounting Ratios)? ▪ Is there a Local Area Network (LAN) capability? ▪ What is the level of integration between stock control, sales, production scheduling, purchasing and production control. ▪ Of interactions that are performed in a week how many are performed via the LAN? (Random sample of 30 interactions: eg. 10 from purchasing, 10 from order fulfillment and 10 production change orders. ▪ … ▪ Does the organization have a Database Management System (DBMS) ▪ If yes, what fields are stored in the database? ▪ What percentage of the data is stored electronically? ▪ What fields in the databases are accessible by your suppliers? ▪ What fields in the databases are accessible by your customers?

Companies within the durable consumer goods sector typically sell their products through channels that include relatively powerful general-purpose retailers, dedicated retailers and/or mail order catalogues. The sector experiences medium product proliferation with high proliferation in some niche markets. The products are price sensitive with only the high-end winning orders on design and flexibility. This trend is changing, however, with the low-end consumer becoming more and more demanding of high-end design at low-end prices. Long waiting periods are becoming increasingly unacceptable and there is at least as much pressure to reduce lead-time as there is to reduce prices.

The company from which this case study is drawn is typical of this sector. It is a medium sized Original Equipment Manufacturer (OEM) assembly factory with one manufacturing site. The company utilizes a batch-flow manufacturing strategy with a product layout. It is a labor and material intensive operation with minimal information technology supported production. It manages a global supply network and distributes chiefly from inventory to a European customer-base.

24.4.2 Data collection

Data was collected over a one and one-half-week period. There was a team of four researchers, two of who were involved full-time. There were four aspects of the fieldwork methodology that facilitated the data collection process and made it a successful one. These were the understanding and support of the main contact, the availability of an on-site office base for the use of the researchers and the focus on collecting hard data in the form of documents. *Support from main contact* was needed to supply initial literature to orientate us to the company and facilitate the data collection exercise. This literature included organizational charts, internal telephone directory, and product brochures. The main contact set up also an on-site office base for the researcher team in advance and introduced the researchers personally to key personnel in each functional area who would provide interviews and data. The *on-site office base* was used as a venue to conduct some of the interviews, as a secure place to log and store data as it was collected, and to display information that the research team needed to share with each other. It also helped to facilitate free discussion of the researchers' preliminary impressions and planning review.

Document collection and analysis was based on three methods of document collection and analysis:

(1) Semi-structured interviews:

- Gained familiarity the organization of the company and the form in which the data was likely to present itself.

- Supplemented the collection of hard data by providing the reasons and history behind *why* things are done in the way that they are.

(2) Document analysis:

- Provided quantitative evidence of claims made in interviews.

- Yielded insight into which metrics the company views as important enough to track and report on.

(3) Observation:

- Allowed the researchers to experience the context of the company on the shop floor level.

- Verified the material process flow maps.

- Created information flow maps.

24.4.3 Analysis of the results

This previous section has presented an overview of the data collection process. A more detailed examination of how the field workbook performed with regards to sub-studies is provided in the following. The results obtained for the two examples of sub studies presented in Table 1 above are discussed below.

Collection of Product Proliferation Data: The aims of this sub-method were to describe the product type, to describe the degree of variety and to index the degree of proliferation. The first two of these aims were achieved, however, the workbook did not assist the development of a process for indexing the degree of proliferation and subsequently we had no such an index on site. The method facilitated the achievement of the aims. Current product lists from one to two years ago for the chosen product or product family was accomplished via the collection of a complete set of current product brochures and a product catalogue from 1997. All of the questions were answered in separate interviews with the main contact person and marketing manager. Most of their responses were confirmed via concurrence and/or by a cross-check with Bill of Materials (BOM's) for each product in the product family of interest.

Collection of Data for Communication and Information Management (C&IM): The aims of this sub-method were to index the degree of integration of its communication and information management systems and to describe the purpose of its communication and information management systems. Samples of customer orders and supplier invoices were collected but not analyzed as planned. Interviews were conducted with the manager and an IT officer. The processes of interviewing and information-flow diagramming revealed worked well to complement each other. We ascertained whether there was an articulated C&IM strategy, whether IT was represented at the strategic level and the use of percentages was helpful in most cases. We were able to discern the C&IM strategy of the company as well as the level of competence in implementing that strategy (see Figure 2). We also extracted the level of use of certain systems and the degree of integration locally and across a wide network.

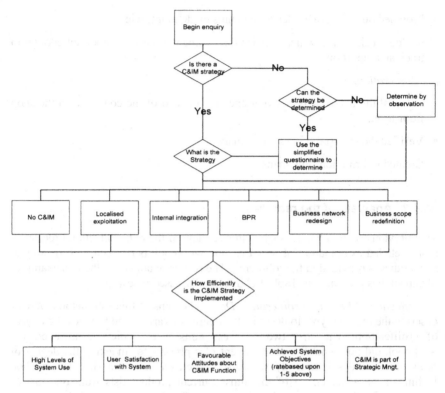

Figure 2: C&IM strategy of the company and the level of implementation competence

Sampling was not mandatory in this instance because the company had the data we needed in a form that was readily accessible. The purchasing manager and the production manager provided answers to most of the questions. Their answers were corroborated by others as well as by a company report on manufacturing stocks. Answers obtained in interviews were substantiated by process flow diagramming done by an independent researcher. We ascertained what types of communication networks existed and their strengths and weakness. These data were challenging to collect and the field workbook was not as facilitative as we had hoped in some areas. For example, we were unable to get specific financial information that would give insight into their levels of investment in C&IM. The field workbook could have been designed to provide alternative methods for gathering data that was not available via the expected channels.

24.4.4 Benefits of the field workbook

The analysis of these two examples has indicated some of the strengths and weaknesses of the field workbook. The process of constructing the workbook yielded advantages in addition to those stemming from the use of workbook itself. Some of the key benefits gained from the process of developing the field workbook include:

- The development process provided an organized forum for deciding upon what would and would not be collected.

- In many cases alternative ways of collecting the sought after data were developed

- The process of developing it also demanded consideration as to how the data would be analyzed.

Some of the key benefits gained from the workbook itself, its design and use in the field include:

- It focused the researchers on the aims at each stage of the exercise

- It facilitated the collection of hard data through lists and logs that acted as stage gates

- It communicated a clear purpose and focus to the interviewees and other company personnel

- It provided alternative paths to reach aims, thus increasing the likelihood of getting the sought after data and also providing triangulation.

- It facilitated preliminary analysis of the collected data exposing gaps before leaving the company site

The benefits of the workbook as a tool for data collection and preliminary analysis are summarized above. Enhancements to the approach will be presented in the next section. And finally, a more streamlined tool dedicated to the collection and extraction of key mass customization data from industrial contexts is presented as the MC Data Acquisition Framework.

24.4.5 Enhancements to field workbook

Upon examination and reflection on the field workbook and its results, the benefit of the field workbook proved to be more as a tool for analysis and less as a tool for data collection. Appreciating the grounded theory approach to research, the original intention was to analyze the data as it was collected. Although there is great value in performing preliminary analysis of the data in real time, the researchers found that in practice, a tool containing both processes was awkward. Paging through a 40-page workbook while interfacing with company personnel proved cumbersome and impractical. The workbook did well in providing a completeness check and in facilitating the preliminary data analysis. However, both of these activities were performed upon returning to the on-site office base. Therefore two separate tools would have sufficed.

The separation of the data collection function from the data analysis activity is likely to produce a more streamlined and efficient overall process. Therefore, a dedicated data collection tool, separate and distinct from the data analysis effort. Focusing on the data collection process itself, the researchers recognized a number of alterations that could enhance its usefulness as a dedicated data collection tool. Three enhancements were proposed:

(1) Communicating which data was required: The workbook needed to be more sensitive to the reality of collecting information from varied sources and in varied forms. One of the most challenging aspects of the methodology was in communicating what data was required. The researchers had clear and focused aims, but translating these into the company documents that contained this information was difficult. This was an issue of communication. To remedy the above, we enhanced the process to include example reports and likely forms in which the data might come. We also require presentations by company personnel on the areas of interest and key processes before beginning the interviews.

(2) Data collection Progress Summary: The field work book included a data log which did well in facilitating the tracking of what data had been collected. However, there was no easy way to assess what data was still outstanding. It would have been beneficial to have a tool that illustrated what had been collected with respect to what data still needing to be collected. A format that enabled easy inspection of the gaps in the data collected would have proved useful. To address this, a tabulated format was introduced. The table includes a column to report on the state of data and the progress made through the collection process.

(3) Alternative routes to the data: Some of the assumptions made regarding the availability and the accessibility of data were wrong. In some cases, data was expected to be unavailable or not in the form needed. When it turned out to be readily accessible, whole sections of the sub-methods were redundant. In other cases, data was assumed to be readily accessible and when it was not, we lacked an alternative route to collecting it. In the worst of these instances, data was missed out. One way to alleviate this from recurring would be to standardize across all sub-studies, the strategy developed for use on the flow diagramming sub-studies. In these processes, flow diagramming for material flows as well as for information flows, we allowed for two possibilities. If the company had process flow diagrams they would be collected and verified. If the company did not have process flow diagrams would they be created from observation. Building this kind of flexibility into the other sub-studies would have proved beneficial. In the next section, we incorporate these enhancements into a mass customization data acquisition framework. This tool is a streamlined and dedicated to the collection and extraction of key mass customization data from industrial contexts. Used in conjunction with the field workbook analysis tool, it is a powerful instrument for researching and profiling the operational characteristics thought to support success with strategies of mass customization.

24.4.6 Mass customization data acquisition framework

The framework presented in this section was developed to enhance the fieldwork methods used to profile the mass customization capability of manufacturing enterprises. The framework, unlike the field workbook, is purpose-built for the process of extracting the sought after data. It is a simple tool that fits on an A3 sheet of paper. At a glance, researchers can get an indication as to the data needed and the possible form it may be in. Source persons and departments are identified and listed conveniently in one place so that repeat visits to the same person for different data can be minimized. Additionally, researchers have a real-time visual picture of the completeness of the data collection process. This framework is currently being used as the data acquisition tool, separate, but complementary to the field workbook, which facilitates preliminary analysis of the data. Both tools are used on-site, but the data acquisition framework is used during interactions with company personnel whereas the field workbook is used at the researchers' base. This separation of the tools has lead to a clearer, more streamlined and more effective fieldwork exercise.

Table 2 presents an abridged version of the table used in practice. In the actual version, each of the columns is expanded to include greater detail, but not to exceed an A3 size sheet. The *first column* incorporates the five factor variables found previously to greatly influence success with mass customization. The two additional areas of interest in this column are 'Change Management and New Process/Technology Diffusion' and 'Production Scheduling, Capacity and Inventory Management' enable the operation to be further understood with regards to how it is structured from a manufacturing systems standpoint as well as from an organizational culture standpoint. The *second column* contains examples of the types and forms in which the data is likely to present itself in the organization's documents and reports. This facilitates the process of communicating what data is required. This column is updated as a common language between the researchers and the company personnel begins to emerge. The *third column* allows for the confirmation and tracking of where the data will be/ was extracted. Review of this information serves as a check on the degree to which the data is being triangulated and corroborated by multiple sources. Finally, the *last column* provides an easy check on the status of the data collection process. Any gaps in the data required can be flagged and tracked, thus providing for greater completeness and minimal unplanned redundancy.

24.4.7 Further enhancements to the field workbook application of findings and future work

Since the first publication of this work [12], further enhancements have been made. The MC Data Acquisition Table described above, improved upon the original field workbook through streamlining the operative size of the tool into a convenient A3 Sheet. Two additional tools are proposed below to streamline the data collection and analysis process even further.

The *MC Focused Site Tour* methodology [13] enables the on-site portion of the data collection exercise to take place in a single day. Where extended on-site interaction is infeasible, this methodology calls for background research into the manufacturer and its industrial sector, as well as pre-visit telephone interviews. The MC Focused Site Tour methodology is currently being tested and validated through field trials.

Table 2: Mass customization data acquisition framework

Key Process/Area of Interest	Example / possible form / doc.	Source	Status?
1. Production Scheduling, Capacity and Inventory Management	MPS process/rules MOQ policies MRP/ERP system scope WIP and distribution stock level report		
2. Production Mix (Product Variety) Management	cycle time batch size product catalogue BOMs		
3. Communication and Information Management	order data gathering process order data transfer process order data storing process WAN, Web		
4. Support from Top level Management	KPIs and why strategic planning summarize financial spend order winners		
5. Supply Chain Management	communication links with suppliers degree of true demand data sharing pareto of lead time by components		
6. New Product Development	process map time-line capturing customer requirements point(s) and extent of customer involvement		
7. Change Mngt and New Process & Learning	production change management process frequency of employee-directed change organization chart key roles & functions		

A second proposed tool, the *MC Competency Profile* [14], provides a template within which to capture the manufacturer's capabilities for analysis (see Figure 3). Adopting a research approach that explicitly builds upon the competency theory of the firm, this proposed methodology captures and ranks a manufacturer's competencies along five key competency dimensions [15]:

- Design and new product introduction (speed and intelligent design with customer-input),
- production flexibility and capacity management,
- supplier management and supply chain agility,
- variety and inventory management,
- communications and information management.

The reader will appreciate that the dimension production flexibility and capacity management has replaced the factor variable support from top-level management. This change highlights the decision taken by the researchers to focus on achieving mass customization operationally with the assumption that a business decision has been taken to pursue mass customization as a competitive strategy. In essence we assume support from top level Management as given and furthermore raise the focus on the factor variable production flexibility and capacity management. Production flexibility and capacity management is defined as the ability to remain economic while responding to/absorbing change and fluctuations in customer demand with regards to the nature, volume and timing of production outputs; included are flexibility of tools, processes, products, routings, volume.

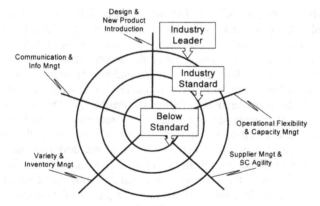

Figure 3: Mass customization competency profile spider diagram

The data collected on these five variables is ranked relative to what is considered standard for that industry. The manufacturer's MC Competency Profile emerges as a result of plotting these rankings. The profile facilitates the comparison of a given manufacturer's key competencies with the key competencies of manufacturers' enjoying success with mass customization. In this way, the profile provides decision support for the practitioner faced with evaluating which of the various MC strategies is most appropriate for a given firm. Through highlighting any competency shortfalls, it is envisaged that the MC Competency Profile can also aid in the development of route maps to mass customization success. This decision support tool is currently being validated statistically [16] and by computer simulation [17].

24.5 Conclusions

This chapter presented the research methodology used to characterize a manufacturing enterprise's mass customization capability. The methodology is a case study bounded by a previously developed theory, which identifies five areas as having particular importance for mass customization success. These areas are effective management of variety; communication and information management; support from top-level management; supply chain management efficiency and agility; and new product development ability. This research did not set out to test the correctness the five factor variables theory. The theory was only used to focus and bind the data collected in the case study. Research into the significance of the five factor variables theory is underway via computer simulation and survey-based investigations. The chapter introduced an instrument in the form of a joint data collection and analysis field workbook. The application of this field workbook to the case of a durable consumer goods manufacturer is critically reviewed. The analysis highlights the fact that a dedicated tool, separate from an analysis aide, would be better suited to the process of collecting data in complex manufacturing contexts. The result is an enhanced data collection tool, namely, the MC Data Acquisition Framework. This tool is separate, but complementary to the previously developed field workbook.

Two further tools are outlined. The MC Focused Site Tour is a methodology which enables the on-site data collection exercise to be completed in a single day. And MC Competency Profile is a dedicated data analysis tool which makes use of a template for capturing the findings during the on-site interactions. It is envisaged that this will supplement the field workbook analysis through visual presentation of the relative ranking of the firms' competencies. These tools are currently being tested and validated in the field. Preliminary feedback is that they are proving useful in efficiently communicating what data is required, in checking that it has been gathered and in analyzing it with a view to discovering the best-fit MC strategy for a given operational context. Through the development of research tools to investigate mass customization within manufacturing firms, it has helped identify success factors for a set of mass customization strategies. It has also proposed conceptual decision support tools to aid in the selection of mass customization strategies and the design of MC systems.

References

[1]   Davis, S.: Future Perfect, Reading 1987.

[2]   Pine II, J. B.: Mass Customization, Boston 1993.

[3]   Kotha, S.: Mass Customization: Implementing the Emerging Paradigm for Competitive Advantage, in: Strategic Management Journal, 16 (1995), pp. 21-42.

[4]   Ahlstrom, P.; Westbrook, R.: Implications of mass customization for operations management: An Exploratory Survey, in: International Journal of Operation and Production Management, 19 (1999) 3, pp. 262-274.

[5]   Spring, M.; Darlymple, J: Product Customization and Manufacturing Strategy, in: International Journal of Operation and Production Management, 20 (2000) 4, pp. 441-467.

[6]   Mchunu, C.; de Alwis, A.; Efstathiou, H. J.: Manufacturers' Ability to Mass Customize: A System Dynamics Modeling Approach in EurOMA, Bath 2001.

[7]   Gunasekaran, A.: Agile Manufacturing: enablers and an implementation framework, in: International Journal of Production Research, 36 (1998) 5, pp. 1223-1247.

[8]   Hart, C. W. L.: Mass Customization: Conceptual Underpinnings, Opportunities and Limits, in: International Journal of Service Industry Management, 14 (1995) 5.

[9]   Efstathiou, H. J. et al.: Assessing the effectiveness of manufacturing information systems, Human Performance in Planning and Scheduling: Fieldwork studies, Methodologies and Research Issues, University of Oxford, 2001.

[10]   Calinescu, A. et al.: Supply Chain Complexity Study: BAE Systems (Plymouth), Graphic (Crediton), University of Oxford and University of Cambridge, 2000.

[11]   Saunders, M.: Research Methodology for Business Students, 1993.

[12]   Mchunu, C.; de Alwis, A.; Efstathiou, J.: Data Acquisition Framework for Facilitating Research in Mass Customization, in: Proceedings of the 1st World Congress on Mass Customization, Hong Kong 2001.

[13]   Mchunu, C.; de Alwis, A.; Efstathiou, J.: How to take a Mass Customization Factory Plant Tour, in: Manufacturing Engineer, 2002.

[14]   Mchunu, C.; de Alwis, A.; Efstathiou, J.: Methodology for establishing a Mass Customization Competency Profile, in: Journal of Operations Management, 2002.

[15]   Prahalad, C. K.; Hamel, G.: The Core Competence of the Corporation in The State of Strategy, in: Harvard Business Review, 1991, pp. 3-15.

[16]   Mchunu C.; de Alwis, A; Efstathiou, J.: Decision Support Framework for Establishing a Best-Fit Mass Customization Strategy, in: International Journal of Operations and Production Management, 2002.

[17]   de Alwis, A.; Mchunu, C.; Efstathiou, J.: Methodology for Testing Mass Customization Strategies by Simulation, in: Proceedings of the International Conference of the Production and Operations Management Society, San Francisco 2002.

Acknowledgements: The authors gratefully acknowledge the support of the Engineering and Physical Sciences Research Council (EPSRC) (Grant No. GR/N11926/01). We acknowledge the contributions of all the members of the Mass Customization for Manufacturing Enterprises Project, Philip Brabazon, Jo Bramham, and Bart MacCarthy. We also acknowledge the contributions and input of our colleagues in the Manufacturing Systems Group at the University of Oxford and the support of our industrial collaborators.

Contact:

Aruna de Alwis
Manufacturing Systems Group, University of Oxford, United Kingdom
E-mail: dealwis@robots.ox.ac.uk

Part VI: Applying Mass Customization to the Fashion Industry

Building a customer centric value chain for apparel and footwear customization

The apparel and the footwear industry are both industries that are forerunners in the application of mass customization. The reason behind this development can be seen in the fact that clothes and footwear offer the potential to address all three possible dimensions of customization: fit (shape, measurements, size), functionality and aesthetic design (taste, forms). Products that require the matching of different physical dimensions or functional requirements often engender a higher price premium than products that are customized just by the possibility of changing colors or design patterns. Clothes and footwear are products that must, first of all, exactly fit their user's measurements. Additionally, customer integration into the aesthetic design of a shoe or a piece of clothes and the adaptation of functional requirements (like the profile of a sole, height of a heel; features of a fabric) are further means of increasing the utility of a product.

Thus, customization in this industry offers a good opportunity to counterbalance additional cost in manufacturing by a higher consumers' willingness to pay. Customization is also favored by more and more suppliers due to the steadily growing pace of change in fashion cycles, high forecasting problems, and multi-channel distribution systems. However, the change has just begun in the fashion industry. Despite various approaches like fast response supply chain systems, the use of digital models in product design, or manufacturing robots substituting the traditionally high level of human labor, the apparel and footwear sector is still dominated by traditional mass (variant) production systems. Thus, making these industries more customer centric is both a great challenge and an immense opportunity. Part VI addresses these challenges and provides a good insight into the various activities needed to transform an industry from a mass production system into a customer centric enterprise.

In Chapter 25 *Bullinger, Wagner, Kürümlüoglu and Bröcker* present the enabling information technologies for process management using the example of the footwear industry. Here, the change from mass production to a made-to-order system forces a complete revision of the processes and IT-systems that support the various phases of the product life cycle. Based on the EuroShoe Project within the 5th Framework Program of the European Community (www.euro-shoe.net), they envision the idea of an Extended User Oriented Shoe Enterprise. (New) appropriate IT-systems have to be selected and implemented. The chapter describes the demands of a transition from mass production towards mass customization in this industry. But not only IT and process design have to change. Mass customization will be only successful if appropriate sales systems exist. For example, customized clothes and shoes cannot be sold exclusively on the internet. The necessity of taking the measurements of each customer demands a direct interaction between seller and buyer. This prerequisite is supported by the demand of many customers for experience shopping or for the opportunity to feel fabrics and materials before the purchase. Thus, mass customization in these industries often requires strong cooperation with retail. This is the theme dealt with in Chapter 26 by *Taylor, Harwood, Wyatt and Rouse.* They present a strategic model for implementing mass customization for clothes on a UK High Street. Although mass customization clothing services have been available in some stores for over 10 years, they have been limited to men's suits. In an exploratory empirical study, the authors identify four stores offering (industrial) made-to-measure clothing and contrast them with four independent (traditional) tailors. The study provides interesting insights into mass customization from a retailer's perspective and may give manufacturers some ideas how to better tune their operations towards the demands of retail.

A special technology that is discussed often in the context of mass customization of fashion items is the use of individualized avatars (virtual mirrors). *Gurzki, Hinderer and Rotter* show how personalization technologies can supplement mass customization in the fashion industry in Chapter 27. The chapter gives an overview of the requirements of business-to-consumer fashion retailing and the available technologies. It develops an approach for an online shopping platform with individualized avatars for animated fashion presentation and integrated natural language text-based customer consulting features. As discussed above in Chapter 1 of this book: personalization is a major enabler for mass customization – not only in the fashion industry.

The last two chapters of Part VI specifically address two important means of becoming more customer centric in the fashion industry. In Chapter 28 *Luximon, Goonetilleke and Tsui* demonstrate how to achieve better fit in the footwear industry. Footwear fitting is generally performed using the two variables of foot length and foot width (or girth), even though feet and shoes are three-dimensional objects. As a result, the matching between feet and footwear are quite variable and can be quite unacceptable even for the same brand of shoes. Footwear fitters speak of 'perfect fit' even though the term 'fit' appears to be nebulous. The authors propose a method of quantifying 'fit' based on 3D tools. The proposed footwear fit quantification can be used to predict potential discomfort and even fit-related comfort if the material properties of the shoe are known. The method can also be used to rank different footwear lasts for any given individual.

However, evaluating the perfect fit of a shoe is one thing, designing and manufacturing it according to customers' demands is another. The foot data has to be translated into a customer specific last and shoe design. Thus, in Chapter 29 *Sacco, Vigano and Paris* evaluate how virtual reality technologies and CAD/CAM enable made-to-measure shoe manufacturing in mass markets. The authors discuss the state of the art of appropriate technology available and provide a glimpse of the future based on a virtual shoe design environment. With this system, a designer draws or modifies the style lines which were created before in the CAD system directly onto a (virtual) shoe model. Designers can fly-through the environment and interact directly with the virtual shoe using immersive interface devices. Such a solution is the starting point for the efficient manufacturing of customer specific shoes in mass markets.

25 Towards the Extended User Oriented Shoe Enterprise

Enabling information technologies for process management of mass customization using the example of the footwear industry

Hans-Jörg Bullinger, Frank Wagner, Mehmet Kürümlüoglu and Andreas Bröcker
Fraunhofer-Institute for Industrial Engineering (IAO), Stuttgart, Germany

Transformation from mass produced goods to mass customized ones is challenging but in some branches inescapable. Such a radical change in nature of production forces a complete revision of the processes and IT-systems that support the various phases of the product life cycle. EUROShoE is the first project putting this change into practice for the European shoe industry. The definition and the modeling of the mechanisms for an Extended Mass Customizing Enterprise (marketing, sales, logistic, production, administration, etc.) play a central role to redefine the processes involved in the product life cycles (design, production, distribution, and dismissal). Here the transformation from manufacturing of mass produced shoes to production of customized (customer oriented) ones, requires a thorough revision of the processes. In connection to the new processes of the Extended User Oriented Shoe Enterprise, (new) appropriate IT-systems have to be selected and implemented. With this in mind, the chapter describes the demands of a transition from mass production towards mass customization.

25.1 Introduction

Customers, whether consumers or companies, do not only want more choices or variety. But, they want exactly what they need - when, where, and how they want it [1]. For many industries, mass customization is one concept to satisfy these customer demands. A study on mass customization in the United Kingdom has shown that the companies who implemented this concept, 21% have increased their market share, 24% decreased the time of response to customers orders, 14% increased their profitability and 5% lowered their manufacturing costs [2]. The shoe industry is one distinct industry sector, where customization promises to be very effective [3].

Since years, the shoe sector within Europe which consists mainly of SMEs with mass production, has a negative trend. After a decline of 4.6% between 1997 and 1998 fell the European production by another 6.1% in 1999 to less than one billion pairs per year for the first time. As a result of this, the employment continued to fall in 1999 (about 12.000 jobs), thus falling to less than 300.000 employees. This represents a 4% drop compared to 1998 [3]. For the European shoe industry, one approach to stop this negative trend and to improve the competitiveness is to capture new market sectors by introducing new products. One high potential market sector is the customized (better fitting/ custom-made) shoes sector, a market which satisfies more and more of the consumer demands. A central role in this connection is the definition and the modeling of the mechanisms of an Extended Mass Customizing Enterprise - marketing, sales, logistic, production, administration, etc.- with a view to restructure the processes involved in the product life cycles - design, production, distribution, and dismissal. Here, the transformation from manufacturing of mass produced shoes to production of customized (customer oriented) ones, requires a thorough revision of the above mentioned processes. In connection to the new processes of the Extended User Oriented Shoe Enterprise appropriate (or new) IT-systems have to be established.

25.2 Mass customization of shoes

Mass customization is a synthesis between mass production and the production of highly specialized and individualized products. It aims at producing high quality specific product for an individual with a similar production cost as of mass produced products and comparatively short delivery time [6]. For (re-)designing a product, process, or business unit, companies have to define which of the four approaches of customization serve their customers best. The approaches are [3]:

- *Collaborative Customization:* conducts a dialogue with individual customers to help them to articulate their needs, to identify the precise offering that fulfills these needs, and make customized products for the customers.

- *Adaptive Customization:* offers a standard, but customizable, product that is designed in such a way so that users can modify it themselves.

- *Cosmetic Customization:* presents a standard product in various types to different customers.

- *Transparent Customization:* provides individual customers with unique goods or services without letting them know explicitly that those products and services have been customized for them.

Collaborative customization is often associated with mass customization and appropriate for enterprises whose customers cannot easily articulate what they want and become frustrated when forced to select out of a vast number of options [3], like e.g. shoe customers. The requirements and the success of mass customiza- tion depends on balance of the three aspects: time-to-market (quick responsiveness – schedule), variety (customization - features), and economy of scale (volume production efficiency - costs). To achieve this balance, three major technical challenges have to be considered [7]:

- Maximizing the reusability of design and process capabilities,

- formulation of a product platform (standardization and modular design),

- integration of the product life cycle.

For mass customization of shoes these technical challenges have to be put into practice. Regarding the shoe market of today, an oversupply of mass produced similar types and forms of shoes of different brands are available but do not completely satisfy customer specific needs. On the one hand, the customer wants a perfect fit and on the other hand a special design for the shoe. In many cases, customers are willing to pay a special price for their unique requests being satisfied in one product. Therefore, a mass customized shoe development approach is envisaged. A more user oriented shoe concept implies a certain level of customization that must be intended not only as style customization but even more important, as fit and comfort customization. Empirical studies on mass customization in the shoe industry show that for customers, fit, quality, and price represent the most important purchase criteria for shoes [8, 9]. These affect the level of individual satisfaction of each single consumer. If the individual satisfaction and the customer's request (custom-made, better fitting, mass production) expressed in consumer studies in Europe are taken as an indicator, we have the following market segmentation [10].

About 65 % of the consumers buy mass produced shoes and accept a very low level of individual satisfaction. The low price is the driving parameter. In countries with a higher purchasing power, customers (approx. 35%) can make a better choice of what they buy. Only a very small minority (approx. 5%) can afford very expensive tailor-made shoes. Mass produced shoes are typically manufactured in low-wage countries, which can keep low prices on standardized large batch production. In contrast to this, better fitting and hand made shoes are mainly produced in industrialized countries with a higher quality production. To close this gap, new concepts and processes are needed which are the future challenges for the European shoe industry. Driving idea is the mass customization of shoes, manufactured for individual customers which best fit the customers' measurements and at the cost of standardized mass-produced shoes with comparably short delivery times.

In the footwear industry concepts and tools already exist for mass customization, e.g. [11, 12, 13]. But, most of these concepts do not integrate the entire product life cycle - design, production, distribution, and dismissal - of a shoe or they are single solutions. The best comparison besides the shoe industry is the apparel industry, e.g. [14], where mass customization is further developed than in the shoe industry. In front of that, the shoe industry needs new concepts, processes, and tools in order to fulfill the market requirements.

25.3   The extended user oriented shoe enterprise

25.3.1   Today's mass production situation of a footwear producer

Today's European shoe enterprises are mostly organized in a traditional way. High prices because of increasing high labor costs are responsible for many problems which have affected shoe manufacturers in the last 10 years. The market pressure (share of European producers in the European market fell from 49% in 1998 to 45% in 1999 [4]) and competition from low-wage countries (e.g. the import from China increased from 17,3% in 1998 to 18,3% in 1999 [4]) created a situation in which the slightest fluctuation in the level of economic activity has major regional and social effects.

One reaction of shoe companies was to shift some of the manufacturing phases (with high labor costs) to other countries (outsourcing). This shifting was mainly in 1998-2000 to Central and Eastern Europe and the Maher countries [4]. The other reaction was the introduction of automatic machines in their manufacturing process capable of reducing the manpower or increasing the productivity. These reactions brought serious problems. On one side, European shoe companies are moving from country to country with their production, struggling to find lower labor costs. Furthermore, outsourcing because of low costs implies higher logistic costs and the loss of control on some important phases of shoe production. Because of that, the shoe manufacturing process today is dispersed among different players like shoe manufacturing companies, subcontractors for specific production phases (e.g. stitching), and suppliers for specific components (e.g. lasts). On the other hand, although machines with remarkable capabilities and significant throughput were developed in the last years, introduction of these machines in the shoe companies has been very slow and only occasionally [10].

Regarding the software tools to plan and manage the activities of the footwear companies, a similar situation can be found. The European footwear companies make use of software applications in the area of accounting, general company management, and in the product design (CAD/CAM). The implementation of ERP systems in shoe companies to manage the manufacturing phases are at a starting point. The reason is that ERP solutions which now are widespread in other industries are not yet customized very well for footwear companies, which in Europe are mainly SME's with 20 to 150 employees [4]. Even when modern tools are present in different functions of the shoe company, they mostly are not aligned with the logic of integrated management and planning of the whole company resources. The situation of the suppliers and subcontractors are even more critical. Only a small number of them are using CAD or ERP systems.

As mentioned above, the solution to increase the competitiveness of European footwear manufacturers and their technology suppliers is to expand to high potential market sectors like customized (better fitting/ custom-made) shoes. To transfer this, the following two aspects must be considered [10]:

- mastering rather than accepting a distributed manufacturing process and

- developing the shoe enterprise into a fast reacting enterprise capable of responding even to individual consumers.

To support the different processes during the product life cycle of the new, customized shoe, the companies need:

- new methods to control and manage the distributed manufacturing process,

- adequate planning and management tools, and

- an appropriate set of manufacturing technologies.

These have to be considered as obligatory prerequisites for the development of shoe making companies from "shoe factory with suppliers and subcontractors" to "Extended Shoe Enterprise".

25.3.2 Mass customization in a footwear company: the extended user oriented shoe enterprise

The Extended User Oriented Shoe Enterprise can be seen as an extended or in some cases as a virtual enterprise. The concept of extended enterprise is applied to an organization, where a dominant enterprise "extends" its boundaries to all or some suppliers whereas the virtual enterprise includes other types of organizations, namely a more democratic structure with a cooperation of equals. That means that an Extended Enterprise is a particular case of virtual enterprises [15].

The integration ("extend" the boundaries of the existing shoe enterprise) of different companies/ actors in an extended (virtual) user oriented (mass customizing) shoe enterprise is connected with the improvement of the performance of distributed organizations and markets. It focuses on the communication of information, co-ordination and optimization of enterprise decisions and processes in order to achieve high levels of productivity flexibility, and quality [cf. 16]. It employs that the development of the Extended User Oriented Shoe Enterprise (Figure 1) has to consider the following aspects: organizational aspects, process modeling, workflow management, handling of distributed manufacturing processes, distribution network management, subcontractor/supplier/partner search, and information exchange.

The organizational aspect is one of the most important aspects for the development of an Extended User Oriented Shoe Enterprise because different actors are involved in the product life cycle and have to be organized. These actors are the shoe manufacturers/ footwear companies, external designers (last makers), suppliers (leather and synthetic material suppliers), component manufacturers, sub-contractors (for cutting and stitching), and last but not least the customers (distribution network).

The processes of the new Extended User Oriented Shoe Enterprise have to be dynamic rather than static. This has to be considered for the process modeling of

- the internal processes of the footwear company,

- the interfaces to external partners, suppliers, subcontractors, etc., and

- the interfaces to market/ customers (distribution network).

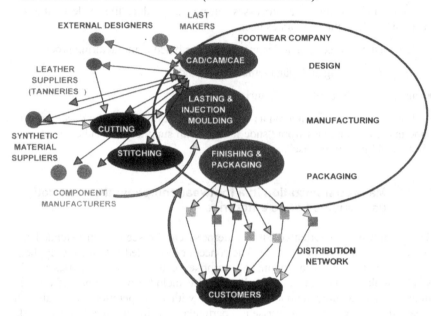

Figure 1: The Extended user oriented Shoe Enterprise

This includes also the workflow management, and handling of distributed manufacturing processes, the offer making/order process management and the selection of subcontractors and suppliers. Another result of the process modeling will be the exchange of information between the different actors. To achieve a good level of integration of the different actors, the integrated and holistic processes have to be supported by appropriate and integrated information and communication technology (ICT). Furthermore, these processes build the basis for ICT requirements that have to be fulfilled for producing mass customized shoes.

25.4   Information technology enablers

Modern IT-systems are one of the main enablers for mass customization [cf. 17], in other words, the enabler of the Extended User Oriented Shoe Enterprise. For this, a broad range of software tools are required to bring customizable products to the marketplace [18]. These tools are able to reduce the increasing complexity of data, produced during the product life cycle of mass customized shoes. Here, the progress of web technologies (internet and intranet) gives the ability to connect the extended or virtual enterprise with different actors, enabling rapid information flow over many regions or even globally. This helps to reduce communication cost and to increase the efficiency of the processes.

The ICT systems for the Extended User Oriented Shoe Enterprise have to be capable of handling

- the extended sales and supply chain (from acquiring customer requests to issuing the relevant design/ production orders),

- the distributed manufacturing process with the footwear company on the one hand, the design centre on the other hand, and

- the network of component manufacturers (coordinating the production and the delivery of all the relevant shoe components produced by suppliers).

This implies a fully integrated IT-environment. As identified, maximizing the reusability of design and the formulating a product platform (standardization and modular design) are critical technical challenges for mass customization [7]. In other words, the product development/engineering is crucial for successfully implementing mass customization. Considering here the extended enterprise with different actors involved, the major implication regarding engineering is a full integrated PDM-, ERP-, and CAD/CAM- environment (shown in Figure 2).

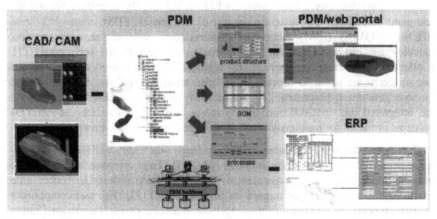

Figure 2: The IT-enablers in the product development

CAD/ CAM-Systems: The formulation of the product platform starts with the use of CAD-systems to design products that maximize the reusability of standard parts and assemblies [cf.19]. On the current market modern solutions for the shoe design are available. These are based on CAD/CAM software packages both with 2D or 3D capabilities. But none of them are adequate to handle the quick adaptation needs and rapid design change capability which arise due to the aim of adapting the shoe design to the specific requirements of each individual consumer.

They also do not have the capability of resting their operations upon a knowledge base where design, engineering, and biomechanical/anthropometrical constraints can be blended and harmonized. Another weak point is the integration with other software packages that are part of the design process [10].

Product configuration systems (PCS): The formulation of the product platform is coupled with the use of product configuration systems [19]. PCS are multifunctional software with logic capabilities to create, maintain, and use electronic product models that allow the complete definition of all possible product options and variation combinations with a minimum of data entries and maintenance. They are used as an interface between the distribution and other domains in the value chain with the objective to support sales and order processing effectively and efficiently [19, 20]. This capability is essential for companies offering customized products satisfying specific customer needs. PCS are often provided as a part of a Computer Aided Selling (CAS) or Sales Force Automation (SFA) module of ERP systems [20]. For modeling complex products, the configuration module of PDM systems are required. The rule-based configuration capabilities of PDM systems provide an important link between sales-based product configurations and production-specific configurations expected by ERP systems.

A *product data management system (PDM)* characterizes the holistic, structured and consistent administration of all data and processes that have to be generated, processed and transferred during the development of new or during the modification of existing products over their complete life cycle [21]. By extending beyond the company boundaries into extended or virtual enterprises, PDM systems like Iman (UGS), ENOVIA (IBM), MatrixOne (Matrix), etc. manages the entire product development process including the external actors via PDM web portal [21]. Collaborative Product Commerce is the highlight. Core functions of PDM systems are an electronic vault for document and file storage, workflow management and product structure management [21]. Additionally, most of the PDM systems provide an integrated link to CAD and ERP systems. This means that the use of PDM systems is crucial for mass customization of shoes in an extended enterprise.

Enterprise Resource Planning (ERP) systems like SAP and BaaN are enterprise-wide information system solutions [22, 23]. They are based on the idea of process integration and management of the information flow in the entire value chain from the point-of-sale to point-of-production [cf. 17]. The main components of ERP systems are manufacturing, supply chain management, finance and human resources.

25.5 Strategic selection of IT-systems: a methodology

For companies (SME's), the evaluation of suitable IT-enablers for mass customization is challenging. Wrong decisions can entail high costs for the company. In order to select and implement a set of appropriate and suitable IT-enablers for mass customization, a strategic methodology for software selection is needed. Here the scenario technique tenders a methodology for the strategic selection of IT systems supporting the new processes of mass customization.

The scenario technique is a concept for the evaluation of standard business software systems [24]. This concept was further developed by the Fraunhofer Institute for Industrial Engineering (IAO), which was further deployed, validated and verified successfully in several projects. Paramount objective is to create an "objective level" to prove the suitability of standard software solutions for the individual company needs. The procedure of software selection is subdivided in the phases (Figure 3): pre-selection, definition phase, detail specification and integration-lab. Finally the contract is signed, which is followed by company-wide roll-out of the IT-system. The pre-selection and design phase are characteristic for the scenario technique, which reduce the operating expense and create a decision base. During the pre-selection phase, the number of suitable IT-systems available on the market is reduced to two to four systems. This pre-selection is based on certain criteria regarding the technique, costs and vendor of the IT-systems (Table 1).

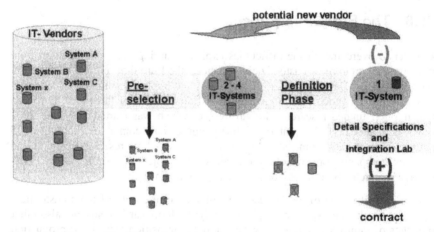

Figure 3: Phases of the IT-system selection

The main instrument for a precise and detailed selection of the 2-4 IT-systems in the design phase is the developed scenarios which are used during the benchmark as a "script" for the presentation of the IT-systems by the vendors. To confirm the collected impressions, which are won during vendors-presentations and to prove the daily use of the software tools, reference customers of the vendors have to be visited. Finally, one suitable IT-system is selected and a pre-contract is signed. The detailed specification and the integration-lab phase are used for a detailed view on the standard business software by installing a test bed within the company. This phase is the preparation of the final decision (contract). It gives an opportunity to revise the decision and to restart this phase with the next vendor in the ranking, if the selected IT-system is not suitable for the company requirements. The contract finishes the selection and starts the company-wide roll-out of the IT-system.

Table 1:   Pre-selection criteria

Technique	Cost	Vendor
IT	*Software*	*Market Potentials*
• platform • architecture • adaptability • integration • technology	• licenses • customizing • increase in functionality • integration	• economic situation • market potentials • partner concept • strategy
application	*operation*	*internal potentials*
• functions • data model • user interface • methods	• maintenance • training • support • consulting	• development • service • training • consulting

25.6   The EuroShoE project

In order to overcome all the difficulties explained during the transformation from mass production to mass customization of shoes, the EuroShoE project has been set up. 'EuroShoE - Development of the processes and implementation of management tools for the Extended User Oriented Shoe Enterprise' is a three-year European Commission funded research 'Growth'-project which was started in March 2001 (www.euro-shoe.net). The interdisciplinary consortium comprises more than 35 partners - companies from the footwear industry as well as research institutes and software companies or technology providers. The project is coordinated by ITIA-CNR, a research lab of the National Research Council of Italy.

The project is one of the largest ever founded in the field of mass customization. The results are expected to give not only the footwear industry but also other branches of industry a major show case how the transition from mass production for anonymous markets towards mass customization in individual markets can be performed. The expected results of the EuroShoE-Project are [4]:

1. A detailed survey of the specific aspects of demand for the customized shoes, a reference model of the business and operational processes of the EUROShoE [25].

2. A set of software tools and procedures to select, configure and integrate IT-systems.

3. A fully implemented IT environment for a test bed EUROShoE.

4. A new generation of foot feature capturing devices and camera based foot scanners.

5. A knowledge based CAD/CAM software for the design of customized shoes.

6. A variety of new conception versatile and multi purpose shoe machines and systems, a physical and virtual (web based) sale environment for selection of customized shoes.

One of the work packages of the EuroshoE-Project focuses on modeling the mechanisms of the extended shoe enterprise. This reference model will cover all the processes/ functions of the extended enterprise with a view to the processes involved in the shoe life cycles. The process oriented modeling activities are based upon an analysis of the structures of typical European footwear companies with their current level of organization. In the first step, the footwear companies (Bally, Calana, ECCO, Frau, Jefar, Lloyd) have been analyzed. Based upon these models and the marketing requirements for mass customized shoes, the reference model of the Extended User Oriented Shoe Enterprise will be created and validated together with the footwear manufactures in the project consortium. This model will be the basis to determine specifications of software systems suitable for an Extended User Oriented Shoe Enterprise.

Following this, selection and implementation of appropriate software tools (ERP, PDM, etc.) to effectively manage complexity of the new Extended User Oriented Shoe Enterprise will be done. For this, a computerized tool-based methodology for selection of the specified software systems will be developed. Due to a continuous change in requirements of the Extended User Oriented Shoe Enterprise, the management and planning systems, such as ERP- or PDM-Systems, etc. have to be continuously adapted during their entire life cycle. This requires a configuration of the software tools according to the enterprise business processes. Here the methods and tools will be developed which support the configuration of the software systems. Finally, a market place will be developed and implemented to support the business to business relationships and material/ component procurement.

For the European shoe industry, mass customization of shoes means a radical change in their way of thinking and working. Starting with the new process of an Extended User Oriented Shoe Enterprise, new methods and tools have to be specified and introduced. The activities of the EuroShoE-Project will put this into practice. These project activities will help the European shoe industry to realize mass customized shoes which will increase firstly the competitiveness of European footwear companies and secondly the customer satisfaction.

References

[1] Pine II, B. J.; Peppers, D.; Rogers, M.: Do you want to keep your customers forever?, in: Harvard Business Review, 1995, pp. 104-114.

[2] Sabljakovic, V.: The New Economy, Electronic Commerce and the Rise of Mass Customization, on www.dbai.tuwien.ac.at/staff/dorn/ Seminare/IM/EC/K19.

[3] Gilmore, J. H.; Pine II, B. J.: The four faces of Mass Customization, in: Harvard Business Review, 75 (1997) 1, pp. 91-101.

[4]   Commission of the European Communities: Report on the promotion of competitiveness and employment in the European footwear industry, Commission staff working document, Brussels 2001.

[5]   EuroShoE: on www.euro-shoe.net.

[6]   Pine II, B. J.: Mass Customization: The new frontier in Business Competition, Boston 1993.

[7]   Tseng, M.M.; Jianxin, J: Mass Customization, in: Handbook of Industrial Engineering, 3rd Ed., 2001, pp. 684-709.

[8]   Kieserling, C.: Übersicht: Mass Customization in der Schuhindustrie, 1999.

[9]   Aumann, F.: Zitex – Industrielle Maßkonfektion – Marktchancen einer Innovation: Das Wichtigste in Kürze zum Forschungsbericht "Untersuchungen des Marktpotentials für Maßkonfektion aus der Sicht der Konsumenten, des Textilhandels und der Bekleidungsindustrie", Muenster 2000.

[10]   EuroShoE Consortium: Description of Work of the EuroShoE-Project, 2001.

[11]   Press Release: Mass Customization in the shoe trade: tecmath uses its scanner technology for the ShoeFit system, on www.tecmath.com.

[12]   eShoes: on www.eshoes.com.

[13]   Custom Made Shoes-Shoes for the 21st Century: on www.digitoe.com/.

[14]   Piller, F.: Fallstudie Mass Customization Levi's PersonalPair: massgeschneiderte Damenjeans – von der Stange, in: Piller, F.: Kundenindividuelle Massenproduktion – Die Wettbewerbsstrategie der Zukunft, München 1998.

[15]   Camarinha-Matos, L. M.; Afsarmanesh, H.; Garita, C.; Lima, C.: Towards An Architecture for Virtual Enterprises, in: Proceedings of the 2nd World Congress on Intelligent Manufacturing Processes & Systems, Budapest 1997.

[16]   UCANet-Project: D1-Survey of study and statistical analysis, on www..democenter.it/ucanet/publicDocs/d1.pdf.

[17]   Piller, F.: Mass Customization and SAP R/3TM – Business Solutions like SAP R/3 as an Enabler of Mass Customization, University of Wuerzburg, 1997.

[18]   A QAD White Paper: Mass Customization – Survival and Growth in the To-Order Sector, on www.qad.com/publications/whitepapers /pdf/mass_customization.pdf.

[19]   Frech, J.; Schwarz, J.; Wagner, F.: Marktstudie – Computer Aided Selling Systeme zur Produktkonfiguration, in: Fraunhofer Institut für Arbeitswirtschaft und Organization IAO, Stuttgart 2000.

[20]   Bourke, R.; Arts, J.; van der Roest, M.: Achieving success with mass customization: The Vital Contribution of Engineering, on www.caenet.com/res/archives/ 9910PDM-ERP.html.

[21]   Bullinger, H. J., Frielingsdorf, H.; Wagner, F. et al.: Marktstudie – Engineering Data Management Systeme, in: Fraunhofer Institut für Arbeitswirtschaft und Organization IAO, Stuttgart 1999.

[22]   Luczak, H.; Eversheim, W.; Stich, V.: Marktspiegel PPS-/ERP-Systeme für den Mittelstand, Forschungsinstitut für Rationalisierung (FIR), RWTH Aachen, 2000.

[23]   Marktrecherche: E-Business-Funktionen in Standard-ERP-Systemen, in: PPS-Management, 1 (2001), pp. 60-70.

[24]   Sonderegger, R. J.; Lanz, T: Konzepte, Evaluation von Standardsoftware mittels Drehbuchtechnik, in: OUTPUT, 1 (1996).

[25]   Piller, F.: The Market for Customized Footwear in Europe: Market Demand and Consumer Preferences, A project report from the EuroShoe Project within the European Fifth Framework Program, Munich and Milan 2002, online available on www.aib.ws.tum.de/piller.

Acknowledgement: The work presented in this chapter was made possible by the highly appreciated support from the European Commission, funding the research project EuroShoE (Growth GRD1-2000-25761, www.euro-shoe.net, Project Coordinator: Prof. Claudio Boer and Sergio Dulio from ITIA-CNR, Milan, Italy) [5]. The authors would like to thank all the project partners and their representatives for their valuable contributions. Special thanks to Bally Shoe Factories Ltd./Caslano for their permission to use some of their graphics of the »digital product concept« for this chapter.

Contacts:

Univ.-Prof. Dr.-Ing. Habil. Prof. eh. Dr. h.c. Hans-Jörg Bullinger
President of the Fraunhofer Gesellschaft, Munich, Germany
E-mail: bullinger@iao.fhg.de

Dipl.Ing. Mehmet Kueruemlueoglu
Fraunhofer-Institute for Industrial Engineering (IAO), Stuttgart, Germany
E-mail: Mehmet.Kueruemlueoglu@iao.fhg.de

26 Implementing a Mass Customized Clothing Service

A strategy model for implementing a mass customized clothing service in a High Street store

Celia P.A. Taylor[1], Ray J. Harwood[1], Jane L. Wyatt[1] and Michael J. Rouse[2]

[1] Department of Textile Design and Production, De Montfort University, Leicester, United Kingdom

[2] Department of Corporate Strategy, De Montfort University, Leicester, United Kingdom

This chapter explores how a mass customized clothing service could successfully be implemented on the UK High St and suggests a strategy model. In the present highly competitive environment clothing retailers need to gain a competitive advantage. Superior customer service was identified as a more effective competitive strategy than price or merchandise, since it is difficult to achieve consistent quality customer service, those that achieve it are not easily copied by competitors. Mass customization is a means of offering superior customer service, customers are supplied with a garment to meet their needs and also enjoy the personal attention this facility requires. System manufactures such as Lectra Systèmes are already collaborating with retailers in Europe to provide garment customization as an in-store service. Although mass customization clothing services have been available in some High St stores in the UK for over 10 years, they have been limited to men's suits. Four High St retailers were identified as having offered the service, however, two of these retailers have ceased the service within the last four years. In contrast the four Independent tailors interviewed offering mass customized clothing are expanding and developing their service. Structured interviews were undertaken in a survey of eight UK retailers. The sample consisted of four High St retailers and four Independent tailors.

26.1 Mass customization in the clothing industry

The UK market [1] for clothing is predicted to become increasingly competitive as consumer demand for clothing is weak due to competition from other retail sectors of consumer durables. Combined with the demographic decrease in the number of 20 to 24 year-olds, the highest spending and most fashion-led consumers, this has resulted in a large number of retailers chasing too few customers. The strong pound sterling and sourcing from abroad have resulted in the market being geared to very tight margins at all stages of the supply chain.

Mass customization is about giving consumers a unique end product when, where and how they want it - at an affordable price. It is characterized by small lot production, timely delivery, competitive cost, and a move away from centralized manufacturing. Mass customization provides a way for a company to obtain or maintain competitive edge in today's international marketplace [2].

There is some ambiguity in the industry concerning just what the service includes. It is widely stated that a mass customization clothing service would allow customers to select from the available fabrics and colors in the range. However, there is some disagreement on whether it includes producing the garment to fit a customers individual body dimensions, or whether this is a made-to-measure service. For this report the authors have assumed that a mass customization service would include producing the garment to provide a better fit using customer body measurements. A mass customization approach is already being successfully employed in the production of corporate wear and military uniforms by some UK companies and in the Centre For 3D Electronic Commerce [3] research project. The aim of this investigation is to identify those key factors that must be considered for successful implementation in the High St.

26.2   Key areas for providing mass customized clothing

For the purpose of this investigation, eight key areas are identified as fundamental for consideration in providing a mass customized clothing service. The strategy model is developed from a retailer/consumer perspective in this case, and illustrates the interrelationships of the components and their contributing factors. In this section the literature and some earlier findings are discussed for each key each area. In the next section, a brief discussion of each area is given based on the results of the interviews with service providers.

(1) Customer: A survey of European consumers [4] found that 66% of consumers go shopping for apparel with a clear idea of what they want, however, only 55% find and buy what they were looking for when they entered the store. The main reasons for not making a purchase were twofold: prices being too high (57%), and not being able to find a style they like (53%). Garments not fitting well and not being able to find their size also frustrated them. Respondents expressed a degree of interest in mass customized clothing, and were beginning to feel "...the individualism and personal fit of custom-made clothing is worth the price and wait." 36% would be willing to pay for certain items to be custom made. 28 % had already bought customized items, and more than half would purchase again [5].

(2) Marketing: According to Gremler et al. [6] the most credible way for a company to acquire new business is word-of-mouth communication between existing and potential customers. However they suggest positive word of mouth does not just result from a customer being happy with service, it is also the interpersonal relationships between employees and customers that encourage positive customer word-of-mouth, and strategies are needed to encourage this

bonding. In a study of individuality in fashion [7], the research implied that differentiation of fashion marketing strategies would be profitable, as customers are attracted by individual treatment and by being surrounded by unique shopping environments and promotional events. Rosen [8] argues that if a company is providing a unique product they need to make customers aware of the service.

(3) Retailer: The advantages of a mass customization service, to the retailer, are claimed by Conrad [9] to be a reduction in inventory, mark downs, and returns, whilst increasing customer satisfaction and customer loyalty. A report by Mintel International Group Limited [1] suggests that fashion trends need to consider the shape and size of typical customers in order to sell products, and that a choice of length in terms of trousers, skirt and dress would maximize sales potential. Fernand [10] argues that fit is a major problem in the British Fashion Industry, and mass customization is one solution to provide better fitting clothes and allow individuality. The consumer's desire for uniqueness is a backlash against uniformity created by globalization [11]. In response to this need designers are beginning to offer one off garments and create more personal shopping environments in their ready-to-wear stores.

(4) Service: Levy and Weitz [12] explain that a customer evaluates retail service by comparing their perceptions of the service they receive with their expectations. They assert that once a retailer has a reputation for service they have a competitive advantage that is difficult for competitors to match. According to Bishop Galiango and Hathcote [13] it is more profitable to retain a customer, than to have to spend money attracting new ones. Their study (of consumer expectations and perceptions of service quality in retail apparel specialty stores) found that although customers indicated merchandise was most important followed by price and then service, it is the latter that provides the best advantage. This finding is supported by the retailer Jaeger who identified exceptional customer service as a means of providing a competitive advantage to get them through the recession in the 1990s and beyond [14]. They recognized the importance of stressing to staff the impression they make on customers, and the value of retaining their customers. Staff provide the customer with an experience in terms of advice, knowledge, and personal attention, so the customer should leave looking and feeling good.

A report by Lyons [15] investigated selling methods in the bespoke tailoring industry and identified that bespoke suits are 'bought unseen'. Therefore the salesperson needs to gain the customers' trust and confidence that they have the ability to create a garment to his or her satisfaction. Lyons claims that bespoke salespeople use a different approach and technique from any other type of salespeople and details the manner of customer service required to sell bespoke products.

(5) Product choice: Increased variety ensures a customer finds a product to meet their need, and satisfies their desire for change. Kahn [16] explains that if consumers are offered a wider variety of goods their desire for change is stimulated by the choice. Increasing variety requires the 'customizer' (retailer) to identify what features need to be customized and which can remain standardized.

An understanding of how customers perceive variety is important so they are offered choices they value. The 'customizer' can then guide and advise the consumer about the product effectively.

(6) Price: A comparison [9] of the cost distribution of manufacturing a casual denim garment by mass production (traditional) and mass customization indicates that mass customized garments are more profitable for the retailer and manufacturer due to greater margins, reduced returns and reduced inventory. Pine [18] argues that those companies that are already mass customizing products and services "...are finding that great variety - even individual customization - can be achieved at prices that approach, and sometimes beat, those of mass producers." A study of the willingness of German consumers to pay a premium price for made-to-measure apparel [19] suggests that 5% was the most accepted premium level, thereafter acceptance dropped rapidly. Only 40% of the respondents were prepared to pay 15% more.

(7) Delivery: Leclerc [17] conducted an investigation to compare consumer attitudes towards spending and saving money, and spending and saving time. The study found that subjects would pay more to avoid waiting for a higher priced good or service than for a lower-priced product. The respondents felt they would rather group losses of time into one day than have them over separate days. The study also revealed that subjects were less likely to make risky decisions with time than with money, since unlike money, time can be saved but not stored. Therefore planning time is important, and subjects claimed that certainty allows them to plan their time better. In terms of mass customization, although a customer may pay a premium, they know that at the end of the agreed waiting time, they will receive a garment to their personal requirements. The customer can therefore plan their purchase, and save many disappointing and unrewarding shopping trips.

(8) Return policy: Jaeger [14] changed their return policy to a 'no fuss' approach as they realized making a big commotion when a customer made a return sent out the wrong message and could result in a customer defecting. Returned mass customized products would be difficult to sell. However, because the customer has had a personal involvement in creating the product he or she is involved in the entire process - they emotionally own the product and are much more likely to keep it and be committed to the purchase [20].

26.3   Case study research

26.3.1 Methodology and sample

There is currently no comprehensive list of mass customized clothing services in the UK, so locating providers was achieved via a number of methods including contacting software providers, apparel directories and web searches. The study sample consisted of eight UK retailers that were identified as offering a customized menswear service. The interviewees were individuals who had been directly and personally involved in managing and often providing the service. The eight retailers consisted of four High St retailers and four independent retailers (tailors). Two of the High Street retailers in the sample had ceased offering the service within the last four years. A survey of the eight retailers involved structured interviews conducted face to face at the retailer's premises. The interviewees were informed that the purpose of the survey was to identify how their customized clothing service was operated. Where possible the interviews were recorded and lasted for approximately one hour. The types of retailers interviewed (Figure 1) were bespoke tailors; bespoke and mass customized retailers; bespoke, mass customized & ready-to-wear retailers; and mass customized and ready-to-wear retailers.

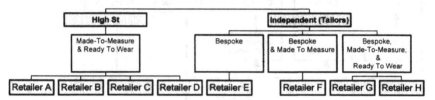

Figure 1: Interview sample

Retailers were asked how they perceived and evaluated the customization service in terms of the most important elements, any problem areas, and how the service might be improved and expanded. Retailers were questioned under the following subject areas:

- How the customization service was developed (identifying customer requirements, target market, determining fabric and style selection, pricing, and promotion).

- How the service was provided (accessing service, taking an order, flexibility, waiting time, refunds and returns).

- How did production take place (manufacturer, translating order, pattern creation, delivery).

- How successful was the service (length of time service operated, profitability, location, expansion options).

- General questions about the service (advantages and disadvantages, body scanning competition).

26.3.2 Key areas in the practice

In the following, the eight key areas for providing a mass customized clothing service are discussed before the background of our own empirical investigation. Figure 2 gives an overview how these factors are connected.

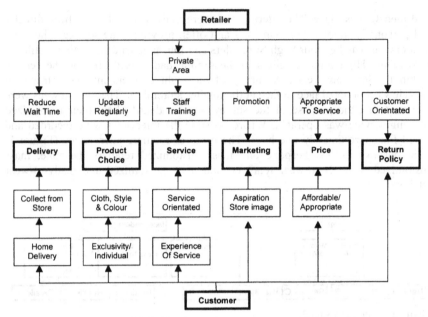

Figure 2: A strategy model for implementing a mass customized clothing service in a High Street store

(1) Customers: The survey suggests that desire for a suit that fits and that is exclusive were the main reasons for purchasing a mass customized garment. Many customers had problems with finding ready-to-wear that fitted correctly. Customers also enjoyed the experience of a customized service. This supports Kurt Salmon Associates [4] finding in their survey of European consumers. According to the retailers interviewed a customers' motive for buying a mass customized garment tended to be exclusivity when purchasing from a Independent tailor, and fit when purchasing from a High St retailer.

(2) Marketing: The Independent retailers seemed to operate more successful marketing strategies than High St retailers. Independent retailer F (see Figure 1) engaged a PR company to generate publicity in newspapers as they are too small to advertise, they also ran promotions. Web sites were also popular with the independents as having only a limited number of shops, they need to tell people that they exist. London based Independent retailers E and G store location often drew people in. In contrast the High St retailers relied on in-store displays as their main promotion, as well as word-of-mouth and introduction by sales staff if customers could not find what they want in ready-to-wear. High St retailer B also mentioned their service in their catalogue. Only High St retailer D advertised periodically, although it was limited. Regional Independent retailer H advertised in the yellow pages, local newspaper and county magazine. All the Independent retailers felt word of mouth was another reason people came to them, while this was the view only half of High St retailers.

A study of retail store image [22] identified a desirable fashion store has a reputation for offering a wide range of clothes, of quality and durability at an acceptable price. They found that the customer experience in the store must back up the marketed image. The retailers felt customers often selected their mass customized service over another competitors service because they aspired to the stores image. For example they may prefer a younger fashionable store image too a classic store image.

(3) Retailer: Retailer's reasons for offering a mass customized service varied. Independent retailers (E-H see Figure 1) offered it as an affordable alternative to bespoke, it offers a higher standard of fit at a reasonable price. High St retailers (A-D) saw it as a service for customers who could not fit into ready-to-wear and also for those who want something more individual, an upper end service. The extent of the problem of customers not finding a fit in retailer ready-to-wear sizing systems is indicated by a study of the American sizing system for women [21]. The study identified that many individuals did not fit into sizing dimensions and were therefore not catered for by the current sizing system. In terms of the environment in which the service was provided, independent retailers offered a much more private and luxurious environment, since they already provided bespoke services they were already geared up for providing a customized service. In contrast High St retailers allocate very limited space within the shop floor, for instance a table for consultations and display.

(4) Service: The amount of training that staff received varied between retailers. For Independent retailers it may involve up to five years training/apprenticeship in tailoring skills, service provision, and measuring. In contrast High St retailer staff training varied between three days to six weeks. High St retailers staff training involved learning about the human form, taking measurements and figurations. The staff also visited the factory to understand the system and how garment's are produced. This meant that the Independent retailers had more highly trained staff who were extremely motivated in their work and had a deeper understanding of tailoring and measuring. High St retailers claimed that the length of training required by the Independent retailers was not necessary to provide the service. However it is arguable that it provides an advantage and a greater degree of flexibility as Independent retailers tailoring staff are able to make minor adjustments that the factory could not achieve. The findings indicated that Independent retailers offered a superior customer service and also enjoy greater success than High St retailers offering mass customization.

This is supported by Bettencourt and Gwinner [23] findings that companies customizing the delivery of their service to the individual customer can provide a potential source of competitive advantage. The firm can customize the service by the way the employee responds to the customer and adapts the service attributes to them. The study found that a firm needs to research customer segments and identify the type of service they want and employ individuals who can perform this adaptive role and train them accordingly. Finally providing an environment in which staff can respond to customer needs empowers employees to customize the service to the customer. A database of customer preferences also helps the customization service [23]. Independent retailer can be more responsive to individual customers, they offer a greater personal service and are extra accommodating as they are more familiar with providing a customized service that is flexible to customers' needs. It is the authors view that they are more confident because of their vast knowledge and experience accumulated from producing bespoke garments. They are not bound by the rules and regulations that operate in a High St store trading environment.

It could be argued that it is contradictory to offer a mass customization service in a High St store environment where there is a pressure to maximize sales. High St retailers need to address the issues required to provide a mass customized service. Currently the store environment is not conducive to the service because it offers limited privacy and luxury. Sales personnel need to be given the appropriate training, and rewarded (e.g., remuneration) in accordance with their skill levels to reduce staff turnover. If the service was provided in a separate small store dedicated to this service and with perhaps the top of the range ready-to-wear, the High St retailer may be able to achieve a similar environment and service as an Independent retailer. It is the authors belief that this could increase the success of the service as a mass customized service provider.

(5) Product Choice: Independent retailers offered a much wider range of fabrics than High St retailers could, as they were already set up with a very wide range of fabric for their bespoke range. Given this limitation within a High St store, a new service could be implanted by offering a product that most people have trouble getting to fit. This could be an essential item in people's wardrobes that they would be prepared to pay extra for the service. However, Kahn [16] explains that increasing consumer choice may require a degree of guidance to be provided by retail staff.

(6) Price: High St retailers prices for a mass customized men's suit started from between £200 - £400 (for retailers at the higher end of the market), Independent retailers starting price was almost twice as much from £500 - £600. The retailers claimed price was based on cost. Goldsmith and Newell's [24] study of the relationship between product innovativeness and price sensitivity of fashionable clothing identified that more innovative customers were less price sensitive. This supports the strategy of price skimming for introducing new products where there is little knowledge about it and there are few alternatives. Independent retailers felt that they would have to modify their prices as competition increased, as although they offered a wider choice of fabrics they were basically selling a similar product at twice the price.

The prices of mass customization services were determined by a number of factors. Firstly it was at a price point that made it more expensive than ready to wear, and an affordable alternative to bespoke. It filled a market gap between the two extremes. The price therefore depended on the cost of the other products the retailer sold, and also the store location. Independent retailers were at liberty to charge twice as much as High St retailers at £585+, since their location, environment, and service were more exclusive and personal than High St retailers.

Although the actual product is very similar, they have the advantage of offering a wider range of fabrics, and are better able to make adjustments if necessary. High St retailers had to price their mass customization garments slightly above their ready to wear to differentiate the service, and give it exclusivity, as well as covering the extra costs of providing the service, while remaining affordable to their customer base.

(7) Delivery: Delivery time for both Independent and High St retailers was on average 4-6 weeks, but could take up to 8 or 10, it could be as short as 3 weeks with the express service at an extra charge. Sanders' [19] study of German customers' acceptance of different delivery times for made-to-measure apparel identified about one week as optimum. Retailers preferred to have customer's collect the garment so they could be fitted. Independent retailers were more willing to post the garment to the customers home, perhaps because they may only have one store and it was not convenient for the customer to return and they had higher profit margins.

(8) Return policy: Generally all retailers felt returns of mass customized garments were unusual. They were also very customer orientated in their returns policy, and were willing to rectify the problem or give a refund if necessary. Retailers H and B felt that if the customers' complaints were unreasonable they would not get a refund. The return policy needs to reassure customers they are going to take home a garment they are happy with and will wear. Duffin [14] argues that giving a full refund with no fuss is more cost effective in the long term as it retains them as a customer and over a lifetime they will spend a lot of money as well as preventing negative word of mouth about the retailer. Several of the retailers interviewed felt a virtual try on system would be advantageous, so customers could see the garment style and fabric on them. There had been occasions when customers did not like the full impact of a fabric or the style when they saw the final garment.

26.4 Conclusions

It appears the High St retailers and Independent retailers have only experimented with mass customization and seem to see it as threat to their other products and services rather than an opportunity. They seem to have overlooked the benefits of filling what Gilmore and Pine [25] describe as the 'satisfaction gap', where there is a difference between the product a company offers, and what a consumer really wants. Providing a mass customized clothing service offers potential savings in terms of satisfied and loyal customers, reduction in inventory and floor space, and the ability to be more responsive to trends.

A High St retailer offering a mass customized service has to give it their full backing in terms of supporting staff providing the service, and not see ready-to-wear sales as the priority, as this damages the service. Although High St retailers claimed it is not currently preferable to have mass customized garments sales exceeding ready-to-wear, the authors believe that if the retailer is satisfying the customer and making a profit from mass customized garments that they should explore the potential, rather than curtail the market.

High St retailers need to train their staff so they are knowledgeable and confident in providing the service. They also need to support the staff in the specialist service they are providing so staff can remain focused and dedicated to service. Retailers should reward staff in accordance with skill levels. It may be very short sighted of a retailer to invest in training a member of staff and not provide rewards (e.g., remuneration) to ensure the trained staff stay.

In terms of marketing there is no advantage in offering a service customers are not aware of, although over-promotion has been mentioned by several retailers as undesirable. The service may lose it exclusive appeal, and for High St retailers ready-to-wear sales, currently more profitable, could decline. Although it is argued that mass customization is ultimately more profitable [9]. Retailers need to identify the garments that customers really want. The service could begin with a staple core product such as trousers and expand into others such as jackets and skirts. The range of styles would need to be appropriate to the market and regular updates would keep loyal customers interested.

The price would need to be higher than ready-to-wear to differentiate the service and also reflect the cost of offering a more personal service. However, it would need to stay accessible to the core customers, if the service was to be offered within the store ready-to-wear range. This restricts the options in terms of fabrics, as offering a wide choice is more costly. If the service was offered exclusively in a separate store then the price could be geared to attract a different market segment. In terms of delivery time retailers thought one week was optimum, although three weeks is minimum at the moment, it is still a long waiting time and retailer felt they could attract more customers if they could reduce the waiting time. The return policy needs to give the customer confidence in the service so that they are not going to get left with something they do not like simply because it was made for them. The customer needs to feel that they are going to get a garment they want, a customer-orientated policy assures problems will be rectified or a refund will be given. Taking into account the above factors, the successful implementation of a mass customized clothing service within a High St environment could potentially increase consumer satisfaction and loyalty, whilst concomitantly strengthening the competitive advantage of the retailer.

References

[1]   Mintel International Group Limited: UK Clothing Market, March 2000.
[2]   Gerber Scientific: What is mass customization?, on www.mass-customization.com.
[3]   Centre for 3D Electronic Commerce: on www.3dcentre.co.uk.
[4]   Kurt Salmon Associates European Consumer Outlook '99.
[5]   Kurt Salmon Associates Consumer Outlook '98: on www.kurtsalmon.com

[6]   Gremler, D. D.; Gwinner, K. P.; Brown; S. W.: Generating positive word-of-mouth communication through customer-employee relationships, in: International Journal of Service Industry Management, 12 (2001) 1, pp. 44-59.

[7]   Park, K.: Individuality on fashion diffusion: Differentiation versus independence, in: Journal of Fashion Marketing and Management, 2 (1997) 4, pp. 352-360.

[8]   Rosen, F.: Enabling Mass Customization, on thinkcustom.org/presentation.htm.

[9]   Conrad, A.: Consumer customization, on www.aimagazine.com/archives/598/.

[10]   Fernand, D: No More Frock Horror, in: The Sunday Times Magazine, March 31th 2002, pp. 32-37.

[11]   Quick, H.: Trend – One of a kind, in: Vogue, 168 (2002), pp. 97-98.

[12]   Levy, M.; Weitz, B. A.: Retailing Management, 3rd Ed., 1998, pp. 570-599

[13]   Bishop Gagliano, K.; Hathcote, J.: Customer Expectations and Perceptions of Service Quality in Retail Apparel Specialty Stores, in: Journal of Services Marketing, 8 (1994) 1, pp. 60-69.

[14]   Duffin, D.: Winning customer ownership – The Jaeger service excellence story, in: Managing Service Quality, 7 (1997) 2, pp. 80-86.

[15]   Lyons, M.: The Theory and Practice of Selling in the Bespoke Tailoring Industry, in: Registered at Stationers Hall, June 6th 1979.

[16]   Kahn, B. E.: Dynamic Relationships With Consumers: High-Variety Strategies, in: Journal of the Academy of Marketing Research, 26 (1998) 1, pp. 45-53.

[17]   Leclerc, F. et al.: Waiting Time and Decision Making: Is Time like Money?, in: Journal of Consumer Research, 22 (1995) 1, pp. 110-119.

[18]   Pine II, B. J.: Mass Customization, Boston 1993.

[19]   Sanders, H.: Financial rewards from mass customization, on www.sanders.ch.

[20]   Gerber Scientific: Mass Customization, on www.mass-customization.com.

[21]   Ashdown, S. P.: Introduction to Sizing and Fit Research, on car.clemson.edu/fit2000/.

[22]   Thompson, K. E.; Ling Chen, Y.: Retail store image: a means-end approach, in: Journal of Marketing Practice: Applied Marketing Science, 4 (1998) 6, pp. 161-173.

[23]   Bettencourt L. A; Gwinner, K.: Customization of the service experience: the role of the frontline employee, in: International Journal of Service Industry Management, 7 (1996) 2, pp. 3-20.

[24]   Goldsmith, R. E; Newell, S. J.: Innovativeness and price sensitivity: managerial, theoretical and methodological issues, in: Journal of Product & Brand Management, 6 (1997) 6, pp. 163-174.

[25]   Gilmore, J. H.; Pine II, B. J.: The four faces of mass customization, in: Harvard Business Review, 75 (1997) 1, pp. 91-101.

Acknowledgements: The authors gratefully acknowledge the support of Lectra Systèmes that made this research possible.

Contacts:

Celia P.A. Taylor and Professor Ray J. Harwood
Department of Textile Design and Production, De Montfort University,
Leicester, United Kingdom
E-mail: Celia.Taylor@students.dmu.ac.uk; rjh@dmu.ac.uk

27 Individualized Avatars and Personalized Customer Consulting

A platform for fashion shopping

Thorsten Gurzki[1], Henning Hinderer[1] and Uwe Rotter[2]
[1] Fraunhofer-Institute for Industrial Engineering (IAO), Stuttgart, Germany
[2] Institut für Arbeitswissenschaft und Technologiemanagement IAT, University of Stuttgart, Germany

Electronic Commerce is growing world-wide. The fashion industry, however, has only experienced minor benefits from this growth. Major problems are the lack of customer consulting and missing possibilities to try on fashion products in online shops with high three-dimensional display quality. This chapter gives an overview of the requirements of business-to-consumer online fashion retailing and the available technologies. It develops an approach for an online fashion shopping platform with individualized avatars for animated fashion presentation and integrated natural language text-based customer consulting features. These technologies are a major enabler for mass customization in the clothing industry.

27.1 Requirements for online fashion retailing

Mass customization based on online technologies requires a high level of integration into the textile process chain. This paper mainly focuses on the interface between suppliers and consumers (Figure 1). It describes the creation of a customer-oriented enhancement of existing E-Commerce solutions especially for the fashion industry. Two heterogeneous system architectures, one focusing three-dimensional garment and customer representation, the other focusing customer consulting services, are combined. The content of this paper is based on the research and development projects FashionMe [1] and ADVICE [2], both funded by the European Union. The platform described is designed in such a way that integration into existing enterprise information technology environments is enabled. The platform will build the foundations for custom-designed online fashion retailing.

Online fashion retailing has special requirements compared to other online retailing sectors. Display quality and consulting offers need to be realized at a high level in order to achieve broad customer acceptance. The main requirements for online fashion retailing are [3]:

- user-friendly offer of additional product information by integration of links;

- enhanced product presentation by means of audio tools;
- improved visualization, 3D representation, animated presentation of garment;
- personalization features: individual product offer, use of personalized avatars;
- individualized online consulting services;
- 24-hours availability of the shop and the consulting services.

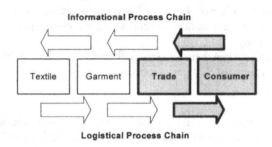

Figure 1:   Part of the textile process chain [5]

In the fashion and textile sector, online shopping has not yet achieved significant economic importance. According to a representative survey conducted in Germany [4], one of the prominent reasons for internet users not to take advantage of online fashion shopping offers is that 60% of the shoppers want to try on the articles before they buy them. 29% mentioned the missing possibility to 'look and feel' the fashion products as a negative characteristic of online fashion shopping. Only 6%, however, say they would miss the real-life conditions of a shopping event in a store. Another main result of the study was that shoppers demand customer consulting services. Checking the fitting of clothes over the internet is still by far not sufficiently possible today [6]. In the following sections, the concept of individualized 3D avatars is introduced, allowing to virtually try on clothing in order to better check the look and the garment characteristics especially when motions are made.

27.2   With visualization to eGarmets

In order to realize 3D garment visualization on an individualized avatar or to use a virtual catwalk displayed with regular web technologies, several components are needed. The person, the garment, the animation and the shop have to be available in transferable file sizes and formats.

Avatar technology: In our context, an avatar can be understood as a digital model of a person. An avatar must have certain prerequisites in order to meet the requirements for the representation of clothing and for walking on a virtual catwalk. The three essential components are a *three-dimensional mesh* representing the shape of the body, *graphical information* that is mapped on this mesh as textures, giving the avatar a realistic appearance, and a virtual skeleton that can be taken as a basis for the definition of motions; this also comprises the definition of joints with certain degrees of freedom as well as adjustable length relations; in the application described here, the H-Anim (Humanoid Animation) standard is used [7].

eGarments: The term eGarments denotes digital, three-dimensional models of real pieces of clothing. Most online product catalogs only consist of two-dimensional pictures. In order to show garment in three dimensions and with animated features, the effort for generating the necessary data material is higher. The easiest way to produce eGarments is to generate data from a CAD program which is used for the design and the cutting construction of the clothing. State-of-the-art of cutting construction is, however, in most cases only two-dimensional. In order to generate a three-dimensional volume model of the garment, these faces must be sewed together virtually and then transferred into a three-dimensional grid model (Figure 2).

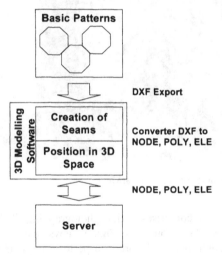

Figure 2: eGarments taken from a CAD system

However, eGarments can also be generated by digitizing the pictures of existing pieces of clothing. This method is employed in cases where, for example, the construction is not computer-based, or when CAD data cannot be transformed into the necessary data format. eGarments are produced in a multistage process. A life–like dummy is equipped with the specific garment and is then scanned in 3D. The basis for this method is that the naked dummy was scanned in the first instance so that the system knows the gages of the dummy. In a second step, the dummy is scanned wearing the garment. By subtracting the known rough model from the dressed model, the necessary geometric data is computed. With this approach, eGarments consist of the offset of the dummy and the garment's surface, and the graphical information which maps the surface and is used as a texture.

27.3 With scanning technology to virtual catwalks

Currently, there is a multitude of 3D scanners on the market that take advantage of different scan technologies. The selection of a certain scanner depends on the specific field of application. Besides the exactness of the measuring results, also aspects such as speed, usability, expenses for mass utilization and interfaces to further data processing systems are decisive criteria [8]. The solution to be described here is a technology developed by a member of the FashionMe project consortium, AvatarMe. The focus of AvatarMe's technology basically lies on the simple and fast generation of internet-enabled, personalized avatars that have a realistic appearance, and not so much on an exact rendering of the actual gages of a person. The scanner is accommodated in a booth that can also be set up in public places (Figure 3).

Figure 3: `Type 6` avatar booth of AvatarMe

The scanning process comprises digital photographing of the model from four different perspectives. Based on these digital views, the avatar is computed by means of existing rough models. In order to perform motions with the model, the avatar is assigned a skeleton as described above. The assignment of all necessary points in the avatar for the individual joints is realized with a simple method: predefined avatars are being used (Figure 4). After size, weight, age and sex of the scanned person have been recorded, the most appropriate avatar is automatically selected from a number of about 60 different predefined body models. Then, the avatar is being personalized by integration of the digital pictures. This procedure provides a first skeleton which can be further refined manually in order to equip fingers with knuckles, for example. By doing so, the avatar does not really comply with the exact body measures of the respective person. The quality, however, is sufficient to walk on the virtual catwalk and get a realistic impression of how the garment would actually look like on one's body. The mesh of such an avatar consists of about 3.800 nodes. This figure can be considered a compromise between the desired accuracy and the minimal amount of data. The procedure is optimized for real-time web animation.

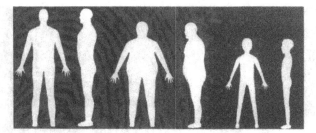

Figure 4: Predefined avatars

The *virtual catwalk* itself is a software that determines the motions of the virtual skeleton. The software animates the avatar according to defined, sex-specific motion sequences, i.e. it can be used with any avatar. The motion sequences contain detailed, time-dependent descriptions of motions for every joint. To generate these motions, two possibilities are used: Either they are generated artificially by means of a 3D editor, or they are digitized by motion capturing of real motions. Motion capturing (e.g., the technology of Vicon [9]) records and digitizes the motions of a human model with all their irregularities. These irregularities provide the realistic impression of the model. By doing so, complex motions can be transformed and edited easily.

The *user interface* design is of crucial importance to the acceptance on the part of the user. The design of the following websites was made in an iterative process under consideration of the results of usability tests according to DIN ISO 9241-10. The interface is divided into a two-dimensional part (Figure 5) and a three-dimensional environment (Figure 6). For representation of the products, samples from the Avatar Booth are used. The desired garments can be put in a shopping cart by drag and drop. With these products selected, the virtual catwalk can be entered.

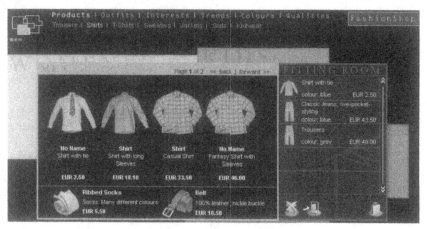

Figure 5: Screenshot of the FashionMe 2D Shop Pilot

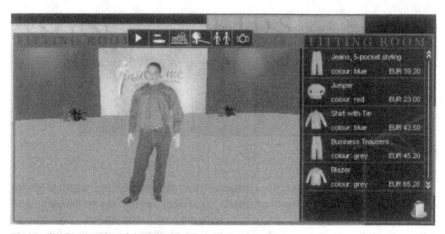

Figure 6: Screenshot of the 3D catwalk

The integration of the various technological components described above requires networking various heterogeneous data sources in a *client/server architecture*. The IT architecture used in the system for setting up an online shop is visualized in Figure 7. On the client side, for the 2D section of the shop a standard Web browser is used. In the 3D section comprising the virtual catwalk, a plug-in (Avatar Animation Engine) is used for visualization which is automatically installed when the user enters the shop for the first time. The avatar is loaded by the Avatar Server, and the files for the selected pieces of clothing are loaded by the eGarment Server. If pieces of garment are selected where the representation is based on a CAD construction, in addition the Cloth Animation Engine must be installed. In this connection, the shop system serves for the representation of conventional product information.

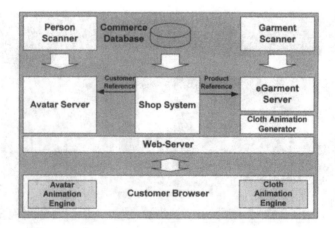

Figure 7: Client/Server Architecture

27.4   Virtual online customer consulting

In order to sell information-intensive products on the internet, automated online consulting features are necessary. Existing natural language text-based virtual sales systems lack intelligent customer consulting and adequate product presentation. The R&D tasks conducted within the ADVICE project aim on improving existing approaches concerning these two aspects.

Virtual sales-assistant systems can be classified roughly into three groups. *(1) Catalog guides* act as a guide through the online offer, reacting on input patterns with a comment and the presentation of a suitable web page or catalog content. *(2) Catalog-search systems* help the customer to retrieve appropriate products – depending on the customer's input – by a case-based reasoning supported (CBR) search in a catalog. The CBR approach allows to identify patterns and to offer similar or related products. *(3) Catalog-extension systems* finally provide additional information on the products and decision support for the buying process. Most systems offer an animated character as an user interface for the customer.

ADVICE acts as a catalog-extension system and is able to provide customer-decision support. This feature allows the use of ADVICE in the fashion sector. The ADVICE approach separates the product knowledge from the presentation knowledge, thereby achieving enhanced suitability of the presentation. A special feature of ADVICE is the operation of the system by speech acts [10]. Incoming natural language customer utterances are analyzed and translated to speech acts, representing the semantics of the utterance, and then transferred to the Dialog Processing Component [11] which is generating the answer. This answer is also represented semantically with XML-based speech acts. The natural language text-answer synthesis of the system is based on these speech acts. Natural language analysis and generation is described in [12]. The generated speech acts are also used for gesture generation of the virtual assistant in the user interface of ADVICE and for the product presentation in the Presentation Manager of the system [13]. The Presentation Manager is located in the Interface Agent (Figure 8), which represents a user session. The Presentation Manager is client-specific. It is available for a broad variety of client devices such as personal computers running the native ADVICE virtual assistant and mobile phones using the Wireless Application Protocol (WAP). Since speech acts include the semantics of the answer, the gestures are generated very accurately according to the natural-language text answer. The core ADVICE system is a multi-agent system. This agent architecture has been adapted to a four tier business system architecture, which fulfills the requirements of business users.

A *virtual character* is interacting with the end user. Two primary goals are leading the project's client development: First, the client is plug-in free to allow any user to use the assistant without having to download client software. And second, the client is able to make advanced human-like visual expressions. These goals have been achieved by creating a JAVA applet viewing a 3D VRML-based model of the assistant character (Figure 9). For the animation, two techniques are used: The motions of the head with its fine facial musculature are modeled directly, whereas the motions of the body are generated with the H-Anim standard. By this, the mimic can be adapted very specifically to any virtual character, that might be cartoon or animal characters as well. The H-Anim standard, too, allows to adapt the body motions (arms, legs) to a multitude of characters. When required, the body motions are loaded dynamically, therefore allowing a high number of possible motions.

The character for an application needs to be carefully designed with respect to the specific requirements (e.g. creditability of the character) of the application and the use in the targeted customer group (refer to [14], [15]). The client is capable of loading websites for which it gets the URLs from the ADVICE Server. By this, the server can present content from external data sources (shop system, Web content management system) in the browser of the user. For the client, predefined questions can be generated in the server application, which are presented to the customer as a sample of possible questions directed to the assistant. By this, also unpracticed users can operate with the system since special data input over the keyboard is reduced.

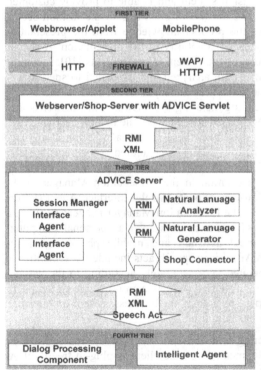

Figure 8: ADVICE System architecture   **Figure 9:** ADVICE client software

27.5 Integrated implementation of fashion visualization and consulting

The results of the projects FashionMe and ADVICE can be combined to an innovative fashion shopping platform. The platform is extended in such a way that shopping processes both on the supplier side and the customer side are supported. Two interfaces exist between the customer and the supplier. First the customer must draw up a personalized avatar using an avatar scanner, then he must enter the shop, where he can go shopping with his avatar. The scanning process can be done in a shopping mall or, in the future, in avatar booths in place of today's photo booths. Alternatively, a personal avatar can be created online without scanning by means of a conventional photo, a selection of standard avatars and manual entry of the person's measures. Since the creation of an avatar is already a personal act of the customer, it is possible to inquire preferences of the customer which later can be used for the online shopping process. In this connection, data protection must be guaranteed, which might require considerable efforts due to the high degree of variation of international data protection regulations. The representation of the avatar is done (as described above) by means of a plug-in.

The JAVA client developed within the ADVICE project is employed for the consulting services. The functionality of the client will be enhanced, so that the client is not only capable of presenting shop-system pages to the customer but also of dressing the avatar according to the natural language utterances of the customer. A further enhancement of the concept will be done for non-broadband connections (e.g., for modems or mobile devices) for which the presentation of the personal, dressed avatar must be realized as a plug-in free picture or sequence of pictures. In such cases, the consulting dialog will be conducted by means of text input and output.

The connection of avatar technology and eGarment technology with shop systems developed in the FashionMe project constitute the core of an integrated fashion platform (Figure 10) consisting of the following components: an Avatar Server, which provides the customer avatars; the eGarment Server, which administers the eGarments; the Shop System, which is responsible for product description and order processes; and the Consulting System, which supports the customer in the process of product selection by means of recommendations. A crucial aspect is the data administration in the back-end systems, so that the E-Commerce components of the platform and other enterprise systems operate on the same up-to-date data inventories. Important interfaces are between the ERP system, CRM system, and Shop System, Avatar Database and Avatar Server, in the eGarment section, and between Fashion Knowledge and Consulting System.

The Shop System accesses customer data in the CRM System and checks the availability of products in the ERP System. Orders are reported directly from the Shop System to the ERP System. Manual editing of an order is not necessary any more. The consulting component taken from the ADVICE project can be easily integrated into the fashion retailing platform. If a customer addresses the assistant as to, for example, suitable beachwear, the system generates an output from the knowledge base of the fashion knowledge component comprising data about beachwear and the personal profile of the customer (size, skin type etc.), presenting the customer one or several products. System-internal communication is conducted (as described before) on XML-based speech acts. Besides generating the natural language answers, the Presentation Manager (Figure 11) analyzes the speech acts and determines which products are to be presented.

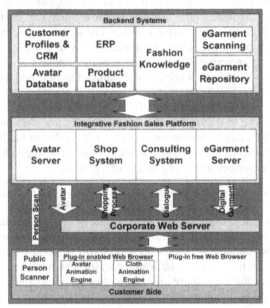

Figure 10: Integrated platform for personalized fashion shopping

The system accesses the Avatar Server over the Content Connector, the eGarment Server and the Shop System, and gets an URL with a reference to a personal avatar dressed with the desired article and a reference to the respective shop page. The Content Connector serves to abstract the details of the underlying content systems, such as the Avatar Server or the Shop System, in order to be able to use the software for a multitude of purposes together with various other systems. The references are embedded into an XML document that is stored in the Virtual Sales Assistant Client. The client transforms the references into a link and makes the browser load the customer's avatar dressed with the respective garment. The Avatar Database is placed in the back-end section in order to enable an enterprise to offer different shops under different brands, if need be, that all operate on a single data inventory. Customers have to place their avatar into the system only once since they can use one avatar in each of the shops. This possibility also enables the use of personal avatars in enterprise networks, where enterprises mutually permit the use of external avatars on their websites.

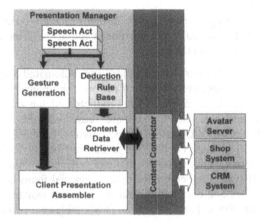

Figure 11: Presentation manager and interface to the avatar server and shop system

The research aspects and software architecture described here are being implemented at various demo business sites. These cases are also subject of user tests to show proof of usability and customer acceptance. The platforms created are supposed to enrich the existing online presentations in order to improve customer relationship, customer care and service. The provision of interfaces for the integration of the discussed architectures into various systems used on-site, e.g. ERP, CBR, assure operability and success of the platform. The FashionMe tool set will be installed in the *Macmoda Online Shops*. Macmoda is the Portuguese and Spanish retail chain of Maconde, one of the five leading European garment producers. The FashionMe architecture can be extended to the presented integrative platform at a later stage of the development.

The system also offers the possibility to use single components in other business cases. At the moment Fraunhofer IAO is working in two initiatives of other business sectors that will take advantage of the modular architecture. A personalizable, multimedia enhanced B-to-C portal for a global manufacturer of cellular phone technologies and a tool portal for craftsmen are to be realized. These projects themselves are part of the e-Business strategy of the companies involved. Therefore, compatibility and cooperation with various heterogeneous software architectures are crucial aspects of the success and acceptance of the projects. The emphasis lies on personalization and individualized customer support in the respective businesses.

27.6 Conclusions

Potentials for further development are currently located in the improvement of the realistic representation of the eGarments, such as truthful representation of the colors, the 'behavior' of the cloth, and the fitting of the garment on the avatars.

The techniques for creating avatars must be further refined in order to accomplish an even more detailed graphical representation while at the same time operating with smaller file size of the avatar.

The combination of fashion visualization and automated natural language consulting with text dialogs aims on solving two main problems of online fashion shopping: the missing visualization of the products and the customers and the missing consulting services. The architecture presented here is based on the projects FashionMe and ADVICE. Both projects involve European companies into their R&D activities. This close cooperation guarantees that the individual systems and the entire integrated architecture will meet business requirements. At the moment, the overall system is being developed. A special focus is to increasingly support sales processes and data administration / content management processes on the side of the suppliers. The platform already enables easy-to-perform online fashion shopping.

References

[1]   FashionMe: on www.fashion-me.com.

[2]   ADVICE: on www.advice.iao.fhg.de.

[3]   Miller, A.; Mueller, A.: FashionMe: The future of Fashion Shopping over the Internet, in: Proceedings of the e-Business and eWork 2000 conference, Madrid 2000.

[4]   GfK Textilmarktforschung: Market Survey Textile & eCommerce: Internet-Shopping for Clothing, 2000.

[5]   Bullinger, H.-J.: Neue Wettbewerbsparadigmen für die Textil- und Bekleidungsindustrie, in: Accurate Response Manufacturing fuer die Textil- und Bekleidungsindustrie, 1998.

[6]   Kraevetz,S.: If That's Me in the E-Dressing Room – Why Doesn't This Fit?, in: The New York Times on the web, September 20th 2000.

[7]   H-Anim Standard Organization: on www.h-anim.org/.

[8]   Rennesson, J.-L.: Automated 3D Body Measurement Systems, in: International Forum for Future Trends in Textiles and Technology, 2000.

[9]   Vicon: on www.metrics.co.uk/animation/.

[10]   Searle, J. R..: Speech Acts: An Essay in the Philosophy of Language, 1969.

[11]   Garcia-Serrano, A.; Henandez, J.; Martinez, P.: Intelligent Assistance to E-Commerce, in: Proceedings of the E-Business and E-Work Conference, Venice 2001.

[12]   Martínez, P.; García-Serrano, A.: The role of knowledge-based technology in language applications development, in: Expert Systems with Applications Journal, 19 (2000) 2, pp. 155-160.

[13]   Gurzki, T.: An Architecture for Integrating Virtual Sales Assistant Technology into E-Business Environments, in: Proceedings of the E-Business and E-Work Conference, Venice 2001.

[14]   Nissler, J.; Machate, J.; Hitzges, A.: How to get the Right Outfit for My Agent? Classification- and Design Methodology for a Virtual Shopping Assistant in a 3D

World, Information, Communication and Cooperation Interfaces, in: ID, 2 (1999), pp. 162-167.

[15] Prabhu, G.; Harel, D.: GUI Design Preference Validation for Japan and China – A Case for KANSEI Engineering?, Human-Computer Interaction: Ergonomics and User Interfaces, in: Bullinger, H.-J.; Ziegler, J. (Eds.): Proceedings of the 8th International Conference on Human-Computer Interaction, 1999, pp. 521-525.

Contacts:

Dipl.-Ing. Thorsten Gurzki
Fraunhofer-Institute for Industrial Engineering (IAO), Stuttgart, Germany
E-mail: thorsten.gurzki@iao.fhg.de

Dipl.-Ing. Dipl.-Geooek. Uwe Rotter
Institut fuer Arbeitswissenschaft und Technologiemanagement IAT,
University of Stuttgart, Germany
E-mail: uwe.rotter@iat.uni-stuttgart.de

28 Footwear Fit Categorization

Ameersing Luximon[1], Ravindra S. Goonetilleke[2] and Kwok-L Tsui[3]
[1] Department of Engineering, American University of Armenia, Yerevan, Armenia
[2] Department of Industrial Engineering and Engineering Management, The Hong Kong University of Science and Technology, Hong Kong
[3] School of Industrial and Systems Engineering, Georgia Institute of Technology, Atlanta, Georgia, USA

There is a growing trend to sell many types of consumer products through the web in order to maintain or enhance a company's competitiveness, and sometimes to establish a niche market. For products such as footwear however, manufacturers are facing quite a challenge to provide consumers with good fitting shoes. Footwear fitting is generally performed using the two variables of foot length and foot width (or girth), even though feet and shoes are three-dimensional objects. As a result, the matching between feet and footwear are quite variable and can be quite unacceptable even with the same brand of shoes. Footwear fitters speak of "perfect fit" and more commonly a "proper" or "correct" shoe fit even though the term "fit" appears to be nebulous. This chapter is an attempt to quantify and categorize footwear fit. Using digital manipulations, the foot shape was "adjusted" to the required heel height. The last and foot were then mapped to each other to determine the level of match and mismatch. The magnitude of the match or mismatch was color-coded and overlaid on the foot surface so that such color maps can be used to determine subjective preferences. The proposed footwear fit quantification can be used to predict potential discomfort and even fit-related comfort, if the material properties of the shoe are known. The method can also be used to rank different footwear lasts for any given individual.

28.1 (Mass) customization of footwear

Research indicates that customer focus can influence today's business [1]. As a result, manufacturers attempt to meet the differing customer preferences through product variety [2]. With growing product variety and opportunities in e-commerce, the old paradigm of mass production becomes sluggish [3], especially, when there is a change of the business paradigm from producer-centered productivity to consumer-centered customization commonly known as mass customization [4]. Mass customization is an attempt to satisfy individual customer needs with near mass production efficiency [5]. By breaking down the product features into components and offering those components to the consumer as choices, customization of the whole or part-product is possible [6].

Historically, there has been a trend to introduce product variety to cater to varying consumer tastes and styles [1], [7]. For example, from 1970 to 1988, the number of running shoe models increased from 5 to over 285 (167 men and 118 women) [1]. In order to keep pace with ever changing customer tastes, thousands of new products are made annually and with each variation, manufacturers attempt to bring products closer to what the customer needs. Even though variety matters to consumers, each product variety may have a differing meaning to different consumers [7]. Product assortments can be beneficial, if variety means different options rather than different products, as finding the required item among a large assortment can be quite frustrating for any consumer. Hence, allowing a customer to choose one product from a "shelf" can be wasteful and can also constrain a customer's ultimate satisfaction even though a store shelf may have great marketing appeal [8]. The difficulty of selecting the right pair of shoes in a shoe store is a classic example. In this case, product variety can be a hindrance rather than a benefit. Hence, the need satisfaction process should be attained not purely through more variety but by manufacturing the "right" products. Mass customization means to generate these "right" products in order to fulfill customers' needs at different levels [6], [9], [10]. The footwear industry can adopt this methodology for the next generation of footwear manufacturing and product design.

Gilmore and Pine [11] have defined four different types of customization: adaptive, cosmetic, transparent, and collaborative. Adaptive and cosmetic approaches provide customization on top of standardized products, and there is no need for the company to learn consumer's specific needs, while transparent and collaborative approaches offer differentiated products and require information on customer preferences [7]. Adaptive customization, is appropriate for sophisticated customers while it is not appropriate for customers who are likely to be confused or frustrated [7]. Consumers need to learn how to adapt the standardized product to fit their needs [7], [11]. Examples include the use different lacing methods in shoes, off the shelf insoles to change or adapt a shoe based on his/her own personal needs, ability to configure a shoe for different applications or activities such as the Nike "Rival", which can alternate between a track shoe and a road shoe [12]. Cosmetic customization refers to customizing the aesthetic aspects of standardized products [11]. For example, Cmax (www.cmax.com) and Nike allow consumers to design their shoes based on a different array of styles, materials and colors [12]. In transparent customization, customer needs are observed without actually involving the customer [11]. For example, shoe wear patterns can be studied and used to improve the outsole [13]. Transparent customization becomes difficult, if not impossible, when preferences are not well defined or observable [7]. In collaborative customization, there is a need to interact with the consumer to identify their needs and preferences [11]. As a result, collaborative customization can create the ideal customized product even though it can be a lengthy process. For example, Digitoe (www.digitoe.com) in their mission to deliver custom footwear uses the following steps:

- Educate the consumer on the custom fitting process.

- Consult customer service personnel for fitting needs and footwear selection.

- Fitting appointment (1-2 hours).

- Incorporate fit information into the *last.*

- Shoe fitting trial and customer approval.

- Selection of material, colors and patterns and finalizing the order.

It seems obvious that the interaction between customer and manufacturer is high in order to achieve collaborative customization. Similarly, orthopedic companies such as Footmaxx (www.footmaxx.com) use computerized gait and pressure technologies to fabricate custom orthotics [1]. Such custom shoe inserts typically cost around US$ 125-200 and take about 1 to 2 weeks to be delivered to one's home. Companies such as Custom Shoe Inserts (www.customshoe-inserts.com), on the other hand, make custom orthotics by using an impression kit to determine the shape, size, and differing foot needs. Collaborative customization aims to satisfy an individual's needs related to fit and comfort and may create the ideal customized product even though it can be time consuming and sometimes frustrating for the consumer [7]. Overall, whatever the chosen level of customization, the existence of a suitable fit-metric for footwear fitting can help, if not enhance customer-fitting needs in relation to footwear.

28.2   Why quantify fit?

The importance of product compatibility for comfort and satisfaction is well known [14], [15]. Today, most consumers select footwear based on length and width measures, even though many studies have shown that these two measures are insufficient for proper fitting [15], [16], [17]. In order for a shoe to fit a person's foot, a good understanding of the foot shape is necessary [15]. A good-fitting shoe should be free of any high pressure points [18], but at the same time should have the right 'feel' and support. A meaningful way to evaluate footwear compatibility would be to determine the dimensional difference between the foot and shoe [15]. If guidelines or standards can be established for these dimensional differences, then footwear fitting can be made much simpler. If a shoe is tight, the pressure or force will produce undue tissue compression making the shoe uncomfortable. When the shoe is loose, there can be foot slippage relative to the shoe resulting in damage or injury to soft tissue. Both these situations are undesirable as they may cause discomfort, pain, or even injury [15]. In order to achieve the right fit, the desired clearance between feet and shoes should be known in addition to supporting the foot at the most appropriate locations.

Since footwear-fit is one of the most important considerations when purchasing footwear [18], it has to be understood when manufacturing footwear. The availability of a fit-metric can be useful for the design and development of the right fitting footwear. With the proliferation of e-commerce, a fit metric will allow people to purchase comfortable footwear with some degree of confidence. Thus, the aim of this chapter is to illustrate a 3D methodology to quantify footwear fit.

Goonetilleke et al [19] proposed a 2-D methodology to calculate the dimensional difference between a foot and shoe last and have discussed how that error can be used as an indicator of the quality of fit. An extension of that method to three dimensions as given here, will allow a complete evaluation of the level of match or mismatch between feet and shoes.

28.3 Methodology to quantify footwear fit

The proposed methodology consists of four steps: (1) 3D scanning and orientation, (2) foot and last alignment, (3) computation of dimensional match or mismatch, and finally (4) the selection of lasts.

(1) 3D Scanning and orientation: The Yeti 3D foot laser scanner (www.vorum.com) from Vorum was used. The left foot of each participant was scanned when standing with equal weight on each foot. The *last* was scanned using the same scanner. After scanning the foot and *last*, the 3-D surface coordinates were extracted [20]. The coordinate system was established by "fixing" the plantar surface of the foot and *last* on the (x,z) plane and the "height" dimension along the y-axis.

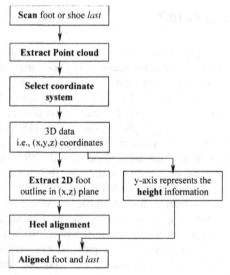

Figure 1: Flow chart for aligning foot and shoe last

(2) Foot and Last Alignment: The 2-D foot outline was extracted from the 3-D scanned data (Figure 1). All points that lay on the (x, z) plane were used to determine the 2-D "shape" information. The points located on the circumference of this 2-D shape were used as the foot outline. The heel centerline of the foot and

last outlines were then obtained using Matlab [19], [20]. The transformations required to orient the heel centerline along z-axis were first determined, after which the combinatorial linear and rotational transformations were applied to the whole foot and Last surface points to generate the 3-D shapes. Figure 2 shows the foot and *last* after alignment. After the foot and Last were matched along the heel centerline, the foot shape was digitally "adjusted" to account for the toe-spring and heel height of the *last* (Figures 3a - 3c). The foot and *last* after matching are shown in Figure 3d.

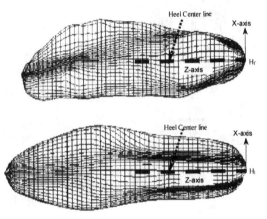

Figure 2: Heel centerline of foot and shoe last (H_f and H_l are the heel points of foot and shoe last respectively)

(3) Computation of dimensional match or mismatch: In order to compute the dimensional differences between foot and shoe last, sections along the z –axis, each spaced at 1mm intervals were used. The dimensional differences at each point were then color-coded for easy interpretation (Figure 4). Positive and negative differences should be distinguished as they have very different implications (Figure 3d). A positive "error" is one where the last surface is outside the boundary of the foot surface, whereas a zero or negative "error" is present otherwise. The concept of sign difference is very important in footwear fitting as a positive error is a loose fit, while a negative error can be categorized as a tight fit.

(4) Selection of Lasts: When the complete foot shape is available, different *lasts* can be selected based on cost, fitting needs and the time available to produce a pair of shoes (Figure 4). The most expensive and time-consuming procedure would be to produce a custom *last* for each individual consumer. For aesthetic customization, the *last* can be selected using foot length and possibly foot width sizing. This method is relatively cheap and less time-consuming since there will be no necessity to make new *lasts*. However, the degree of fit may not be satisfactory as foot length and foot width may be insufficient to generate the 3-D shape of foot [16], [17]. A different set of measures [20] may allow a closer match between *last* and feet if it is possible to obtain the characterizing parameters for each of the two three-dimensional shapes.

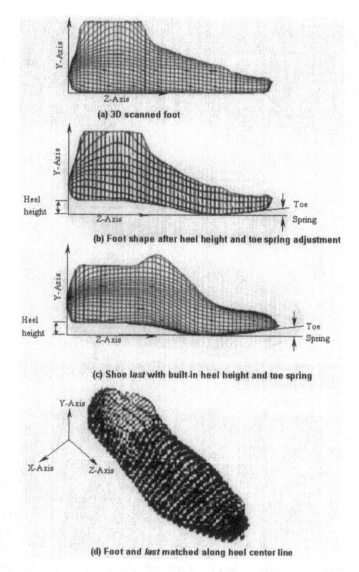

Figure 3: Foot and shoe last matching

28.4   Discussion and conclusions

Currently, most of the mass customization in the footwear industry is in the aesthetic domain, with some sizing customization. This method may require several fitting trials before the preferred fit for footwear can be attained. For groups that are unable to give their subjective opinions (example, subjects who have no sensation in their feet and children who cannot express the degree of fit), a fitting trial will be somewhat meaningless. With the proliferation of e-commerce, footwear purchase through internet can be greatly enhanced if a fit metric is present. Thus it is vital that the proper "clearance" between the foot and shoe be present for the foot to function as needed.

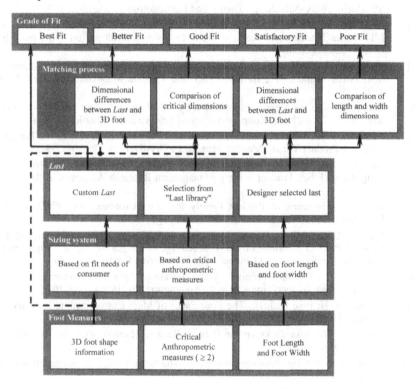

Figure 4: Last selection to achieve different degrees of fit

In this chapter, we proposed a method based on dimensional differences to quantify footwear fit. Important elements of this approach involve the orientation and alignment of foot and shoe, and shape adjustment to take care of shoe characteristics such as toe-spring and heel height. One weakness of this method is neglecting the shape change with bending at the metatarsophalangeal joint. However, the method allows the designer or developer to have some understanding of areas that may cause fit problems for users so that design modifications may be performed even without fitting trials. The method can be further enhanced through biomechanics, perception, and sensation studies [15]. Since the discomfort and pressure tolerance level on the surface of the foot may be different, there may exist a need to scale the allowable dimensional differences to account for similar sensations. Further study is needed to generate desired guidelines for the clearances between feet and shoes.

References

[1]   Cox, M. W.; Alm, R.: The Right Stuff – America's Move to Mass Customization, in: Annual Report of Federal Reserve Bank of Dallas, 1998.

[2]   Ishii, K.; Juengel, C.; Eubanks, C.: Design for Product Variety: Key to Product Line Structuring, in:ASME Design Theory and Methodology Conference, 1995.

[3]   Anderson, D. M.: Mass Customization, the proactive management of variety, in: MIT Sloan Management Review: Build-to-Order Consulting, on www.build-to-order-consulting.com.

[4]   Anderson, D. M.; Pine II, B. J.: Agile Product Development for Mass Customization: Niche Market, JIT, Built-to-Order and Flexible Manufacturing. New York 1997.

[5]   Jiao, J.: Design for Mass Customization by Developing Product Family Architecture, Ph.D. Thesis, Hong Kong University of Science and Technology, 1998.

[6]   Kahn, B., 1998, Variety: From the Consumer's Perspective, in: Ho, T.-H.; Tang, C. S. (Eds.): Product Variety Management Research Advances, 1998, pp. 19-38.

[7]   Lancaster, K: 1998, Markets and Product Variety Management, in: Ho, T.-H.; Tang, C. S. (Eds.): Product Variety Management Research Advances, 1998, pp. 1-18.

[8]   Du, X.: Architecture of Product Family for Mass Customization, Ph.D. Thesis, Hong Kong University of Science and Technology, 2000.

[9]   Martin, M. V.; Ishii, K.: Design for Variety: A Methodology for Understanding the Costs of Product Proliferation, in: Proceedings of the ASME Design Engineering Technical Conferences and Computers in Engineering Conference, 1996.

[10]   Spencer, J. E.: Robotics Technology and Advent of Agile Manufacturing Systems in the Footwear Industry, in: Assembly Automation, 16 (1996) 3, pp. 10-15.

[11]   Gilmore, J. H; Pine II, J. B.: The Four Faces of Mass Customization, in: Harvard Business Review, 75 (1997) 1, pp. 91-101.

[12]   Friedman, W.: Nike Picks up the Pace in Race to Harness Web, in: Advertising Age, 71 (2000) 10, p. 4.

[13]   Cavanagh, P. R.: The Running Shoe Book. Mountain View, CA: Anderson World, 1980.

[14]   Cheskin, M. P.: The Complete Handbook of Athletic Footwear, New York 1987.

[15]   Goonetilleke, R. S.; Luximon, A.: Designing for Comfort: A Footwear Application, in: Proceedings of the Computer-Aided Ergonomics and Safety Conference, Maui 2001.

[16]   Goonetilleke, R. S.; Ho, C.-F.; So, R. H. Y.: Foot Anthropometry in Hong Kong, in: Proceedings of the ASEAN 97 Conference, Kuala Lumpur 1997, pp. 81-88.

[17]   Goonetilleke, R. S.; Luximon, A.: Foot Flare and Foot Axis, in: Human Factors, 41 (1999), pp. 596-606.

[18]   Quimby, H. R.: The Story of Lasts, in: National Shoe Manufacturers Association, New York 1944.

[19]   Goonetilleke, R. S.; Luximon A.; Tsui, K.-L.: The Quality of Footwear Fit: What We Know, Don't Know and Should Know, in: Proceedings of the Human Factors and Ergonomics Society Conference, San Diego 2000, pp. 515-518.

[20]   Luximon, A.: Foot Shape Evaluation for Footwear Fitting, Ph.D. Thesis, University of Science and Technology, Hong Kong 2001.

Acknowledgment: This work has been supported by the Research Grants Council of the Hong Kong Special Administrative Region, China (Project No. HKUST 6074/99E).

Contacts:

Dr. Ravindra S. Goonetilleke
Department of Industrial Engineering and Engineering Management,
Hong Kong University of Science and Technology, Hong Kong
E-mail: ravindra@ust.hk

Kwok-L. Tsui
School of Industrial and Systems Engineering.
Georgia Institute of Technology, Atlanta, Georgia, USA
E-mail: ktsui@isye.gatech.edu

29 Virtual Reality and CAD/CAM for Customized Shoe Manufacturing

How virtual reality and CAD/CAM enable custom shoe manufacturing in mass markets

Marco Sacco[1], Giampaolo P. Viganò[1] and Ian Paris[2]
[1] ITIA-CNR Institute of Industrial Technologies and Automation, National Research Council, Milano, Italy
[2] CSM3D International Limited, Somerset, England

This chapter examines some issues involved in producing custom footwear on a mass-market basis. The starting position for custom shoes is the measurement of feet. Traditional methods can be replaced nowadays more and more by modern measuring technology based on 3D scanners. While discussing foot measurements, the chapter will also address common misconceptions about foot measuring, shoe sizes, better fitting footwear and comfort. However, getting exact feet measures is just the necessary, but not the commensurate condition for customized footwear. The feet data have to be translated into a customer specific last and shoe design. In this context, we will outline the advances in current CAD/CAM software to facilitate the production of custom footwear satisfying the needs of mass-market consumers economically. Nowadays, shoe design is mainly done by hand. However, when providing custom footwear this is too time consuming and too vulnerable to errors when designs are interpreted in manufacturing. To solve this problem, the VRSHOE system was developed. A designer draws (creates) or modifies in this virtual shoe design environment directly on a (virtual) shoe model the style lines that were created before in the CAD system. Designers can fly-through the environment and interact directly with the shoe model using immersive interface devices. The user interface, the environment and the results obtained in this project will be presented.

29.1 Measurement of feet – the first step to custom footwear

"When making good fitting shoes it helps that leather stretches and that feet are rubbery"- Anonymous

The human foot is a living machine that consists of a complex but efficient arrangement of bones, joints, ligaments, tendons, muscle, fatty tissues and skin. It is a remarkable piece of engineering that is able to cope with the fact that your heel can strike the ground with a force equal to twice your body weight.

Considering that the average man walks 2,500 miles every year the foot has a lot of work to do. No wonder that providing better fitting shoes is an important task. This chapter focuses on issues involved in designing custom footwear on a mass-market basis. The starting position is the measurement of feet, and, thus, the discussion of common misconceptions about foot measuring, shoe sizes, better fitting footwear and comfort. The basic shoemaking process is not an exact engineering discipline: leather stretches, feet shapes change continually. In addition, wooden shoe lasts – the production utility that is responsible for the exact fit – does not look like a foot, the shoe upper is not a true development of a last but is forced to fit a last. Finally, shoes change their shape while being worn. All these factors make exact measurement of feet difficult.

29.1.1   Foot / last measures and sizes

The measurement of feet has never been easy [9, 10]. Firstly, the foot has not an exact size, it changes its shape under different conditions (weight on, weight off, outside a shoe, inside a shoe). The foot also changes shape during the course of a day. In some cases people have been quoted as being as much as a whole shoe size from getting up in the morning to going to bed at night. In addition, the foot alters during a person's journey from child to adult to senior citizen. It is also possible that sudden changes in the social activities of an individual also affect foot shape and size. Paradoxically, measurement techniques of feet have changed only little over the last 100 years and still predominantly rely on tape measure or contact devices that simply measure lengths, widths or girths. Fundamentally this creates a problem as feet are not always measured in the correct position, the tape is not always held at the same tension. Two people are unlikely to be able to measure the same foot in exactly the same way. Finally, this manual process takes rather long, especially if it has to be applied in large scale at retailers offering custom footwear.

The sizes of lasts and shoes are defined by the traditional system of providing a number to indicate size and a letter to indicate the width fitting. An example of a common statement made by the average man in the street is that "My feet are size 8C". Firstly, it is unlikely that both feet are the same size and secondly, this statement is relatively meaningless. Last and shoe sizes are only an indication to approximate dimensions. The size naming system is only a title, a size 8C from different companies – or even for different shoes produced by the same company – may very well have different dimensions. The focus over a number of years has been on theses "numerical" measures. But in fact the most important measures are shape, position of major landmark features of a foot, and only finally dimensions. Last and shoe sizes are only an approximation to their dimensions today.

Attempts have been made to establish a standard for feet and last measurements. There are many well-known foot measurements. However, these methods have always encountered some difficulties. feet and last measurements have evolved over a number of years as different people have searched for the "magic measurements", i.e. one that are easy and quick to measure and provide the answer to "why does this shoe do not fit?" But when trying to obtain clearly defined mathematical rules to define these measurements a number of issues have to be considered:

- Some measures are mathematically poorly defined and will not produce accurate measurements for all shapes of feet.

- Other measures are not based on the result but are based on the traditional manual methods used to obtain these measures.

- Some measures are based on anatomical features that cannot be obtained from the shape of the feet alone.

29.1.2 Foot measurement systems for better fitting footwear

"After 130 years of Shoemaking why is it that we still fit test even the simplest of women's court shoes?"– Tony Darvill, founder of Shoemaster.

The fit of a shoe is dependent on a whole range of factors. The size and shape of the foot and last are obviously important even if it is obvious that lasts do not look like feet and feet do not have the same shape inside a shoe as outside the shoe. Accurate measurements of feet would appear to be a necessity, as would an understanding of the relationship between foot and last measures. The physical characteristics of a foot can also affect fit, for example a hard bony foot will have different requirements compared to a soft fleshy foot. If the relationship between foot and last were the only factor affecting the quality of fit then the problem would be relatively easy to deal with. However, we also have to contend with the fact that different types of shoes will have different fitting properties. This is further complicated by the fact the type of materials used to construct the shoe will also affect its fit.

Unfortunately, there are also subjective factors that affect a customer's judgment of whether a shoe is a good or bad fit. These include the acceptance of the look of the shoe, the "comfort" of a shoe, a preference for tight or loose shoes etc. In reality, there are many factors that affect fit. Therefore there is no-one in the shoe industry who will claim to always be able to make a good fitting shoe at the first attempt.

Modern technology may provide a solution allowing a more complex measurement of feet compared to the traditional systems. Especially modern scanning technology could be able to match not just a particular foot to a standard size but to deliver personal foot sizes for each individual. The footwear industry has been long searching for the "best in field foot scanner" that would allow us to use 3D-computer techniques to analyze the shape of a foot and calculate whatever measurements are required by the people that know what to do with the information. Our criteria for a foot scanner was as follows:

- A full 3D computer model of a foot should be generated, thus allowing whatever foot measurements are required to be taken.

- The accuracy of the foot model should be accurate enough to make better fitting, custom shoes.

- The scanner should produce results that are not affected by whoever scanned the foot and what time of day (or week) that the scan was taken.

- The foot scan should take the smallest amount of time, therefore reducing errors caused by foot movement.

- It should be possible to scan a foot in a variety of positions.

- We should be able to scan a foot whatever shape and no matter how big or small it is.

- If required, not only the foot but also the "leg" could be scanned in one operation.

Our investigations led us to conclude that digital camera technology using stereoscopic matching software would meet our requirements best. The next challenge was to find the best supplier of this technology and then apply it to the specific application of foot scanning. The benefit we expected to achieve from such a foot scanner is that it would be possible to record more information, with greater accuracy and more consistence compared to traditional manual or current technology based techniques. This would give us a solid platform to answer the real question "what is it that makes a good fitting shoe"? The end result of this process has been the launch of the *"FotoFit"* foot scanning system (provided by Shoemaster). FotoFit is designed specifically for shoe retailing, custom footwear and orthopedic footwear sectors and is claimed to be the first full 3D scanning device of it's kind in the market. Feet can be scanned with or without hosiery. The scanner is in fact a modeling device as it creates a 3D virtual model of a foot using digital images of a foot (Figure 1). This model can easily be measured and compared with other feet or shoe lasts.

Not only can the scanning system take measurements automatically. It also creates a 3D photograph of the foot that can be rotated in real time to view the foot from any direction. The scanning experience of a customer is now complete as they can see their feet captured within a computer terminal in retail. They can even be sent home with a copy of their foot model for use at a home computer. FotoFit can also be used to educate customers about foot shapes and good fitting shoes. In this context, the following activities are possible:

- *Foot shape changes under different conditions:* A foot can be scanned under different conditions i.e. standing, sitting, flat on the floor, standing on a heel-piece etc. Comparisons of measures and foot shape under these different conditions are easy to carry out.

- *Daily foot changes:* A foot can be scanned at morning, noon and night to show and quantify the natural changes mentioned above.

- *Foot changes with time:* A foot can be scanned today, tomorrow, weekly, monthly, quarterly, yearly etc. to monitor changes in a customer's feet. Additionally, in a medical environment where corrective footwear is prescribed changes in a foot shape can be easily monitored and shown to the patient. A foot scan can become the shoe industry's equivalent of a dental check-up.

- *Evolutionary foot changes:* Foot surveys can be done more easily as capturing data is quick and simple. Foot surveys could become a continual process instead of an activity that is carried out infrequently.

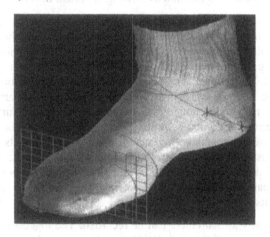

Figure 1: A foot scan with sock, showing typical measures

29.2 CAD/CAM software for customized footwear

However, getting exact feet measurements is just a necessary, but not the commensurate condition for customized footwear. The feet data have to be translated into a customer specific last and shoe design. In this context, modern CAD/CAM software plays a major role to facilitate the production of custom footwear in mass markets. We will use the example of the Shoemaster CAD/CAM system to illustrate the role of such a system for the design process of custom, better fitting footwear. This system has at its core the philosophy that all components of a shoe need to be developed from a common source of data that would take into account the anatomy and dimension of the foot shape. This idea, developed at a time (1960s) when the PC was still a dream, positions Shoemaster as a central system that integrates all parts of design, development and manufacture, allowing all people involved in this to work more closely together. Additionally, the system "enables" companies to gain benefits from associated technology that may have been developed in isolations e.g. leather cutting machinery, last milling machines, foot scanners etc. Shoemaster can be regarded as the glue that holds people, the equipment and the process together [3, 11].

The Shoemaster system is extensively used for the production of "normal". mass market footwear. In applying the system to custom or orthopedic footwear we are able to offer a semi-automatic adjustment of shoe components to allow shoes to be modified to an individuals requirements whilst using bulk manufacture techniques and processes. We start with the acquisition of data. Shoemaster records customer (or patient) information which not only includes foot measurement but also other facts that need to be taken into account when attempting to produce good fitting or the appropriate corrective footwear. Data can be recorded each time a customer is examined allowing historical data to analyzed easily.

As mentioned before, foot measuring devices alone do not produce good fitting shoes! What they do provide is a consistent and accurate method to measure a foot at a given point in time under certain conditions and to provide more information in a shorter period of time. The potential errors caused by human measurements are removed. 3D last data is taken into the computer via various manual digitizers or automatic scanners. This data is used to create a 3D surface model of the wooden last that can be shown in wire-frame or as a shaded surface. Feet and lasts are measured by mathematically defining a number of points on the surface and calculating measurements between them. These measurements are stored within a library of customer and last information along with other facts and figures e.g. type of foot disorder, category of last etc. Profiles and shapes are also stored so that a comparison of shape can also be made for Last's and Feet of similar measurements.

However, any measurement can be recorded. The biggest problem has been to get the industry or different individuals to agree on the *correct* measurements. Therefore we have taken the policy of providing a system that produces consistent measurements whose method of calculation can be changed by the operator as appropriate. Many lasts and customers can be stored in the database. Search rules can be set up to search for lasts with given characteristics (within a user defined tolerance). These search criteria are stored for future use and allow different criteria to be set up for different foot disorders or characteristics. Any last that has been selected can be modified to produce a new 3D model. This can be done in relation to the foot shape as the foot can be shown in "transparent mode" inside the last. Once the last shape has been defined the 3D data is transferred to a standard engineering package, these develop NC tooling paths to drive computer controlled lathes or turning machines.

Shoe designs are created by drawing lines onto the surface of the last or by digitizing designs that have been drawn by hand. A specification for thickness, color and texture of leather is added to this geometry to allow a 3D visualization of the Shoe to be produced and viewed in any direction. Soles that have been previously created can be automatically adjusted to fit any last, so that a visualization of the complete shoe can be easily created. The system allows a computer flattening of the last to be created. This also creates a 2D form of the Shoe design. This 2D geometry is used as a basis for the creation of the shoe upper patterns. As Shoemaster is a completely integrated system then any change made in 2D, possibly for production reasons, can be immediately viewed in real time within 3D therefore ensuring that the integrity of the 3D construction and design is maintained. The 2D geometry of shoe upper patterns is transferred to 2D cutting devices to allow templates or leather pieces to be produced directly.

One of the major inefficiencies when producing custom footwear is today that basic pattern engineering is reproduced each time a last shape is altered, even if the basic shoe is the same design. With as system like Shoemaster, shoe designs are engineered once. Any shoe design can be easily transferred to any new last shape with the associated patterns being automatically adjusted and a 3D visualization of the new shoe being reproduced within a few seconds. The potential for allowing customers to view a shoe design on their last, prior to it's production, is an exciting prospect. An important point to remember is that if the new Shoe design is not quite correct then it can be easily modified and due to the integrated nature of Shoemaster any pieces or tooling associated with that alteration are automatically updated and the 3D view of the shoe is adjusted accordingly. An additional feature is the ability to view two shoes at the same time allowing the "balance of the design" in a pair of shoes to be easily evaluated.

29.3   VRSHOE: A new process to design footwear

29.3.1 System approach

In a shoe factory today modeling and styling process are achieved in two different ways: manual and semiautomatic (i.e. CAD aided) [8]. In this section we want to discuss a new process that has been developed in order to integrate the semiautomatic method with a Virtual Reality Environment (VRE) to give the whole design process more immersion and interaction, thus supporting the designer's work. With the *VRSHOE SYSTEM*, developed jointly be the authors of this chapter, new last models can be built starting from already available digital models. Also new basic models of lasts can be based on existing knowledge easily. The new shoe models, once created and verified in the VRE, are sent to the CAD system where the 3D model is flattened to obtain the shell and to step into the engineering phase of shoe design. This process is accomplished in two separate steps:

(1) Immersive visualization: In this step the VRE imports the last model and style lines from the CAD or from a database where the digitized lasts are stored and gives the designer a 3D immersive view of the footwear. In order to achieve the import phase, an appropriate geometric file format (such as VRML or IGES) has to be used. In the VRE the user can fly-through and move around the shoe model. The 3D stereoscopic effect is implemented using stereoscopic glasses (shutter-glasses like) and a stereo projector, while the interaction device is a sensorized pen and a sensorized last. All the VRE has been implemented in the Paradigm Simulation software VEGA.

(2) Styling in virtual reality: The VRE allows a designer to draw (create) or to modify, directly on the shoe model, the style lines previously created in the CAD (Shoemaster). In this step the user can fly-through the environment and interact directly with the model using immersive interface devices; the interface tools for the user interaction are implemented in the VRE as 3D tool-objects. The "virtual board" has all the buttons and menu for the function and tool calls (such as a virtual pen for the styling).

In the future two more features will be added: the possibility to create the last model from scratch in the VRE and a virtual mannequin for aesthetic tests. It will be possible to link also a Virtual Environment where it will be present a "Virtual Mannequin": the stylist at the end of the design phase will be testing the style of the new footwear using the mannequin [2].

The whole system integrates a 3D CAD process oriented software (CSM3D Shoemaster) and a virtual environment (Figure 2). Thus it is possible to design the footwear and get immersive feedback. One of the main aspects is the exchange of data in order to represent and modify the shoe geometry in both environments. To achieve this result a database of lasts of footwear has been created starting from 3D models exported from the Shoemaster CAD (a translation from the Shoemaster format must be performed to obtain a file that can be used in the Virtual Environment). The application loads a geometric model from the lasts database and shows to the user a three-dimensional view of the product. Therefore the user can design the styles in an immersive environment and then export them for importing into CAD.

For the interaction devices we used an immersive prototype architecture (CrystalEYES 3D glasses, Pholemus trackers, a Pholemus stylus and a BARCO projector for 3D projection on wide screen) with the aim to investigate the possibilities offered by latest technology. This provides also the possibility for the user to work in a desktop mode eventually using only a standard mouse and the 3D glasses. The position tracking system has greatly improved the user interaction in the VRE. The user leans the stylus against the last in his hand, he can feel the contact with the last and see the "virtual pencil" that follows its profile while drawing the styles. The virtual last follows the movements of the real one, the virtual pencil is positioned exactly according to the position of its sensor (6 degrees of freedom). In this way user's senses are projected in the VRE. This kind of haptic feedback is called "tactile augmentation" [4] Physical objects are used to obtain the haptic feedback in VREs [5]. To interact with the 3D interface (menus, buttons, etc.) the lack of haptic feedback on the input device led to develop an alternative solution: change of color or brightness and sound effects. Sounds were introduced also to notify menu animations an other significant events.

The CAD system for shoes/lasts models design is Shoemaster. A critical factor of the system is represented by the integration with CAD environment; the two systems must be able to communicate with each other. Two possibilities are available: real-time connection or suitable import-export converters. While in the first case an integrated system is obtained, where the communication channel masks the data transfer, in the second it's the user to perform, through batch converters, the data exchange. The real time converter increases the complexity of files management and links the virtual environment to the specific CAD system. The decision to adopt a converter in a standard format (VRML) gives the potential to link VRShoe with different software tools and support web visualization. The application is written in C++ using Vega functions to control the Virtual Environment. Vega has a menu based environment (LynX) that makes easy and user friendly the creation of the Virtual Reality Environment and provides a wide collection of functions that allow runtime operations on every element in the Virtual Environment and on peripherals connected to the system. The LynX package is used for the Virtual Environment set-up: it's substantially an editor, with a window based interface, where menus are used to set Initial conditions for several features of the Environment being developed. Through LynX menus and dialog boxes, objects, lights, sounds, colors, points of view and all the environment features can be created.

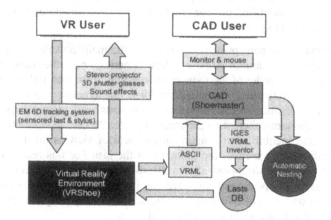

Figure 2: Architecture of VRShoe

29.3.2 Interface development

One of the prominent features of VEShoe is its user interface. Much effort was spend in the development phase to design a user interface that supports the traditional way of working of a footwear designer. The environment interface is structured by morphological components in order to achieve a correct interpreta-tion and monitoring of the product usability logic [6]. Such components can be

- sign, relatively to the graphical icons study and the navigation indexes choice;

- ergonomic, referred to the relation between user, working tools, environments;

- perceptive, visual and haptic - or sensorial corporeal;

- chromatic, relatively to interaction between field and objects;

- technological, related to technologies developed for the VRShoe environment.

The screen background of the system is configured spherical and reticular shaped, concave towards the user position. The virtual space enclosed between user and background constitutes the traversable operating field, containing the working tools and the user itself. The choice of the chromatic component is that of a dark and cool color, that amplifies the perceptive distance of the background without disturbing the work in first plane.

The environment is organized in *tool "containers"* which appearance in the field determined by the user, according to a working methodology that re-propose the real procedure. Initially, the field is empty and only on the right side just one index object is visible. Such object activates the appearance of the main working console that shows the five functional groups of the operative modes so ordered: draw, modify, shape typology of shoe, color, saving. The main console enters in field accompanied by an acoustic signal, translating towards the centre of the field of view enough to show the five push-buttons for the operative modes. Each operative mode is identified by five graphical icons that semantically synthesizes the groups of tools (Figure 3a).

The chromatic and perceptive components of the main console are memory of the wooden containers and of work and draw benches, not in the formal proposal but in the functional order of the tools contained and in the material visual choice of the surfaces. Activating the single buttons, an acoustic signal follows the appearance of 3D solids containing up to seven icons and chromatically grouped tools. The shapes are coherent to the main console. Particular attention has been paid to the fundamental need to maintain the field of view as much as possible free, to comfortably maneuver the pointer and the last. For this reason the main console automatically disappear after a stand-by time without selection of any button; to make it appear again the user has just to select a small white button always visible on the right. The solids, instead, can be temporally cancelled or put aside in view at the side of the operating field.

The main menu consists of five push-buttons activating the related menus. After one of these buttons is pressed the selected menu opens with a combined animation of rotation, translation and scale change, while playing a sound effect. Each menu can be moved grabbing it with the pointer and dragging into the desired position. When the pointer touches a button this is highlighted (a sound is associated to this event) and pushing the key on the Polhemus Stylus the button is activated, it is animated and performs the selected function. All the animations are accompanied by sound effects that enhance the degree of immersion focusing the user attention on what is happening in the Virtual Environment. The virtual foot object allows the tracking and the visualization of the drawn styles in real time in the virtual environment. It has been represented like the real wooden model that the user holds in his hand. The virtual foot can be moved at pleasure and eventually can become transparent to allow the symmetric copy option (Figure 3b). Fundamental points in the interface study were analysis of the potential user's needs in the virtual environment and ergonomic and functional usability of the VRShoe project. The idea is to create an environment to support technologically complex work making it intuitive and immediate to use (Figure 3c).

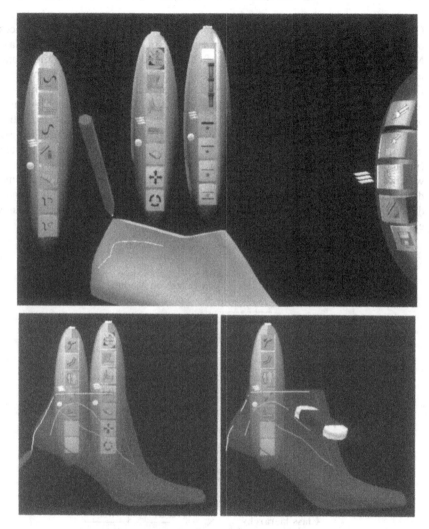

Figure 3: The VRShoe user interface: (a) Tools menu (top), (b) Last transparency (left), (c) Erasing a line in the VRShoe (right)

To achieve this results, the 3D graphic interface was developed in two parallel steps. First a software module was developed aimed only to the creation of an interface and the interaction with it. Second, the set of graphical objects that make up the interface, their movements and the interaction mode were designed.

(1) Development of software modules: The first step in the interface development was the creation of a collection of C++ classes describing the features and the behavior of the whole components in the environment. Starting from a set of general purpose C++ classes based on Vega, other new classes (derived from those still available) were built. Figure 4 shows them in their hierarchy tree. The

GUI class manages events generated from tools handled from the user (instances of the class Tool) and several three-dimensional menus (instances of the class Menu3D). VRShoe creates an instance of GUI and adds to it the main menu and others menu. The structure and the behavior of these menus are stored in a text file; such file is read at the application start-up and can be updated by the user immersed within the VRE (using a virtual keyboard). Furthermore each instance of the class Menu3D can build by itself its structure directly from the geometric model (loaded by Vega during the initialization phase): for each node of the 3D model hierarchy defined as DOF (degree of freedom) a new button is created (instance of the class Button 3D). Each button selects an item of the menu to which it belongs. The resulting action is executed by the main application. Some default buttons (open, close, icons, etc.) have their own default behavior already defined. Each button has two attributes describing the action to perform and eventual parameters needed.

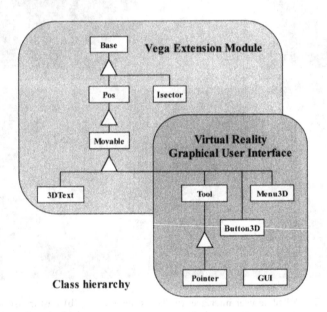

Figure 4: Interface Architecture

(2) Development of graphical objects: The configuration text file for menus was created in an automatic way due to the "streamable" property of Menu3D and Button3D classes, that is they have the capability to read and to write their configuration in a file. Also the virtual keyboard was created as an instance of Menu3D (it's a menu whose buttons are its keys). Therefore only two menus are created: the main menu and the virtual keyboard. Creating the main menu also its buttons and submenus are automatically created (in a way transparent to the application), thus each menu builds its buttons and its submenus by itself, recursively. The menu configuration can be modified by hand (with any text

editor) and also by the user through the 3D interface (the user can move around menus and save their new position). Moreover the menu configuration file stores some parameters for the animation, the list of the items and the submenus, and so on. One of the most important features of this interface module is the independence from a specific application. In practice, it can be reused as is for all the VR applications in which the user interacts with the VRE from a fixed position.

29.4 Conclusions

The question "what makes a particular shoe fit well" is rather complex and no exact answer exists. However, through the use of technology we can improve our knowledge of the factors that effect fit and define guidelines that would allow us to produce various types of footwear that should be very close to a good, personal fit. The most important thing is to involve the customer. Only be integrating customers into finding the solution customer satisfaction can be achieved.

By using the FotoFit System presented in this chapter it is possible to gain full confidence by customers at the earliest stage. This will be achieved by using a measuring method that is quicker, more accurate and more consistent than traditional methods that are prone to human error. Modern scanning technologies provide a solid platform from which the process of achieving better fitting footwear can be started. Fotofit can revolutionize shoe retailing and manufacturing alike. The quality and type of information that is now available will enable forward thinking companies to offer a service and product quality that has not been possible in the past.

VRShoes was born as a demonstrator of the usefulness of virtual reality in design and creative production in the footwear sector. The system provides an immersive environment where a user can interact in a natural way with the objects surrounding him (through a virtual pen and a sensorized last), which are projected on a wide screen. Using 3D glasses, also a stereoscopic vision can be obtained. Much attention has been paid also to the design of the user interface in order to enrich the easiness of use and the perception of immersion sense. The system carries the following major benefits:

Reducing time to market: Reducing the number of operations a designer currently must complete to reach a new footwear model leads to enormous time savings during the design process. This is in particular the result of eliminating the acquisition of style lines by a digitizer and the consequent adjustment in the CAD environment, avoiding also the imprecision of drawing the lines twice (when they are designed and when they are acquired).

Material saving and thus environment safeguard: It's not more necessary to place a casing on the last to be able to design its style lines. The lines can be drawn now virtually, thus neither ruining the last nor demanding casings or other tools. The economic saving can not be evaluated yet as the system just exists in form of a pre-industrial prototype using rather expensive devices. However, the path has been traced.

Easiness and naturalness: The virtual environment fully reflects the operating way of traditional designers (essentially manual and intuitive), but provides in addition the advantages of modern CAD tools to speed up particular operations. Also, storing many data provides good examples/bases for future models.

Among future developments is the development of a virtual wearer that, at the moment of style definition, will be able to aid the designer to figure the use of the footwear to be developed. This "virtual wearer" subsystem should visualize the new shoe model dressed, eventually even while walking around virtually. Such an environment will be able to study especially the behavior of footwear during its use – allowing another important steps towards the design of better fitter, customer centric footwear.

References

[1]   Avai, A.; Dulio, S.; Sacco, M.; Boër, C. R.: Agile, One Pair Shoe Production System, in: CENIM Workshop, Nanjing 1999.

[2]   Sacco, M.; Dulio, S.; Zhi Hang, L.; Mottura, S.; Boër, C. R.: Concurrent Design Tools for Shoe and its Manufacturing Process, in: CENIM Workshop, Karlsruhe 1999.

[3]   Shoemaster: on www.shoemaster.co.uk/.

[4]   Human Interface Technology Lab - Tactile Augmentation: on www.hitl. washington.edu/research/tactile

[5]   Sowizral, H.: Interacting with Virtual Environments Using Augmented Virtual Tools, in: Proceeding of SPIE, Stereoscopic Displays and Virtual Reality Systems, 2177, 1994

[6]   Bowman, D.; Hodges, L.: User Interface Constraints for Immersive Virtual Environment Applications, Graphics, Visualization, and Usability Center, in: Technical Report 1995

[7]   Ulieru, M; Goldsmith, P.: A Fuzzy-Logic Based Computer Aided Sport Shoe Design System, in: 17th International Congress on Biomechanics, Calgary 1999.

[8]   Sacco, M.; Viganò, G. P.; Boër,C. R.: A Virtual Environment for Shoe Design – Engineering and Manufacturing, in: Prime 2001, Progress in Innovative Manufacturing Engineering, Sestri Levante 2001.

[9]   Clarks: Manual of Shoemaking, 6th Ed., 1989.

[10]   Rossi, W. A.; Tennant, R.: Professional Shoe Fitting, in: Pedorthic Footwear Association, USA 1999.

[11]   FotoFit: on www.fotofitscanner.co.uk/

Acknowledgments: We acknowledge Arch. Calabi Daniela of Tplane S.r.l for her contribution in the user interface development. The VRShoe research was funded by MURST within the project SPI-6 – Innovative Production System Program.

Contacts:

Dr. Marco Sacco
ITIA-CNR Institute of Industrial Technologies and Automation,
National Research Council, Milano, Italy
E-mail: m.sacco@itia.cnr.it

Ian Paris
CSM3D International Limited, Somerset, England
E-mail: ian@shoemaster.cix.co.uk

Part VII: New Directions

Future challenges for building the customer centric enterprise

Despite all the pages filled before with research on how to make enterprises more customer centric, there are still many open questions. Thus, the last part of the book discusses the future of mass customization, customer integration, and personalization. *Piller and Tseng* will comment on fields for further research needed to develop new processes, tools and programs for integrating the customer into value creating activities, both on the technological and the operational process side. They envision a future in which mass customization becomes an integral part of business operations, co-existing side by side with mass production. To this end, six areas are identified which require special attention when implementing a mass customization system n practice. These are also fields where more research is needed most.

30 New Directions for Mass Customization

Setting an agenda for future research and practice in mass customization, personalization, and customer integration

Frank T. Piller[1] and Mitchell M. Tseng[2]
[1] TUM Business School, Department of General and Industrial Management, Technische Universitaet Muenchen, Germany
[2] Department of Industrial Engineering & Engineering Management, The Hong Kong University of Science & Technology, Hong Kong

In the closing chapter of this book we would like to share with our readers our view about the future of the customer centric enterprise and the enabling strategies of mass customization, customer integration, and personalization. We will also comment on fields for further research necessary for the development of new processes, tools and programs for integrating the customer into value creating activities, both on the technological and the operational process side. We identify six areas where – from our perspective and within our field of knowledge – more research is needed most. These are also fields which we think require special attention when implementing a mass customization system: (1) Issues concerning the design of products and product architectures. (2) Consumer behavior with customer interaction tools. (3) Drivers of customer satisfaction and dissatisfaction with custom made products. (4) Impact of integrating a user and customer into value creation on knowledge management and information management. (5) Capability analysis and systems engineering for concurrency in value chains. (6) Measurement of value contribution in mass customization systems.

30.1 Converging mass customization into main stream operation

Mass customization has been receiving increasing popularity in business and academia. Specific features that differentiate mass customization from mass production have been quietly accepted in many spheres of production management. For instance, it has become prevailing practice in many industries today that the orders companies received and orders they sent out are smaller, more frequent and rushed. This differs from the common practice of large backlogs and long lead-time. The ramification of universally accepted "just in time" inventory practice goes beyond promoting efficiency. It also raises economic volatility from a macro perspective. However, the ability to hold orders to the last minute is only one capability of the advent of readily available computing and communication capability through the forms of agile manufacturing, focused factories, flexible manufacturing, and computer integrated manufacturing. The other side of strategic advantages, particularly leveraging the pervasive connectivity with customers has equal, if not more significant impacts in the years to come. Namely, customers will be at the center of value creation and are becoming active players in the value chain. In other words, instead of mass producing products and services for an anonymous market, each single customer can participate in the value chain and can be treated as an individual.

While customer integration offers plenty of new options for profit generation, the model also demands new ways of interacting with customers and users. New models for value creation derive not only from customization, but also from new business models aimed at enhancing the service content of manufacturers and retailers. Additionally, approaches to enhance the eco-friendliness and sustainability in operation management accentuate the need to reduce waste, particularly, waste in producing products that no customers need. The previous chapters of this book have identified and discussed many different approaches towards customer centric manufacturing and interaction strategies. From these principles, new research requirements derive.

The objective of this closing chapter is to give indications for topics on an agenda for future direction in mass customization. While the previous chapters have answered questions about the design, implementation and operation of a mass customization system, this chapter will raise new questions. How can we arrive at new processes, tools and programs for integrating the customer into value creating activities, both on the technological and the operational process side? Though customer involvement in value creation has been well accepted in the service industry, the unique aspects of reflecting individuality in value creation is still a vast open area to be studied in many academic disciplines and many industries alike. Many questions are still open – or perhaps have not even been asked? In the following, we will describe six fields where – from our perspective and within our field of knowledge – more research is still needed:

- Design of products and product architectures for mass customization,

- Consumer behavior and customer interaction tools,

- Drivers of customer satisfaction and dissatisfaction with custom made products,

- Customer knowledge management and user driven innovation,

- Capability analysis for manufacturing & logistics and systems engineering for concurrency in value chains,

- Measurement of the value contribution of individualization: does mass customization pay – und under which conditions?

These six issues are also the fields that demand special attention when implementing a mass customization system. The reason for many failures of mass customization pioneers can be traced back to these areas as well. Mass customization is characterized by a high intensity of information exchange compared to mass production [1, 2, 3]. While modern flexible manufacturing technologies have existed for more than a decade, systems capable of handling the increasing intensity of information and interaction with the customers have only existed for a few years. New internet technologies in particular can be seen as a main enabler. The significance of information handling capabilities is based on the process of elicitation [4], the mechanism for interacting with the customer and obtaining specific information in order to define and translate the customer's needs and desires into a concrete product specification. In the meantime customers have experienced difficulties in finding out what their budget will allow them to do, and providers have failed to recognize these problems. In order to perform these processes efficiently in both directions, systems sufficient to cover the arising coordination costs have to be available. Thus we feel that new research on mass customization is especially needed in fields connected with customer interaction and integration. Our (subjective) selection of future research demands and emphases for the implementation of a mass customization system in business should therefore be seen in this light. However, as part five of this book has shown, research on manufacturing systems for mass customization continues to make valuable contributions to the further development of our field.

30.2 Designing products and product architectures for mass customization

Product design can 'make or break' any manufacturing enterprise. Effective product design can enable customers to achieve what they want, and at the same time provide them with a wonderful experience. It will also ensure ease of provision and of fulfillment and most importantly, it makes scale of economy possible at mass production level. The entire value chain starts with a good design. However, traditional design focus has been limited to developing and producing at a point product design level. Little attention has been given either to developing a family of products with underlined product architecture, or to developing products which enable the consumer to participate in the understanding, selection, and negotiation. Consequently customization has often become an "after thought" from the provider point of view and hence a limited option. One distinctive principle of mass customization is a two stage process of product development [5]. While product architectures and the range of possible variety can be pre-determined during a product family design stage, a second design and development stage takes place in close interaction between the customer and the supplier. Based on what has been learnt from the second stage, the product family can be upgraded, which in turn leads to capability enhancement on the part of the providers. Zipkin [4] calls this process the elicitation of a mass customization system, the mechanism for interacting with the customer and obtaining specific information in order to define and translate the customer's needs and desires into a concrete product specification.

For stage one of product design for mass customization, the product family approach has been recognized as an effective means of accommodating an increasing product variety across diverse market niches, while still being able to achieve economies of scale [6]. In addition to leveraging the costs of delivering variety, product family design can reduce development risks by reusing proven elements in a firm's activities and offerings [7, 8]. The backdrop of a product family is a well-planned architecture – the conceptual structure and overall logical organization of generating a family of products – which provides a generic umbrella to capture and utilize commonality. Each new product is instantiated within this architecture and extends so as to anchor future designs to a common product line structure. The rationale of such a product family architecture lies in not only unburdening the knowledge base from keeping variant forms of the same solution, but also in modeling the design process of a class of products that can easily be varied according to the individual customers' requirements within a coherent framework. However, while the basic principles of product family design are understood and well documented in literature, in the context of mass customization many questions remain unanswered.

In addition, product family design for mass customization also has to take stage two into account: the co-design process by a customer. The design options, their combination possibilities, and, in particular, their contribution to fulfilling a desired need have to be understood by (un-trained, inexperienced) customers configuring their own products. In order to understand this process better and to develop an appropriate design of configuration processes simultaneous to the basic product design phase (stage one), more knowledge on the ways in which customers perform their design activities is required.

Issues that need to be addressed for implementing Mass Customization are also rich areas that call for further research. These issues span both stages of the mass customization design process:

- What is a product family? What are the underlying structures? How do we represent these structures?

- How can a product family best represent the capability of an enterprise? How can they be matched to an existing set of resources and enterprise capabilities?

- Can different players (providers, customers, suppliers, production engineers, etc.) communicate well with the same product family structure?

- How do we evolve a product family as products which move with changes in customers' requirements, product technology and enterprise capability?

- How can we move from designing a point product to designing a product family? Should there be changes in design methodology?

- How can (basic) product design and the design of the configuration processes be connected? What are the interfaces between both development processes, and what are the success factors of a good design-by-customer?

- How can services and (material) products be combined into customized bundles?

- How do customers and users perform their design activities? Which factors motivate them and drive user satisfaction (this question is also covered in the following section)?

- Last, but not least, how can different parties work together with trust and integrity ?

30.3 Consumer behavior and interaction tools

An area that demands special attention during the implementation of a Mass Customization system is the front end of customer interaction. This is also a field that necessitates further research. Designing the front end addresses issues in consumer behavior in business systems based on customer integration and mass customization. While most mass customization approaches implemented in practice are based on offering a huge number of varieties and choices, in comparison with a mass production system, there is still only very little understanding about the perception of choice and the joy or burden of configuration experienced by customers. For instance, many mass customization companies offer an astronomical number of options, some of them are in trillions. The vast majority of options do not generate customer contentment; instead, they cause a great deal of confusion and may even turn customers away. The lack of understanding also applies to areas such as the evaluation of different customization options within particular product categories. Knowledge generated by research in this field has to be used to improve customer-company interfaces (configuration engines) and to design product and process architectures that maximize the return from customer interaction in terms of willingness to pay (reflecting the value perception).

Customers often have no clear knowledge of what solution might correspond to their needs. At times, these needs are not apparent to the customers themselves. As a result, customers may experience uncertainty or even perplexity during the design process. Uneasiness could also be spawned by the supplier's behavior too. The newer and more complex the individualization possibilities are, the more information gaps increase. A customer orders from the supplier and often pays in advance for a product she can only evaluate in virtual form, and has to wait days or even weeks to receive it. These uncertainties can attribute additional and most likely, hidden, transaction costs. One of the supplier's most important tasks is to ensure that the customer's expenditure is kept as low as possible, while the benefit/value perceived by the customer has to be clearly delineated.

Interaction systems for mass customization and service management are the premier instrument for reducing these costs [9]. Known as configurators, choice boards, design systems, toolkits, or co-design-platforms, these systems are responsible for guiding the user through the configuration process. While the term "configurator" or "configuration system" is quoted quite often in literature, it is used for the most part in a technical sense addressing a software tool. The success of such an interaction system is, however, by no means only defined by its technological capabilities, but also by its integration in the entire business models and whole sales environment, its ability to allow for learning by doing, to provide experience and process satisfaction, and its integration into the brand concept. Tools for user integration in a mass customization system have to contain much more than arithmetic algorithms to combine modular components. Toolkits for customer integration are also costly to develop, implement, operate, and change (investments for recent web based toolkits for mass customization start at 100,000 US$, and most companies often end up investing at least ten times this sum, see [10]. Hence, the programming of such a tool is both a risky and important investment for a company.

One would expect a rich source of research literature and ample empirical insights in this apparently important issue. There is a fair amount of literature on the technical aspects of product configurators and how to integrate them with the other elements of a mass customization system (e.g. [11, 12]). But from a literature review [13] we could hardly identify any empirical analysis on the actual interaction patterns of customers with toolkits for mass customization. We still lack understanding of how users actually interact with extant systems for customer integration and configuration, i.e. how they proceed while designing a product, and which patterns are visible in the discovery of one's own needs [14, 15].

Thus to build a successful front end for customer interaction in a mass customization system, one has to pay attention to the following issues which are, of course, also important questions for researchers to address:

- What are the motives and incentives for integrating users into value creation? Which factors drive consumers to spend time (and high opportunity costs) on configurators? How can companies implement these mechanisms to create fruitful business models?

- Do users have a relatively clear perception of the intended outcome of the design process? How "targeted" as opposed to a pure trial and error procedure is the design process? How many variants are explored and changed before making a final decision?

- What tools and interfaces (configurators) have to be developed – and to what extent are their success factors of use in order to integrate the customer with regard to consumer co-design? What does this approach mean for the design profession?

- How must configurators and interaction tools be designed in order to overcome an information overload? Do users follow specific patterns while interacting on a mass customization web site?

- Can we observe "learning effects" of users interacting on a mass customization web site during the course of configuration?

- How do different levels of the decoupling point influence customer integration? What are the effects of different decoupling points on the performance of a mass customization system?

- What are the technological approaches for configuration systems, and how can these systems be integrated into large ERP systems? How can configurators and customer relationship management systems be integrated?

- Which standards and data exchange formats allow shared knowledge and use of these configuration tools along an integrated supply chain within different independent partners?

30.4 Drivers of customer satisfaction and dissatisfaction with custom made products

The issues identified in the proceeding sections lead to an important question: How satisfied are customers of mass customization products and what are the drivers of their satisfaction? The importance of this question is evident [13]. Supposedly, only users who have a particular minimum level of satisfaction with the configuration toolkit will finalize the design process and purchase the product, recommend the site to their acquaintances, and come back themselves – always assuming that the satisfaction with the product designed is sufficiently high. Research in customer satisfaction confirms the importance of this construct [16]. It also seems conceivable that the satisfaction with the process has a large impact upon the satisfaction with the product in mass customization [17].

First, it has been shown that the perceived quality of the product itself and that of the entire shopping experience are closely related [18, 19]. Manufacturers therefore often strive for shelf-space in high-level outlets. In a mass customization system, the physical store is often dominated in the virtual environment. The experience and customer expectations are transmitted through intensive interactions on internet or retail based systems to achieve customization. The customer's trust in such a system is very much dependent on the quality of service delivered to the customers. Companies have to be very sensitive in building up trust and reliability in order to reduce the risk seen by prospective customers. Secondly, and even more important, is the fact that in a mass customization system the individual product is the direct result of the process. A mass customizer is offering a solution capability, not a product. A felicitous and successful configuration process will therefore have an impact on both process and product satisfaction.

A major concern in designing the configurator is the number of options available to customers. It is often mistaken that the number is a major driver for customer satisfaction. In fact, excess variety may result in external complexity

[20, 21]. The number of choices can overwhelm customers during product and service configuration as well as during product use if the shoe is equipped with lots of technical support tools (like sensors etc.). The number of choices on typical mass customization sites exceeds these well-known decision problems by far. In fact, one has to convert the choice numbers into a familiar area to get an adequate understanding of how many choices the customer has. For example, if one wanted to build a shop large enough to display all variants of Cmax.com sport shoes (circa $3*1021$) the surface of the whole earth would be insufficient – in fact one would need 7,000 planets the size of the earth, each completely covered with a shop. The burden of choice may simply lead to information overload, resulting from the limitations of the human capacity to process information. So what is the optimal degree of variety and extent of customization possibilities – from the customer's perspective and not (only) from the manufacturer's perspective?

Consequently, building systems that can provide transactions for mass customization need to address the following aspects:

- Who are the customers? What are the expected satisfactions? Is your mass customization configurator compatible with these satisfactions?

- Which user characteristics cause the satisfaction differences likely to be observed? Are there great differences between different customer groups? Which factors cause these differences?

- What is the interrelation between the buying process and product satisfaction?

- Which usability characteristics of a toolkit cause the satisfaction differences likely to be observed?

- What is the optimum amount of customization and customer integration? Which factors explain the value perception of customer integration by consumers? What is an appropriate number of choices from the user's perceptive?

- Do different process designs and experiences of toolkits make it possible to handle different degrees of variety from the user's perceptive?

- To what extent is the role of a more active designer rather than a more passive chooser desirable?

30.5 Customer knowledge management and user driven innovation

By pursuing (mass) customization and interacting with each customer individually companies gain access to a new source of value that is based on the aggregation of customer knowledge which enables them to better perceive market information. Integrating the customer into configuration (customer co-design) allows access to so-called "sticky local information" [22]. Sticky information arises when the costs of information exchange between two different actors are higher than processing this information within one unit. They originate in location specific costs like technological and organizational activities of decoding, transmitting and diffusing the information. We can argue that often customer specific information is sticky in that sense. Tastes, design patterns, and even functionalities are rather subjective and difficult to describe objectively. By integrating the customer into the design of a product or service and transferring customer needs and wishes into customized products, a company gains access to the sticky information and can transfer it to explicit knowledge. By aggregating this knowledge, the company can generate better market research information and more accurate forecasting concerning customer needs.

This is especially true when the firm operates mass production for anonymous markets alongside the customized business – which is the case for many manufacturers and retailers [23, 24]. For the portion of business that is manufactured on stock, the customization segment provides panel-like market research information without the common panel effects biasing the results. The information gained here can be used to improve forecasting and sales planning of products made to inventory. Additionally, new product development and continuous improvement can benefit substantially from such user information. Customers and users provide direct input for new product development – without the usual biases of market research and prototype testing.

Note that the knowledge generated here relates to all aspects of use, design, performance, fit, and further aspects. It also provides important input for increasing the sustainability of value creation along the whole value chain. Customer knowledge makes it possible to design products which better fit the customers' needs and to better forecast volumes of specific models, thus, reducing waste of unwanted and ill-fitting products. This information results from the sheer possibility of interacting. The information gains are not based exclusively on actual sales or customization but just on tracking the information gained by interacting with the users on a configuration tool and within a service enriched sales process.

However, in order to realize the potentials described before, many questions still have to be answered. They include:

- How can knowledge resulting from the intense integration of the customer into the processes of configuration and product design be generated and used?

- How can this knowledge be integrated into innovation processes and market information provided by a company's experts?

- How can tools and instruments be developed to transform the knowledge gained during the integration process into new competencies? How can this knowledge be saved and stored in the knowledge management systems?

- How can the tools and principles from the area of collective learning and knowledge management be transferred to a company that is no longer involved in the manufacturing of products but in interacting with each single customer?

30.6   Capability analysis and systems engineering for concurrency in value chains

Mass customization has been made possible due to the recent advancement in capabilities of modern manufacturing technology such as flexible manufacturing systems or modular product structures. It significantly reduces the fixed cost in comparison with the variable cost and hence variety can be provided without significant increases in cost. This was the starting point of mass customization, and most early research in the field was based on this (e.g. [4, 23, 25, 26, 27, 28, 29]). However, flexible manufacturing and corresponding planning systems are necessary but insufficient for successful mass customization. These systems have to be supplemented by information technologies capable of handling the information flows and transaction costs connected with mass customization. The advent of pervasive connectivity resulting from a rapid increase in internet usage provides the necessary and affordable connection among all parties in a mass customization system.

Thus, the successful introduction of mass customization depends on the extension of the scope of concurrent engineering beyond the traditional boundary of design and manufacturing to include customer interaction, marketing, service and recovery. It has been suggested that concurrent enterprises are very much in line with the idea of the real time economy where the customer is central to the value creation. Mass customization requires a dynamic network organization especially when the goods to be customized have a higher degree of complexity and of customizability, as Pine et al. [30] stated: "Mass Customization (...) requires a dynamic network of relatively autonomous operating units. Each module is typically a specific process or task, like making a given component, a distinctive welding method, or performing a credit check. The modules, which may include outside suppliers and vendors, typically do not interact or come together in the same sequence every time. Rather, the combination of how and when they interact to make a product or provide a service is constantly changing in response to what each customer wants and needs." Process modules (specific processes or tasks) have to interact with each other or come together in a different sequence for every product or service demanded by customers. With mass customization, a firm no longer offers products or services, but the capability of interacting with a customer in order to co-produce an individualized solution [24, 31]. Mass customization elevates the contribution of each participating element from activity to capability. Customers' needs drive the entire value chain. But how is this value chain composed and organized?

Large enterprises like Procter&Gamble (customized cosmetic program Reflect.com), Adidas (MiAdidas sport shoes with customized fit and functionality), or Lego (customized assembly of construction kits in the factory) and a huge number of small start-up companies demonstrate that new forms of customer interaction are already a reality. However, these examples also show that new forms of organizational structures may be needed. Procter&Gamble placed all mass customization specific operations into a specific unit, separated from the standard mass production operations. However, this unit also acts as a learning field for the whole company. Other companies like Adidas on the other hand try to integrate the customized operation into the mass production manufacturing and sales system, creating a hybrid organization structure. What is the better approach for each different situation?

In order to assist companies to build the best fitting structure for a specific customization task and the corresponding value chain, it is important to foster research on how the situation and structure variables of mass customization are composed (following the contingency theory). What are the factors which define a specific customization task and distinguish it from other tasks? And how can variables describing different modes of organizational structures and value chain constellations be matched with these specific tasks resulting in appropriate organizational capabilities? Research on the nature of these capabilities is also needed. Firms have to be able to create "dynamic capabilities" [32, 33] which enable them to adapt, integrate, and reconfigure the firm's skills and competences in order to react more sufficiently to new customers' requests (in the short term) or to adopt a changing business environment (in the long term). However, there is still a lot of research needed in order to transfer the tools and principles from this area to a company that is no longer based on the manufacturing of products but on interaction with each single customer. One question of prime importance is, for example, what enables a firm to decouple the knowledge gained in the integration process in order to deepen and re-direct market-driven and technological competencies in dynamic environments? [32].

To this end, issues in this field that have yet to be highlighted in building mass customization systems include:

- What methodology is available to describe capabilities and to share them in an extended value chain network?

- How can modular process models be created and configured in order to integrate the capabilities of different firms in order to fulfill a specific customer's order? To do so, the idea of inter-organizational cooperation and virtual companies has to be developed much further. The benefits of integrating suppliers into customized manufacturing and capability development are well described in theory, but not applied in practice. Here, further research is needed to establish scalable and transferable coordination schemes in the logistics domain.

- How do we collaborate across boundaries of organization, culture, and geography? How do we incorporate the pervasive connectivity of the internet to encourage the participation of all parties including consumers, suppliers, service providers and others?

- What are the effective ways for participating parties to express their needs without getting bogged down with tedious input and constantly reflect their needs and changes in requirements?

- Where do we price the urgency? The trade off of delayed satisfaction and increase in volume efficiency is obvious from the provider perspective but may not be the case for consumers.

- How can tools and principles from the area of collective learning and knowledge management be transferred to a company that is no longer based on the manufacturing of products but on interacting with each single customer? How can tools and instruments be established to transfer the knowledge gained during the integration process into new competencies?

30.7 Measuring the value contribution of individualization: does mass customization pay – and under which conditions?

Mass customization and customer integration are connected with new cost and profit structures that can be summarized with the term "economies of mass customization" [34]. This opens the field to an important area of research: the quantitative empirical evaluation of mass customization. Taking the growing interest in mass customization on the one hand and the dominance of quantitative empirical research in many academic disciplines on the other hand into account it seems surprising that there is almost no large-scale empirical research about mass customization (see [13] for an overview). Empirical evaluation of mass customization and customer integration is still dominated by case studies and small samples, or by very broad approaches including all kinds of new supply chain systems. With the advent of a broader implementation of mass customization enough data could be available in the near future to embark upon research which evaluates the success and performance factor of mass customization quantitatively. However, qualitative research is not useless and does still provide interesting insights. In order to cover a field characterized by a heterogeneous population of firms and strong growth rates only the combination of research focusing on a large group of cases quantitatively with qualitative in-depth studies of some exceptional examples seems sufficient. The identification of promising practices can help not only to set the starting point for more detailed research, but should also work as a show case of what is already possible. In addition, analyzing "worst practices" and company failures may help managers to prevent errors.

Empirical evaluation needs measures and performance indicators to report the different outcomes of a mass customization system. This task is imperative because the current accounting systems are not designed for assessing the true economical benefits from the total value chain point of view. Thus, managers fail to consider mass customization because of an inability to provide convincing economical justification. Given that the focus has been shifted from controlling cost to value creation, however, our accounting and controlling systems are still dominated by the calculation of "product costs" [35]. Savings and additional costs resulting from different degrees of interaction with the customers are not covered by most industrial reporting / accounting systems in practice. Activity-based accounting and the balanced score card approach provide some first solutions. However, approved ratios for calculating the value of a customer relationship, for example, or parameters for evaluating the extent of the market research information gained by aggregated customer knowledge are still missing.

In addition, the value contribution of mass customization has to be evaluated from the customers' perspective, too. For users, the decision to buy individualized products is basically the result of a simple economic equation [5, 36]: if the (expected) returns exceed the (expected) costs the likelihood that they employ mass customization will increase. Mass customization related costs for a customer are, for example, the price of the product (or the price premium if the individualized product has a higher price than a standard offering), and the drawbacks of the user's integration into value creation during the configuration process we discussed earlier (such as risk, information overload, time and effort required, demand for trust, delivery time etc.). Returns for customers are twofold: firstly possible rewards from the design process such as flow experience or satisfaction with the fulfillment of a co-design task, and secondly the value of customization, i.e. the increment of utility customers gain from a product that meets their needs better than the best standard product attainable [13]. As the latter might be more enduring, this points to the utmost significance of the value of individualization. Only if the increment in the customer's value or utility suffices, are they likely to design their own products via mass customization sites and be willing to pay a price premium. In fact, a number of small scale field studies indicated that consumers are willing to pay price premiums for individualized solutions (e.g. [37]). This important hypothesis, however, needs to be tested in large-scale research.

From a manufacturer's point of view, price premiums are not the only motive for employing mass customization solutions. The chance of sustainable differentiation from its competitors is also of high importance. Today's market heterogeneity, increasing variety, steadily declining product life cycles, decreasing customer loyalty, and the escalating price competition in many branches of industry are the main motivators for firms going into mass customization. Thus, the sheer willingness of consumers to interact within a mass customization system and to try a toolkit for mass customization is obligatory. In other words: mass customization will be a perpetual phenomenon only if, respectively only in markets where the value of individualization exceeds a minimum level. To our knowledge there have been hardly any attempts to explicitly measure (i.e. quantify in economic terms) the users' need for individualization or to quantify the value of customization from a user's perspective (see as a theoretical exception [38]; and empirically [39]). Thus, research on the economic value of getting an individualized product or service is an issue of vital importance. Only if enough customers value the advantages of customized goods highly enough, is this system likely to become a mass phenomenon, too.

The following questions need to be answered when considering a mass customization business. They are also important questions for further research:

- How does performance differ in different industries, and how can these differences be explained? How are profit mechanisms and cost structures changing in different branches of industry, how does industry structure influence the extent of economies of integration?

- What are the effects of the different levels of the decoupling point (the point where the customer is integrated into value creation) on the performance of a mass customization system? What are the success factors of mass customization from a financial perspective?

- How highly do actual and potential customers value individualization? Are customized products objectively or merely subjectively individual? Which factors have an impact on this valuation? What options of customization (fit, functionality, design) are valued most and in which context? What options of comfort and eco-friendliness are valued most and in which context?

- How can methods and controlling tools be designed to evaluate the economical effects of "design-production-service-end-of-life" value chain? How can modern approaches like the balanced score card or target costing be modified and implemented to evaluate the value of the economies of mass customization?

- Customer information has become a very valuable asset in customization. Conceivably, there is intrinsic value in customer data, which then raises the question as to who owns consumer data? No law until now, for example, has given any indication as to who is the owner of a 3D-body scan. Here, basic research is urgently needed.

30.8 Conclusion

In this chapter we have discussed some important issues for implementing a mass customization business. Considering mass customization as a new style of conducting business, these issues are also the very questions that require further development by mass customization researchers. By discussing these issues, we hope to stimulate further discussions and research. We also hope to draw the attention of managers considering mass customization to important success factors. Research into the impact of customization on other functions such as design, marketing, or customer interaction is just beginning. We agree with Wind and Rangaswamy [40] that until now a lot attention has been paid to the impact of mass customization on manufacturing. Knowledge about patterns, procedures and problems experienced by consumers configuring goods and services is limited, and best practices are yet to be identified and put in wider application in business. There are still many interesting and fruitful research opportunities in these and other related fields.

Finally, we would like to conclude with one major piece of advice: The findings, patterns, and suggestions for action found in this book, in other recent writings and discussions on mass customization are far from a well established science. They are even further away from being applicable as generic strategic patterns. Rather, they should be considered as potential starting points for a mass customization concept. After all, mass customization research and practice have to be customized, too.

References

[1] Duray, R. et al.: Approaches to Mass Customization: Configurations and Empirical Validation, in: Journal of Operations Managements, 18 (2000), pp. 605-625.

[2] Lee, C.-H.; Barua, A.; Whinston, A.: The Complementarity of Mass Customization and Electronic Commerce, in: Economics of Innovation & New Technology, 9 (2000), pp. 81-110.

[3] Reichwald, R.; Piller, F. T.; Moeslein, K.: Information as a critical success factor for mass customization, in: Proceedings of the ASAC-IFSAM 2000 Conference, Montreal 2000.

[4] Zipkin, P.: The Limits of Mass Customization, in: Sloan Management Review, 42 (2001), pp. 81-87.

[5] Du, X.; Tseng, M. M.: Characterizing Customer Value for Product Customization, in: Proceedings of the 1999 ASME Design Engineering Technical Conference, Las Vegas 1999.

[6] Tseng, M. M.; Jiao, J.: Mass Customization, in: Salvendy, G. (Ed.): Handbook of Industrial Engineering, 3rd Ed., New York 2001, pp. 684-709.

[7] Ulrich, K.: The Role of Product Architecture in The Manufacturing Firm, in: Research Policy, 24 (1995), pp. 419-440.

[8] Sawhney, M. S.: Leveraged High-Variety Strategies: From Portfolio Thinking to Platform Thinking, in: Journal of the Academy of Marketing Science, 26 (1998) 1, pp.54-61.

[9] von Hippel, E.: Perspective: User Toolkits for Innovation, in: The Journal of Product Innovation Management, 18 (2001), pp. 247-257.

[10] Rogoll, T.; Piller, F.T.: Configuration Systems for Mass Customization and Customer Interaction: A Study and Comparison of 24 Configuration Systems, ThinkConsult 2003, Munich 2003.

[11] Bourke, R.: Product Configurators: Key Enabler for Mass Customization - An Overview, in: Midrange Enterprise, 8/2000.

[12] Weston, R.: Web Automation, in: PC Week, 32 (1997), p. 76.

[13] Franke, N.; Piller, F.T.: Key research issues in user interaction with configuration toolkits, in: International Journal of Technology Management, 14 (2003).

[14] Park, C. W.; Mothersbaugh, D. L.; Feick, L.: Consumer knowledge assessment, in: Journal of Consumer Research, 21 (1994), pp. 71-82.

[15] Stabell, C. B.; Fjeldstad, O. D.: Configuring value for competitive advantage: on chains, shops, and networks, in: Strategic Management Journal, 19 (1998), pp. 413-437.

[16] Johnson, M. D.; Gustafsson, A.: Improving customer satisfaction, loyalty, and profit: an integrated measurement and management system, San Francisco 2000.

[17] Riemer, K.; Totz, C.: The Many Faces of Personalization – an Integrative Economic Overview, in: Tseng, M. M.; Piller, F. T. (Eds.): Proceedings of the World Congress on Mass Customization and Personalization MCPC 2001, Hong Kong 2001 *(see also Chapter 3 of this book)*.

[18] Anderson, E.; Sullivan, M.: The Accedents and Consequences of Customer Satisfaction for Firms, in: Marketing Science, 12 (1993), pp. 125-143.

[19] Patterson, P.; Johnson, L.; Spreng, R.: Modeling the determinants of customer Satisfaction for Business-to-Business Professional Services, in: Journal of the Academy of Marketing Science, 25 (1997), pp. 4-17.

[20] Friesen, G. B.: Co-creation: When 1 and 1 make 11, in: Consulting to Management, 12 (2001) 1, pp. 28-31.

[21] Huffman, C.; Kahn, B.: Variety for Sale: Mass Customization or Mass Confusion, in: Journal of Retailing, 74 (1998), pp. 491-513.

[22] von Hippel, E.: Economics of Product Development by Users: The Impact of "Sticky" Local Information, in: Management Science, 44 (1998) 5, pp. 629-644.

[23] Kotha, S.: Mass customization: Implementing the emerging paradigm for competitive advantage, in: Strategic Management Journal, 16 (special issue 1995), pp. 21-42.

[24] Piller, F. T.: Mass Customization, 3rd Ed., Wiesbaden 2003.

[25] Ahlström, P.; Westbrook, R.: Implications of mass customization for operations management: an exploratory survey, in: International Journal of Operations & Production Management, 19 (1999), pp. 262-274.

[26] Anderson, D. M.: Agile product development for mass customization, Chicago 1997.

[27] Kotha, S.: From Mass Production to Mass Customization: The Case of the National Industrial Bicycle Company of Japan, in: European Management Journal, 14 (1996), pp. 442-450.

[28] Pine II, B. J.: Mass Customization, Boston 1993.

[29] Victor, B.; Boynton, A. C.: Invented Here, Boston 1998.

[30] Pine II, B. J. et al: Mass customizing products and services, in: Planning Review, 21 (1993) 4, pp. 6-13.

[31] Piller, F. T.; Reichwald, R.; Schaller, Ch.: Building customer loyalty with collaboration nets, forthcoming in: Mills, Q. et al. (Eds.): Collaborative Customer Relationship Management, Cambridge 2002.

[32] Fowler, S. et al: Beyond products: new strategic imperatives for developing competencies in dynamic environments, in: Journal of Engineering and Technology Management, 17 (2000), pp. 357-377.

[33] Teece, D. J. et al. : Dynamic capabilities and strategic management, in: Strategic Management Journal, 18 (1997), pp. 509-533.

[34] Piller, F. T.; Moeslein, K.: From economies of scale towards economies of customer integration: value creation in mass customization based electronic commerce, in: International Journal of Electronic Commerce, forthcoming 2003.

[35] Darlington, J.: Lean thinking and mass customization: The relationship between production and costs, in: Management Accounting, 77 (1999) 10, pp. 18-21.

[36] Chamberlin, E. H.: The Theory of Monopolistic Competition, 8th Ed., Cambridge 1962.

[37] EuroShoe Consortium: The Market for Customized Footwear in Europe: Market Demand and Consumer's Preferences, in: Piller, F. T. (Ed.): A project report from the EuroShoe Project within the European Fifth Framework Program, Munich and Milan 2002.

[38]   Dewan, R.; Jing, B.; Seidmann, A.: Adoption of Internet-based product customization and pricing strategies, in: Journal of Management Information Systems, 17 (2000) 2, pp. 9-28.

[39]   Franke, N.; von Hippel, E.: Satisfying Heterogeneous User Needs via Innovation Toolkits: The Case of Apache Security Software, in: Research Policy 2002.

[40]   Wind, J.; Rangaswamy, A.: Customerization. Journal of Interactive Marketing, 15 (2001) 1, pp. 13-32.

Acknowledgments: This research was funded partly by grants from the European Commission (GIRD-CT-2000-00343). The authors gratefully acknowledge the input of Nikolaus Franke, Vienna University of Economics and Business Administration, to this chapter in regard to toolkits for user interaction and customer behavior.

Contacts:

Dr. Frank T. Piller
TUM Business School, Technische Universität München, Munich, Germany
E-Mail: piller@ws.tum.de

Prof. Mitchell M. Tseng, Ph.D.
Department of Industrial Engineering & Engineering Management,
The Hong Kong University of Science & Technology
E-Mail: tseng@ust.hk